# Current Developments in Mathematics 2015

*edited by*

David Jerison
Mark Kisin
Paul Seidel
Richard Stanley
Horng-Tzer Yau
Shing-Tung Yau

International Press
www.intlpress.com

Current Developments in Mathematics, 2015





ISBN 978-1-57146-331-9

20 19 18 17 16 1 2 3 4 5 6 7 8 9

Typeset using the LaTeX system.

# Contents

Current Developments in Mathematics, 2015

# Kähler-Einstein metrics and algebraic geometry

Simon Donaldson

ABSTRACT. This paper is a survey of some recent developments in the area described by the title, and follows the lines of the author's lecture in the 2015 Harvard Current Developments in Mathematics meeting. The main focus of the paper is on the Yau conjecture relating the existence of Kähler-Einstein metrics on Fano manifolds to K-stability. We discuss four different proofs of this, by different authors, which have appeared over the past few years. These involve an interesting variety of approaches and draw on techniques from different fields.

CONTENTS



## 1. Introduction

General existence questions involving the Ricci curvature of compact Kähler manifolds go back at least to work of Calabi in the 1950's [**11**], [**12**]. We begin by recalling some very basic notions in Kähler geometry.

- All the Kähler metrics in a given cohomology class can be described in terms of some fixed reference metric $\omega_0$ and a potential function, that is

(1) $$\omega = \omega_0 + i\partial\overline{\partial}\phi.$$

- A hermitian holomorphic line bundle over a complex manifold has a unique *Chern connection* compatible with both structures. A Hermitian metric on the anticanonical line bundle $K_X^{-1} = \Lambda^n TX$ is the same as a volume form on the manifold. When this volume form is derived from a Kähler metric the curvature of the Chern connection can be identified with the Ricci tensor. (In general in this article we will not distinguish between metrics and Ricci tensors regarded as symmetric tensors or $(1,1)$-forms.)

- If the Kähler class is a $2\pi$ times an integral class a metric can be regarded as the curvature of the Chern connection on a holomorphic line bundle. The Kähler potential parametrising metrics in (1) has a geometrical meaning as a change in the Hermitian metric on the line bundle: $|\ | = e^{\phi}|\ |_0$.

One of the questions initiated by Calabi was that of prescribing the volume form of a Kähler metric in a fixed cohomology class. By the $\partial\overline{\partial}$-lemma this is the same as prescribing the Ricci tensor (as a closed $(1,1)$-form in the class $c_1(X)$). This Calabi conjecture was established by Yau in 1976 [**50**]. In particular when $c_1(X) = 0$ this gives the existence of *Calabi-Yau metrics*, with vanishing Ricci curvature. Another question raised by Calabi involved *Kähler-Einstein metrics* with "cosmological constant" $\lambda$, so Ricci $= \lambda\omega$. The case $\lambda = 0$ is the Calabi-Yau case as above and if $\lambda$ is non-zero we may assume that it is $\pm 1$. The Kähler-Einstein condition can then be expressed as saying that we require that the Hermitian metric on the holomorphic line bundle $K_X^{-1}$ given by the volume form of $\omega$ realises $\pm\omega$ as the curvature form of its Chern connection. Explicitly, in complex dimension $n$ the equation to be solved for a Kähler potential $\phi$ is

$$(\omega_0 + i\partial\overline{\partial}\phi)^n = \omega_0^n \exp(\pm\phi + h_0) \tag{2}$$

where $h_0$ is the solution of $\partial\overline{\partial}h_0 + \omega_0 = \pm\text{Ricci}(\omega_0)$ given by the $\partial\overline{\partial}$-lemma.

In the case when $\lambda = -1$ there is a straightforward existence theorem, established by Aubin and Yau. In fact the proof is significantly simpler than that when $\lambda = 0$. The point is that the nonlinearity from the exponential in (2) occurs with a favourable sign. The condition on the Kähler class means that the manifold $X$ is of general type and the metrics are generalisations of metrics of constant curvature $-1$ on Riemann surfaces of genus 2 or more. The harder case is when $\lambda = 1$. Then the condition on the Kähler class means that $X$ is a *Fano manifold.* A result of Matsushima [**39**], also from the 1950's, shows that existence can fail in this case. Matsushima showed that if there is a solution the holomorphic automorphism group of $X$ is reductive (the complexification of a compact Lie group). This means that Fano manifolds, such as the projective plane blown up in one or two points, with non-reductive automorphism groups cannot support such a metric. The same Calabi-Aubin-Yau scheme which proved existence in the cases $\lambda \leq 0$—the "continuity method" (see 4.2 below)—can be set up in the positive case, but must break down for these manifolds. More precisely, the difference arises because of the absence of a $C^0$-estimate for the Kähler potential due to the unfavourable sign in the equation (2). On the other hand there are many cases where the existence of a solution has been established, using arguments exploiting detailed geometric features of the manifolds and the theory of "log canonical thresholds". The question which arises is to characterise in terms of the complex geometry of the manifold $X$ exactly when a solution exists. In the early 1990's Yau conjectured that the appropriate criterion should be in terms of the *stability* of the manifold $X$ and, after two decades of work by many mathematicians, this is now known to the case. The precise formulation is in terms of a algebro-geometric notion of *K-stability* and the statement is that a Fano manifold admits a Kähler-Einstein metric if and only if it is K-stable.

One of the attractive features of this problem is the range of techniques which can be brought to bear. By its nature, the statement involves an interaction between algebraic geometry and complex differential geometry and, as we shall see, there are important connections with global Riemannian geometry, with pluripotential theory in complex analysis and with nonlinear PDE. Four different proofs of the main result have appeared up to the time of writing.

(1) Deformation of cone singularities. (Chen, Donaldson and Sun [**19**]);
(2) The continuity method (Datar, Szekelyhidi [**21**]);
(3) Proof via Kahler-Ricci flow (Chen, Sun, Wang [**17**]);
(4) Proof by the variational method (Berman, Boucksom, Jonsson [**6**]- the statement proved here is slightly different).

The purpose of this article is to survey these recent developments.

## 2. K-stability

The notion of "stability", in this context, arose from the study of moduli problems in algebraic geometry and geometric invariant theory. It is not usually possible to give a good structure to a set of all isomorphism classes of algebro-geometric objects. For example, generic 4-tuples of points in the projective line up to the action of projective transformations are classified by the cross-ratio, but there is no satisfactory way to define the cross-ratio if three or more points coincide. The idea is that the isomorphism classes of a suitable restricted class of stable objects do form a good space. There is a general circle of ideas relating the algebraic approach to these questions to metric structures and differential geometry involving the notion of a moment map and the "equality of symplectic and complex quotients". In particular there is a large literature, going back to the early 1980's, developing these notions in the framework of gauge theories and results such as the existence of Hermitian Yang-Mills connections on stable vector bundles [**48**]. But the author recently wrote a survey which emphasised this side of the story [**29**], so we will not go into it here beyond noting that the Kähler-Einstein problem fits naturally into wider questions of the existence of constant scalar curvature and extremal Kähler metrics; questions which remain largely open.

Typically, stability is defined by a *numerical criterion* on *degenerations* of the objects in question. In our situation, we consider polarised varieties $(X, L)$, so $L$ is an ample line bundle over $X$ and the sections of $L$ embed $X$

as a projective variety in some projective space. The relevant degenerations are *test configurations* which are defined as follows. For an integer $m > 0$ a test configuration of exponent $m$ for $(X, L)$ is a flat family of schemes $\pi : \mathcal{X} \to \mathbf{C}$, with a relatively ample line bundle $\mathcal{L} \to \mathcal{X}$ and a $\mathbf{C}^*$-action on $\mathcal{X}, \mathcal{L}$ covering the standard action on $\mathbf{C}$. For $t \in \mathbf{C}$ we write $X_t$ for the scheme-theoretic fibre $\pi^{-1}(t)$ and $L_t$ for the restriction of $\mathcal{L}$ to $X_t$. We require that for all non-zero $t$ the pair $(X_t, L_t)$ is isomorphic to $(X, L^m)$. (Note that, due to the $\mathbf{C}^*$-action, it suffices to know this for *some* non-zero $t$.) For technical reasons we also suppose that the total space $\mathcal{X}$ is normal.

The numerical criterion, in our situation, is provided by the Futaki invariant. In its general form, [**27**], this is defined for any $n$-dimensional projective scheme $Z$ with $\mathbf{C}^*$-action and $\mathbf{C}^*$-equivariant ample line bundle $\Lambda \to Z$, as follows. For each integer $k \geq 0$ we have a vector space $H^0(Z; \Lambda^k)$ with an induced $\mathbf{C}^*$ action. Write $d(k)$ for the dimension of the space and $w(k)$ for the sum of the weights of the action. For large $k$, $d(k)$ is given by a Hilbert polynomial which has degree exactly $n$ (since $\Lambda$ is ample), while $w(k)$ is given by a polynomial of degree at most $n + 1$. Thus $d(k)/kw(k)$ has an expansion, for large $k$:

$$\frac{d(k)}{kw(k)} = F_0 + k^{-1} F_1 + \dots,$$

and the Futaki invariant is defined to be the co-efficient $F_1$. In our situation, we define the Futaki invariant $\mathrm{Fut}(\mathcal{X})$ of a test configuration of exponent $m$ to be $m^{-1}$ times the Futaki invariant of the central fibre, with line bundle $L_0$ and the induced $\mathbf{C}^*$-action.

With these definitions in place we can state the main definition of this section.

Definition 1. *A polarised variety $(X, L)$ is K-semistable if for any test configuration $\mathcal{X}$ we have* $\mathrm{Fut}(\mathcal{X}) \geq 0$. *It is K-stable if equality holds only when $\mathcal{X}$ is a product $X \times \mathbf{C}$.*

Note that in the last clause we allow the $\mathbf{C}^*$-action on $X \times \mathbf{C}$ to be induced from a non-trivial action on $X$. What we have called K-stability is often called K-polystability in the literature. The precise statement of the result mentioned in the previous section, verifying Yau's conjecture, is

Theorem 1. *A Fano manifold $X$ admits a Kahler-Einstein metric if and only if $(X, K_X^{-1})$ is K-stable.*

Here the "only if" is usually regarded as the easier direction and is due to Berman [**3**], following related results of various authors The uniqueness of the metric, modulo holomorphic automorphisms, is a relatively old result of Bando and Mabuchi [**2**]. We will not say more about these results here but focus on the "if" direction.

To give some background to the technical aspects of the proofs sketched in Section 4 below we will now try to explain why Theorem 1 is plausible. First we go back to the definition of the Futaki invariant of $(Z, \Lambda)$ in the case when $Z$ is a manifold, which was in fact the original context for Futaki's definition [**31**]. Choose a Kähler metric $\omega$ on $Z$ in the class $c_1(\Lambda)$ preserved by the action of $S^1 \subset \mathbf{C}^*$. Viewing $\omega$ as a symplectic structure, this action is generated by a Hamiltonian function $H$ on $Z$. Then the Futaki invariant can be given by a differential geometric formula

$$\int_Z (R - \hat{R}) H \frac{\omega^n}{n!}, \tag{3}$$

where $R$ is the scalar curvature of $\omega$ and $\hat{R}$ is the average value of $R$ over $Z$. This formula can be derived from the equivariant Riemann-Roch theorem and can also be understood in terms of the asymptotic geometry of sections of $L^k$ as $k \to \infty$, in the vein of quasi-classical asymptotics in quantisation theory. What this formula shows immediately is that if $\omega$ can be chosen to have constant scalar curvature—in particular if it is a Kähler-Einstein metric—then the Futaki invariant vanishes. This given another way, different from the Matsushima theorem, of ruling out Kähler-Einstein metrics on 1 or 2 point blow-ups of $\mathbf{CP}^2$. The definition of K-stability employs the Futaki invariant in a more subtle way; it is not just the automorphisms of $X$ which need to be considered but of the degenerations. The *Mabuchi functional* gives a way to understand this phenomenon. This is a functional $\mathcal{F}$ on the space $\mathcal{H}$ of Kähler metrics in a given cohomology class on a manifold $X$ defined via its first variation

$$\delta\mathcal{F} = \int_X (R - \hat{R}) \delta\phi \frac{\omega_\phi^n}{n!}. \tag{4}$$

Here $\delta\phi$ is an infinitesimal variation in the Kähler potential and one shows that such a functional $\mathcal{F}$ is well-defined, up to the addition of an arbitrary constant. By construction a critical point of $\mathcal{F}$ is exactly a constant scalar curvature metrics which, in the setting of Theorem 1 can be shown to be Kähler-Einstein. (We mention here that there is another functional, the *Ding functional* which has many similar properties to the Mabuchi functional and plays an important part in many developments. This is only defined for manifolds polarised by $K^{\pm 1}$.)

There are three possibilities:

- $\mathcal{F}$ is bounded below on $\mathcal{H}$ and attains its infimum;
- $\mathcal{F}$ is bounded below but does not attain its infimum;
- $\mathcal{F}$ is not bounded below.

An extension of Theorem 1 is the statement that these three possibilities correspond to $X$ being respectively $K$-stable, $K$-semistable (but not $K$-stable) and not $K$-semistable.

Now suppose that $\mathcal{X}$ is a test configuration for $(X, K_K^{-1})$ and, for simplicity, that the total space is smooth. Choose a Kähler metric on this total space, invariant under $S^1 \subset \mathbf{C}^*$. Pulling back by the $\mathbf{C}^*$-action the restrictions of this metric to the fibres $X_t$ for non-zero $t$ can be regarded as a family of metrics $\omega(t)$ on the fixed manifold $X$ parametrised by $t \in \mathbf{C}^*$ but these metrics have no limit, among metrics on $X$, as $t \to 0$. It is natural to think of this limit as a "point at infinity" in the space $\mathcal{H}$ of Kähler metrics on $X$. As we discuss further in Section 4.4 below, the role of the Futaki invariant is to determine the asymptotic behaviour as $t \to 0$ of the Mabuchi functional in such families obtained from test configurations. Theorem 1 can be understood roughly as saying that if there is no minimum of $\mathcal{F}$ this can be detected by studying the asymptotics at points at infinity of this kind (derived from algebro-geometric data).

Apart from its intrinsic interest—in giving an algebro-geometric criterion for the solubility of a PDE—Theorem 1 result also has implications for the construction of compactified moduli spaces of Fano manifolds, see work of Odaka [**41**], Spotti, Sun and Yao [**44**] and Wang and Zu [**49**]. For moduli questions, the notion of K-stability is also relevant in the negative case of varieties with ample canonical bundle. In this case Odaka [**40**] showed that a variety is K-stable if and only if has semi log canonical singularities, which is equivalent to stability in the sense of Alexeev, Kóllar and Shepherd-Barron. Berman and Guenancia showed that for such varieties this is equivalent to the existence of a Kähler-Einstein metric [**8**] (with a suitable definition in the singular case). Compactifying moduli spaces of manifolds involves adding points corresponding to singular varieties and it is interesting to relate the behaviour of the Kähler-Einstein metrics to the algebraic geometry of the singularities. There is much recent progress in this direction, see [**32**] for example.

## 3. Riemannian convergence theory and projective embeddings

In this section we will discuss some ideas which play an important role in three of the four proofs considered in Section 4 below. The general context can be explained as follows. In solving a PDE problem *compactness*—the ability to take limits in some kind of approximating scheme—is usually crucial. On the other hand in our problem we need to exhibit the obstruction to solving the problem (the existence of a Kähler-Einstein metric) as an algebro-geometric object (a test configuration with non-positive Futaki invariant). In the framework of Ricci curvature in Riemannian geometry there is a well-developed convergence theory of *Gromov-Hausdorff limits*; thus, in the Kähler situation, we would like to relate these limits to algebraic geometry and that is the topic of this section.

We begin by recalling some of the main results from the Riemannian theory of manifolds with a lower bound on the Ricci curvature. The foundation of the theory is the link between the Ricci curvature and volume expressed by the Bishop comparison theorem. For simplicity we just consider the case of an $m$-dimensional Riemannian manifold $M$ with Ricci $\geq 0$. Then Bishop's theorem states that for each $p \in M$ the volume ratio

$$v_p(r) = \frac{\mathrm{Vol}(B_p(r))}{\Omega_m r^m}, \tag{5}$$

is a weakly decreasing function of $r$. Here $B_p(r) \subset M$ is the metric ball of radius $r$, and we introduce the normalising constant $\Omega_m$—the volume of the unit ball in $\mathbf{R}^m$—so that $v_p(r)$ tends to 1 as $r$ tends to 0. If $M$ is compact with total volume $V$ and diameter $\leq D$ it follows that

$$\text{Vol}(B_p(r)) \geq \kappa r^m, \tag{6}$$

with $\kappa = V/D^m$. Recall that the *Gromov-Hausdorff distance* between two compact metric spaces $A, B$ is defined as the infimum of the numbers $\delta$ such that there is a metric on the disjoint union $A \sqcup B$ which extends the given metrics on $A, B$ and such that both $A, B$ are $\delta$-dense in $A \sqcup B$. If $(M_i, g_i)$ is a sequence of compact Riemannian $m$-manifolds with Ricci $\geq 0$, $\text{Vol}(M_i) = V$, $\text{diam}(M_i) \leq D$ then Gromov's compactness theorem asserts that there is a subsequence which converges in the sense of this Gromov-Hausdorff distance to some limiting metric space $(Z, d_Z)$. (More generally, the same result applies if we have any fixed lower bound on the Ricci curvatures.) The proof is an elementary argument based on ball-packing considerations and the lower bound (6). It is sometimes convenient to express this Gromov-Hausdorff convergence in terms of a natural topology on the disjoint union

$$\mathcal{M} = Z \cup \bigsqcup_i M_i.$$

Thus for $q \in Z$ it makes sense to talk about points $p_i \in M_i$ which are close to $q$.

Results of Anderson [**1**], Cheeger-Colding [**13**] and Cheeger-Colding-Tian [**15**] give finer information about such "non-collapsed" Gromov-Hausdorff limits. (Here the non-collapsing refers to the volume lower bound, which rules out the collapse of the sequence of $m$-dimensional manifolds to some lower dimensional space.) The notion of Gromov-Hausdorff convergence can be extended to sequences of spaces with base points: the metric balls of any fixed radius centred at the base points are required to converge as above. For each point $q \in Z$ and sequence of real numbers $\lambda_j \to \infty$ we consider the sequence of based metric spaces $(q, Z, \lambda_i d_Z)$. After perhaps passing to a subsequence we have a based Gromov-Hausdorff limit which is a metric cone, a *tangent cone* of $Z$ at $q$. The regular set $R \subset Z$ is defined to be the subset where some tangent cone is $\mathbf{R}^m$ and the complement $Z \setminus R$ is the singular set $S$. If the manifolds $M_i$ satisfy a fixed bound on the Ricci curvature $|\text{Ricci}| \leq \Lambda$ then more can be said. Anderson showed that there are fixed $\delta_m, \kappa_m$, depending only the dimension, such that if $p \in M_i$ and $r \leq \Lambda^{-1/2}$ then if the if the volume ratio $v_p(r)$ is greater than $1 - \delta_m$ there are harmonic co-ordinates on the sub-ball $B_p(\kappa r)$ in which the metric satisfies $C^{1,\nu}$ estimates. Together with results of Cheeger and Colding this shows that the regular set is open in $Z$ and the Riemannian metrics converge to a $C^{1,\nu}$-Riemannian metric $g_\infty$ on $R$. (This means that if $U$ is a pre-compact open set in $R$ there are $C^{2,\nu}$ diffeomorphisms $\chi_i : U \to M_i$ which, regarded as maps into $\mathcal{M}$, converge to the inclusion $U \to Z$ and such that the $\chi_i^*(g_i)$

converge in $C^{1,\nu}$ to $g_\infty$.) The singular set $S$ has Hausdorff codimension at least 4. (In the general Riemannian context, this is a recent result of Cheeger and Naber [**14**], but in the Kähler case which will be our concern it goes back to Cheeger, Colding and Tian [**15**].) The corresponding statements apply to tangent cones: each has a smooth Ricci-flat metric outside a closed singular set of codimension at least 4.

We want to relate these ideas to algebraic geometry and in this section we will focus on the case considered in [**30**]. Thus we suppose that $(X_i, \omega_i)$ are Kähler manifolds of complex dimension $n$ with fixed volumes $V$, diameters $\leq D$ and $|\text{Ricci}| \leq \Lambda$ and with a Gromov-Hausdorff limit $Z$, as above. We suppose that these are polarised manifolds, so that $\omega_i$ is the curvature of a Hermitian holomorphic line bundle $L_i \to X_i$. The main result is that $Z$ can be endowed with the structure of an algebraic variety. More precisely, we allow passage to a subsequence and form the disjoint union $\mathcal{M}$ as above. Then there is a continuous map $I : \mathcal{M} \to \mathbf{CP}^N$ with the two properties.

- There is some fixed $k$ such that for sufficiently large $i$, the restriction of $I$ to $X_i$ is an embedding defined by the holomorphic sections of $L_i^k \to X_i$;
- The restriction of $I$ to $Z$ is a homeomorphism to its image, which is a normal projective variety in $\mathbf{CP}^N$.

This result can be seen as an extension of the Kodaira embedding theorem to singular limit spaces and the proof extends some the ideas in one approach to the Kodaira theorem (an approach which seems to be well-known to experts but does not feature in standard textbooks). Suppose that $L \to X$ is a holomorphic line bundle over a compact complex manifold and $\sigma_0$ is a holomorphic section of $L$ over an open subset $V \subset X$. Let $\beta$ be a cut-off function with compact support in $V$, extended by 0 over $X$ and with $\beta = 1$ on some interior region $V_0 \subset V$. Then we can regard $\sigma = \beta\sigma_0$ as a smooth section of $L$ over $X$ in an obvious way. This will not be a holomorphic section because we have a term coming from the cut-off function

$$\overline{\partial}\sigma = (\overline{\partial}\beta)\sigma_0,$$

but if we have Hermitian metrics on $L$ and $X$ we can project $\sigma$ to the space of holomorphic sections using the $L^2$ inner product on sections of $L$, arriving at a holomorphic section $s$. Of course in this generality the construction need not be useful—the projection $s$ could be 0. The idea is that under suitable hypotheses we can arrange that $s$ is not zero and is very close to the original section $\sigma_0$ over $V_0$. The key is to estimate the error term $\eta = \sigma - s$, which is given by a Hodge Theory formula

$$\eta = \overline{\partial}^* \Delta^{-1} \overline{\partial}\sigma,$$

where $\Delta$ is the $\overline{\partial}$-Laplacian on the $L$-valued $(0,1)$ forms. Of course this formula only makes sense if $\Delta$ is invertible, i.e. if the cohomology group

$H^1(X; L)$ is zero. Then we have

$$\|\eta\|_{L^2}^2 = \langle \overline{\partial}^* \Delta^{-1} \overline{\partial}\sigma, \overline{\partial}^* \Delta^{-1} \overline{\partial}\sigma \rangle = \langle \overline{\partial}\overline{\partial}^* \Delta^{-1} \overline{\partial}\sigma, \Delta^{-1} \overline{\partial}\sigma \rangle = \langle \overline{\partial}\sigma, \Delta^{-1} \overline{\partial}\sigma \rangle,$$

and so

$$\|\eta\|_{L^2}^2 \leq \|\Delta^{-1}\| \, \|\overline{\partial}\sigma\|^2,$$

where $\|\Delta^{-1}\|$ is the $L^2$-operator norm. Now suppose that $L$ is a positive line bundle and the metric on $X$ is the Kähler metric $\omega$ given by the curvature of $L$. There is then a formula of Weitzenböck type

$$\Delta = P^* P + \mathrm{Ric} + 1, \tag{7}$$

where $P$ is the $(0,1)$-component of the covariant derivative on $L$-valued $(0,1)$ forms. But all we use is that $P^*P$ is a non-negative operator, so if $\mathrm{Ric} \geq -1/2$ (say) then $\Delta \geq 1/2$ and $\Delta^{-1}$ is defined and has $L^2$-operator norm at most 2. So we get

$$\|\eta\|_{L^2} \leq \sqrt{2} \|\overline{\partial}\sigma\|_{L^2}.$$

So if we can arrange that $\overline{\partial}\sigma$ is small, in $L^2$ norm compared with $\sigma$, then we get a non-trivial holomorphic section $s$. By construction, $\eta$ is holomorphic over $V_0$ and the $L^2$ norm of $\eta$ controls all derivatives there, by the usual elliptic estimates, and we can hope to show that $s$ is a small perturbation of $\sigma_0$ in any $C^r$ norm. The whole approach extends easily to a case when the original section $\sigma_0$ is not exactly holomorphic but approximately so, measured in terms of bounds on the $L^2$ norm of $\overline{\partial}\sigma_0$.

To prove the usual Kodaira embedding theorem in this way we first make the simple observation that changing $L$ to $L^k$ for some $k > 1$ corresponds to rescaling the metric by a factor $k$—i.e. scaling lengths of paths by a factor $\sqrt{k}$. Under this rescaling the Ricci curvature is multiplied by $k^{-1}$ so if we start with any positive line bundle and take a suitable power we can arrange that the condition $\mathrm{Ric} \geq -1/2$ holds after rescaling. We take $U \subset X$ to be a small co-ordinate ball centred on a point $p \in X$. The flat model is given by the trivial line bundle $\Lambda$ over $\mathbf{C}^n$ with a Hermitian structure corresponding to the Euclidean metric on $\mathbf{C}^n$. In this structure the trivialising section $\tau$ of $\Lambda$ has Gaussian norm

$$|\tau(z)| = e^{-|z|^2/4},$$

which decays very rapidly at infinity. After rescaling, the geometry of the manifold $X$ in the small ball $U$ is close to the flat model and we get an approximately holomorphic section $\sigma_0$ modelled on $\tau$. By suitable choice of parameters one arranges a cut-off function $\beta$ with the support of $\nabla\beta$ contained in the region where $\sigma_0$ is very small, so that $\|\overline{\partial}\sigma\|$ is also very small. Here the "suitable choice of parameters" involves choosing $k$ large. The upshot is that for a suitable $k$ and for each $p \in X$ we construct a holomorphic section of $L^k$ "peaked" around $p$ and in particular not vanishing at $p$. Thus the sections of $L^k$ define a holomorphic map of $X$ into a projective space and one can go further to show that this map is an embedding, once

$k$ is sufficiently large. (Note that the formula (7) is the same as that used in the proof of the Kodaira-Nakano vanishing theorem, which is invoked in the usual proof of the embedding theorem via blow-ups.)

Returning to our main discussion, we do not want to do analysis directly on the Gromov-Hausdorff limit $Z$ but instead establish uniform estimates on the converging sequence $X_i$. Consider any polarised manifold $(X, L)$ with Kähler metric given by the curvature of $L$. We endow $H^0(X, L)$ with the standard $L^2$ metric and for $x$ in $X$ we consider the evaluation map

$$\mathrm{ev}_x : H^0(X, L) \to L_x.$$

We define $\rho_L(x)$ to be square of the norm of this map so the statement that $\rho_L(x) > 0$ for all $x \in X$ is the same as saying that the sections of $L$ define a map $\tau : X \to \mathbf{P}(H^0(X, L)^*) = \mathbf{P}$. More generally, a lower bound on the $\rho_L(x)$ gives metric control of this map. The operator norm of

$$d\tau_x : TX_x \to T\mathbf{P}_{\tau(x)}$$

is $\rho_L(x)^{-1/2} \max(|(\nabla s)_x|)$, where the maximum is taken over holomorphic sections $s$ of $L$ with $L^2$ norm 1 vanishing at $x$. In our situation, with Ricci curvature and diameter bounds, there is a well-known upper bound on $|\nabla s|$, so a strictly positive lower bound on the $\rho_L(x)$ gives a Lipschitz bound on the map $\tau$. Replacing $L$ by $L^k$ we see that the crucial point is to find some $k$ and $b > 0$ so that for all $i$ and for all $x$ in $X_i$ we have

$$\rho_{L^k}(x) \geq b. \tag{8}$$

(Such a bound is sometimes referred to as a *partial $C^0$-estimate.*) It is straightforward to reduce to the case when the dimension of $H^0(X_i, L_i^k)$ is independent of $i$, so that we can regard $\tau_i : X_i \to \mathbf{P}$ as mapping into a fixed projective space. If the bound (8) holds then, after possibly taking a subsequence, we can pass to the Gromov-Hausdorff limit and define a continuous map $\tau_\infty : Z \to \mathbf{P}$ with image a projective variety. Further arguments then show that after perhaps increasing $k$ this map $\tau_\infty$ is an homeomorphism from $Z$ to a normal projective variety.

The central issue then is to establish the lower bound (8), as has been emphasised in many places by Tian. Let $q$ be a point in $Z$ and $C(Y)$ be some tangent cone to $Z$ at $q$. If $q$ is a smooth point then $C(Y)$ is $\mathbf{C}^n$ and we have the model Gaussian section as discussed above. In general we always have a Hermitian line bundle $\Lambda$ over the regular part of $C(Y)$ with a holomorphic section $\tau$ satisfying exactly the same Gaussian decay with respect to the distance to the vertex of the cone. We choose a suitable open subset $U$ of the regular part of the cone. It follows from the definitions that for large $i$ there are diffeomorphisms $\chi_i : U \to X_i$ which are approximately holomorphic isometries with respect to rescalings of the metrics $k\omega_i$ for some suitable large $k$. (Here the approximation can be made as close as we like by taking $k$ large.) Then are then two main technical points to address.

- We want to have lifts $\tilde{\chi}_i : \Lambda \to \chi_i^*(L^k)$ to approximate isomorphisms of Hermitian line bundles over $U$.
- We want a suitable cut-off function $\beta$ on $U$ with $|\overline{\partial}\beta\tau|$ small in $L^2$.

Given these, we can transport the section $\beta\tau$ to an approximately holomorphic section of $L_i^k$ over $\chi_i(U)$and follow the projection procedure to get a holomorphic section $s$ of $L_i$ modelled on $\tau$. The derivative bounds on $s$ give a lower bound on $|s|$ over all points in $X_i$ close to $q$ and an elementary covering argument establishes the bound (8).

The first technical point involves considerations of the holonomy of the connection on $\Lambda^* \otimes \chi_i^*(L_i^k)$, which has very small curvature by construction—this is straightforward if $U$ is simply connected. The second technical point involves the singular set in $C(Y)$, and in particular the fact that this has Hausdorff codimension strictly greater than 2 (see 4.1 below).

## 4. Four proofs

We now come to the core of this survey in which we discuss four different proofs of the equivalence between stability and Kähler-Einstein metrics. In total these proofs run to many hundreds of pages so it is impossible to give any kind of thorough account of them here. All we can do is to explain general strategies and some salient points in the arguments.

**4.1. The proof by cone singularities.** (Note that the announcement [**18**] contains an outline of this proof.) Given a Fano manifold $X$ we fix some suitable $m \geq 1$ and a smooth divisor $D$ in $|-mK_X|$. For $0 < \beta \leq 1$ we can define a class of Kähler metrics on $X$ with cone singularity of angle $2\pi\beta$ along $D$ and extend the whole theory to this case. (When $\beta = r^{-1}$ for an integer $r$ these are orbifold metrics, and hence well-established. There are also close analogies with the theory of parabolic structures and singular Hermitian Yang-Mills connections as developed in [**9**] for example.) We can define a modified Mabuchi functional

$$\delta\mathcal{F}_\beta = \int_X (R - \hat{R})\delta\phi + (1-\beta)\int_D \delta\phi - c\int_X \delta\phi \tag{9}$$

where $c = (mc_1(X)^n)^{-1}$ is the ratio of the volume of D to the volume of X, so that the right hand side vanishes when $\delta\phi$ is a constant. Roughly speaking, we have a family of functionals $\mathcal{F}_\beta$ with critical points the Kähler-Einstein metrics with cone angle $\beta$ and the strategy of the proof is to follow a family of such critical points as $\beta$ increases. We want to show that either the family continues up to $\beta = 1$, which gives our desired Kähler-Einstein metric on $X$, or that the critical point moves off to infinity and that this yields a test configuration violating the K-stability condition.

To begin we need to show that a solution exists for *some* $\beta$. Take $m > 1$ and $\beta = r^{-1}$ so that we are in the orbifold case. If $r > m/m - 1$ then characteristic class arguments show that the we are in the situation of negative

Ricci curvature and the desired solution follows from a straightforward orbifold extension of the standard Aubin-Yau theory. Next one shows that the set of $\beta$ for which a solution exists is *open*. This can be achieved using a suitable linear elliptic theory on manifolds with cone singularities **[28]**.

Suppose then that there is some $\beta_0 \in (0, 1]$ such that a solution $\omega_\beta$ exists for $\beta < \beta_0$ but that there is no solution for $\beta = \beta_0$. We need to extend the theory sketched in Section 3, for smooth metrics with bounded Ricci curvature, to Kähler-Einstein metrics with cone singularities. To begin we show that a metric with cone singularity can be approximated in the Gromov-Hausdorff sense by smooth metrics with Ricci curvature bounded below (**[19]**, Part I). Then the Cheeger-Colding theory implies that there is a subsequence $\beta_i$ increasing to $\beta_\infty$ and a Gromov-Hausdorff limit $Z$ of the $(X, \omega_{\beta_i})$ and $Z$ has metric tangent cones at each point. We call a tangent cone $C(Y)$ "good" if the regular set $C(Y_{\rm reg})$ is open, the metric is induced by a smooth Kähler metric there and for each compact subset $K \subset Y_{\rm reg}$ and each $\eta > 0$ there is a cut-off function $\gamma$ of compact support in $Y_{\rm reg}$, equal to 1 on $K$ and with

$$\|\nabla\gamma\|_{L^2} \leq \eta. \tag{10}$$

The main technical result is that all tangent cones to $Z$ are good. Given this, an extension of the arguments outlined in Section 3 above show that $Z$ is naturally a normal projective variety, carrying a singular Kähler-Einstein metric $\omega_\infty$. Moreover if we write $X_i$ for the metric space $(X, \omega_{\beta_i})$ and $D_i \subset X_i$ for the divisor $D$ then there is a divisor $\Delta \subset Z$ such that the pairs $(X_i, D_i)$ converge to $(Z, \Delta)$.

The new feature in this case, which leads to the difficulty in proving that tangent cones are "good", is the possibility of codimension 2 singular sets. This is the critical dimension with respect to the cut-off control (10). If $\psi$ is a compactly-supported function on $\mathbf{R}^m$ and if $\psi_\lambda$ is the rescaled function $\psi_\lambda(x) = \psi(\lambda^{-1}x)$ then

$$\|\nabla\psi_\lambda\|_{L^2} = O(\lambda^{(m-2)/2})$$

which tends to 0 with $\lambda$ if $m > 2$. This allows one to construct cut-off functions with derivative arbitrarily small in $L^2$ adapted to a compact set $A$ of Hausdorff codimension strictly greater than 2. In the codimension 2 case one needs appropriate control of the volume of the $\lambda$-neighbourhood $N_\lambda(A)$:

$$\mathrm{Vol}\ N_\lambda(A)) \leq C\lambda^2.$$

This is equivalent to the notion of *Minkoswki codimension* $\geq 2$ i.e. for any $r$ the set $A$ can be covered by $O(r^{-2})$ balls of radius $r$.

To complete the proof of the Theorem one wants to show that if indeed the family of solutions breaks down at some $\beta_\infty$ as considered above then there is test configuration with central fibre $Z$ and with non-positive Futaki invariant, so the original manifold $X$ is not K-stable. To this end one can extend the whole theory of stability and test configurations to pairs consisting of a variety and divisor, with a real parameter $\beta$. There is a modified Futaki invariant $\mathrm{Fut}_\beta$ which compares with the usual formula (3) (in the smooth case) just as the modified Mabuchi functional (9) compares with (4). One wants to construct a test configuration $(\mathcal{X}, \mathcal{D})$ for the pair $(X, D)$ with central fibre $(Z, \Delta)$ and show that

$$\mathrm{Fut}_{\beta_\infty}(\mathcal{X}, \mathcal{D}) = 0. \tag{11}$$

The Futaki invariant $\mathrm{Fut}_\beta$ depends linearly on $\beta$ so the fact that $(X, D)$ is stable for small $\beta$ implies that $\mathrm{Fut}(\mathcal{X}) = \mathrm{Fut}_1(\mathcal{X}, \mathcal{D}) \leq 0$.

Let $G$ be the automorphism group of the pair $(Z, \Delta)$—a complex Lie group. The existence of this test configuration $(\mathcal{X}, \mathcal{D})$ follows from general principles once it is established that $G$ is *reductive*; the complexification of a compact subgroup $K \subset G$. Using projective embeddings in some $\mathbf{CP}^N$, the pairs correspond to points $[X, D]$, $[Z, \Delta]$ in a suitable Hilbert scheme $\mathbf{S}$ which in turn is embedded in some large projective space $\mathbf{P}$. The group $PGL(N+1, \mathbf{C})$ acts on $\mathbf{S}$ and $\mathbf{P}$ and what we know from the convergence discussion above is that $[Z, \Delta]$ is in the closure of the orbit of $[X, D]$. The group $G$ can be identified with the stabiliser of $[Z, \Delta]$ in $PGL(N+1, \mathbf{C})$. If $G$ is reductive a version of the Luna slice theorem gives a slice for the action of $PGL(N+1, \mathbf{C})$ at $[Z, \Delta]$ (one takes a $G$-invariant complement to the action of $G$ on the tangent space of $\mathbf{P}$ at $[Z, \Delta]$). A well-known result of Hilbert and Mumford, applied to the $G$ action on this slice, shows that there is a 1-parameter subgroup $\mathbf{C}^* \subset G$ such that $[Z, \Delta]$ is in the closure of the $\mathbf{C}^*$-orbit of $[X, D]$ and this is equivalent to the desired test configuration.

The reductivity of $G$ is an extension of the Matsushima result to the case of pairs and singular varieties. Likewise the vanishing of the Futaki invariant (11) is an extension of the simple case considered in Section 2 above, for manifolds of constant scalar curvature. For the proofs one has to work with the singular metric $\omega_\infty$ using techniques from pluripotential theory.

**4.2. The proof by the continuity method.** In the continuity method one fixes a positive $(1,1)$ form $\alpha$ representing $c_1(X)$ and tries to solve the family of equations for $\omega_s$, with parameter $s \in [0, 1]$:

$$\mathrm{Ricci}(\omega_s) = (1 - s)\alpha + s\omega_s. \tag{12}$$

Yau's solution of the Calabi conjecture shows that there is a solution for $s = 0$ and it is well-known that the set of parameters for which a solution exists is open; if the solution can be continued up to $s = 1$ we have a Kähler Einstien metric, so the problem is to prove closedness (under the stability hypothesis). Thus we suppose that for some $S < 1$ there are solutions for $s < S$ but none for $s = S$. The set-up is very similar to that of cone singularities above, indeed the latter can be regarded as a variant of the continuity method, replacing the smooth form $\alpha$ with the current defined by a divisor.

If one knew that the fixed form $\alpha$ was bounded with respect to $\omega_s$ then the $\omega_s$ would have bounded Ricci curvature and the results discussed in Section 3 above would apply immediately to give a limiting metric on a normal projective variety. So the major difficulty is that we do not have such a bound. The Ricci curvature of the $\omega_s$ is positive so the fundamentals of the Cheeger-Colding theory apply and we obtain a Gromov-Hausdorff limit $Z$ (more precisely, a limit of some sequence $(X, \omega_{s_i})$ with $s_i$ increasing to $S$). But this theory does not ensure that the regular set in $Z$ is open, or give the good convergence properties over the regular set exploited in the argument of Section 3. This is one of the main problems overcome by Székelyhidi in [**46**].

To explain some of Székelyhidi's arguments, we restrict attention to the case when $S < 1$. Consider first a unit ball $B$ centred at a point $p$ in a Kähler manifold, with metric $\omega$, and a vector-valued holomorphic function $f : B \to \mathbf{C}^m$. Suppose that the Ricci curvature is bounded in $B$, say $|\mathrm{Ric}| \leq 4$, and that the pair $(\omega, f)$ satisfy the equation

$$\mathrm{Ric}(\omega) = \sigma\omega + f^*(\Omega_{\mathrm{Euc}}), \tag{13}$$

for some $\sigma \in [0, 1]$ where $\Omega_{\mathrm{Euc}}$ is the standard Euclidean Kähler form on $\mathbf{C}^m$. We claim that for any $\epsilon > 0$ we can find a $\delta$ (independent of $\sigma$) such that for any such $\omega$, if the volume ratio $v_p(1)$ exceeds $(1 - \delta)$ then in fact

$$|\mathrm{Ric}(p)| \leq \epsilon.$$

First, the Ricci curvature is non-negative so by the Bishop inequality we can pass to a smaller ball with centre $p$ (rescaled) and preserve the volume bound. By the results of Anderson we may as well assume that the metric on $B$ is $C^{1,\nu}$-close to the Euclidean metric in harmonic co-ordinates. By a suitable version of the Newlander-Nirenberg integrability theorem we can also suppose that these co-ordinates are actually holomorphic. The bound on the Ricci curvature and the equation (13) means that $|\nabla f|^2 \leq 2n$ and since $f$ is holomorphic we get interior bounds on all higher derivatives of $f$ and hence on the Ricci tensor, in these holomorphic co-ordinates. Thus if $|\mathrm{Ricci}(p)| > \epsilon$ we will have $|\mathrm{Ricci}| > \epsilon/2$ over a ball of definite size centred at $p$. The $C^{1,\nu}$ bound on the metric tensor gives $C^{,\nu}$ bounds on the Christoffel symbols. From this Székelyhidi shows that there is some unit tangent vector $v$ at $p$ such that, in geodesic polar-coordinates centred at $p$, there is a definite lower bound on $\mathrm{Ric}(\frac{\partial}{\partial r}, \frac{\partial}{\partial r})$ at all points close to $p$ and along geodesics starting from $p$ at a sufficiently small angle to $v$. Then the proof of the Bishop inequality shows that this Ricci curvature *reduces* the volume of $B$ by a definite amount (compared with the Euclidean ball) determined by $\epsilon$. So $\mathrm{Vol}(B) \leq (1 - \delta(\epsilon))\Omega_{2n}$, say. Choosing $\delta = \delta(\epsilon)$ we get a contradiction to the hypothesis that $|\mathrm{Ric}(p)| \geq \epsilon$ and the claim is established.

Clearly the result extends (with a suitable $\delta(\epsilon)$) to the case when $\Omega_{\mathrm{Euc}}$ is replaced by any smooth positive $(1,1)$-form $A$ defined over a suitable neighbourhood of the image of $f$. In the case at hand one can cover $X$ by a finite number of holomorphic co-ordinate charts. Working near a given point in $X$ such a chart yields the holomorphic map $f$ above (with $m = n$) and we take the form $A$ corresponding to $(1-s)\alpha$ in this chart. If $s \leq S < 1$ we get a $\delta(\epsilon)$ such that the discussion above applies to any rescaling of a small ball in $(X, \omega_s)$.

Now let $q$ be a point in the regular set of the Gromov-Hausdorff limit $Z$ and let $p_i \in (X, \omega_{s_i})$ be a sequence converging to $q$ in the sense of the Gromov-Hausdorff convergence. Let $B_i$ be the unit ball obtained by rescaling a small ball (of fixed radius $\rho$, independent of $i$) about $p_i$. For any given $\delta > 0$ we can suppose that for all subballs $\tilde{B} \subset B_i$ (not necessarily centred at $p_i$) the "volume defect" of $\tilde{B}$ is less than $\delta$. We choose $\delta$ as above, for some $\epsilon < 1$. Now let

$$M = \max_{x \in B_i} \left( |\text{Ricci}(x)|\, d(x, \partial B_i)^2 \right).$$

A standard line of argument shows that in fact $M \leq 4$. For if not let $\tilde{p} \in B_i$ be a point where the maximum is attained and $\tilde{B}$ be the ball of radius $\tilde{d}/\sqrt{M}$ centred at $\tilde{p}$, where $\tilde{d}$ is the distance from $\tilde{p}$ to the boundary of $B_i$. Rescaling $\tilde{B}$ to unit size we get a ball to which the previous results apply. After rescaling those results give $|\text{Ricci}(\tilde{p}) \leq \epsilon d_0^{-2} M < d_0^{-2} M$ which is a contradiction to the choice of $\tilde{x}$. Thus $M \leq 4$ and in particular $|\text{Ricci}(p_i)| \leq 4\rho^{-2}$. The conclusion is that Székelyhidi is able to show that, when $S < 1$ the Ricci curvature is bounded near points in the regular set in $Z$. It follows that the regular set is open and carries a $C^{1,\nu}$ Kähler metric. Going further, he extends the discussion to tangent cones and shows that these are all "good" in the sense discussed in 4.1 above. Note that in this situation the singular set can have real codimension 2, different from the simpler situation considered in Section 3. (In fact results of C. Li [**35**] in the toric case suggest that the limit as $s \to S$ will develop cone singularities along a divisor.)

In [**46**], Székelyhidi established the partial $C^0$ estimate along the continuity method and used this to show that another notion of stability, introduced by S. Paul, implies the existence of a Kähler-Einstein metric. The corresponding result for $K$-stability was established in the subsequent paper [**21**] of Datar and Székelyhidi. The output from the convergence theory is that if the continuity method breaks down at $S \in (0,1]$ there is a limiting projective variety $Z$, a singular Kähler metric $\omega_S$ and a closed non-negative $(1,1)$ current $\alpha_\Psi$ on $Z$ satisfying the equation

$$\text{Ricci}(\omega_S) = (1-S)\omega_S + S\alpha_\Psi.$$

The possible presence of singularities means that this equation needs to be interpreted. The current $\alpha_\Psi$ is locally written as $i\partial\overline{\partial}\psi$ where $\psi$ is an $L^1$ purisubharmonic function. Globally, $\Psi$ is a singular Hermitian metric on the anticanonical bundle. Datar and Székelyhidi set up a theory for pairs consisting of a variety and a $(1,1)$-current, analogous to the theory for pairs (variety, divisor) discussed above. The new feature is that their space of pairs is infinite dimensional. They are able to carry through a similar strategy to that outlined in 4.1 above by approximating $(1,1)$-currents by those defined by divisors. The form $\alpha$ on $X$ can be taken to be the restriction of the Fubini-Study metric under an embedding $X \subset \mathbf{P}$, then there is a integral geometry formula

$$\alpha = \int_{\mathbf{P}^*} [H \cap X] d\mu(H),$$

where $\mathbf{P}^*$ is the dual projective space parametrising hyperplanes $H \subset \mathbf{P}$, $\mu$ is the standard measure on $\mathbf{P}^*$ and $[H \cap X]$ is the current of the divisor $H \cap X$ in $X$. Replacing the integral by a finite sum gives the approximation procedure which is the starting point for these arguments.

The results of Datar and Székelyhidi go further than the statement of Theorem 1 in two directions. First they prove an analogous result for solutions of the *Kähler-Ricci soliton* equation. Recall that this equation is

$$\text{Ricci}(\omega) - \omega = L_v \omega, \tag{14}$$

where $v$ is a holomorphic vector field and $L_v$ is the Lie derivative. Such metrics are the appropriate analogues of Kähler-Einstein metrics on Fano manifolds with non-vanishing Futaki invariant and represent fixed points of the Kähler-Ricci flow (modulo holomorphic diffeomorphisms), which we discuss further below.

In another direction, Datar and Székelyhidi's proof is compatible with group actions so they prove that to test K-stability of a Fano manifold $X$ it suffices to consider test configurations with an additional compatible action of $\text{Aut}(X)$. This is important because an outstanding defect of the general theory is that it is very hard to verify K-stability of a polarised variety. The problem becomes more tractable for manifolds with large symmetry groups. Toric manifolds, with a complex torus action having an dense orbit, can be described in terms of polytopes. Both Kähler metrics invariant under the action of the corresponding real torus and toric test configurations can be described by convex functions on the polytope and the stability condition is relatively explicit. However in the toric Fano case the existence problem for Kähler-Einstein metrics and Kähler-Ricci solitons was completely settled by Wang and Zhu in 2004 [**49**], and no interesting phenomena arise, from the point of view of stability. Ilten and Süss [**33**] consider $n$-dimensional varieties with an action of an $(n-1)$-dimensional complex torus and develop a combinatorial description of these. In this way they are able to produce new examples, of manifolds which are K-stable, and the theorem of Datar and Szekelyhidi gives corresponding explicit new results about the existence of Kähler-Einstein metrics. In a similar vein, Delcroix studied group compactifications [**24**]; that is he considered a manifold $X$ which contains a complex reductive Lie group $G$ as a dense open subset and such that both left and right translations on $G$ extend to $X$. These can be described by polytopes in the Lie algebra of a maximal compact real torus in $G$ and Delcroix extends the arguments of Wang and Zhu to find an explicit condition for the existence of a Kähler-Einstein metric. This work of Delcroix was essentially self-contained and did not invoke general existence results such as Theorem 1. A subsequent paper [**25**] later showed that his condition emerges from an analysis of equivariant degenerations of $X$ and extended the results to the larger class of spherical varieties, using the theorem of Datar and Székelyhidi.

**4.3. The proof by Ricci flow.** If $X$ is a Fano mainfold the relevant version of the Ricci flow is the evolution equation

$$\frac{\partial \omega_t}{\partial t} = \omega_t - \text{Ricci}(\omega_t), \tag{15}$$

for a one-parameter family of metrics $\omega_t$ in the class $c_1(X)$. This can be expressed in terms of the Kähler potential. For each $t$ there is a unique $h_t$ such that

$$\omega_t - \text{Ricci}(\omega_t) = i\partial\overline{\partial}h_t,$$

normalised so that $\max_X h_t = 0$ and we can write $\omega_t = \omega_0 + i\partial\overline{\partial}\phi_t$ where $\phi_t$ evolves by:

$$\frac{\partial \phi_t}{\partial t} = h_t. \tag{16}$$

It has been known for many years that this equation has a solution for all $t \in [0, \infty)$, starting with any initial condition. The main result of the paper of Chen, Sun and Wang [**17**] is that this flow converges as $t \to \infty$ to a "weak" Kahler-Ricci soliton metric $\omega_\infty$ on a normal projective variety $X_\infty$ (in fact a "Q-Fano variety"). That is to say there is an algebraic torus action on $X_\infty$ an element $\xi$ in the Lie algebra of this torus and a metric on the regular part of $X_\infty$ (locally, in $X_\infty$, given by bounded potential) and which satisfies the equation (12) with respect to the holomorphic vector field generated by $\xi$.

In the case when the limiting metric is weak Kähler-Einstein but $X_\infty$ is not isomorphic to $X$ the same arguments as in 4.1 above, using the reductivity of the automorphism group, show that there is a test configuration for $X$ with central fibre $X_\infty$ and Futaki invariant zero. When the limit is a genuine Kahler-Ricci soliton the statement of [**17**] is slightly more subtle. They show that there is a destabilising test configuration for $X$, with central fibre $\overline{X}$ and strictly negative Futaki invariant, and a further degeneration (which might be trivial) of $\overline{X}$ to $X_\infty$. The upshot then is a trichotomy:

- $X$ is K-stable: the Ricci flow converges to a Kahler-Einstein metric on $X$;
- $X$ is K-semistable but not $K$-stable, the limit is Kähler-Einstein (possibly singular) but $X_\infty$ is not isomorphic to $X$;
- $X$ is not K-semistable: the limit is a genuine Ricci soliton and $X_\infty$ is not isomorphic to $X$.

The foundation for these results is the subsequential convergence of the flow established by Chen and Wang in the earlier paper [**20**]. That is, they prove that for any sequence $t_i \to \infty$ there is a subsequence $i'$ such that $(X, \omega_{t_{i'}})$ converges (in the same sense as in the previous subsections) to a weak Kähler-Ricci soliton metric on a Q-Fano variety. A fundamental difficulty in proving this is that it is not known that the Ricci curvature is bounded, either above or below, along the flow. This prevents a direct application of the Cheeger-Colding convergence theory. Results from the previous literature on Kähler-Ricci flow, including those of Perelman elaborated and extended by Sesum and Tian [**43**], yield three important pieces of information.

(1) The scalar curvature $R_t$ is bounded along the flow: $|R_t| \leq C_1$
(2) The potential $h_t$ is bounded along the flow: $|h_t| \leq C_2$
(3) There is a uniform Sobolev inequality:

$$\|f\|_{L^{2n/n-1}} \leq C_3 \left(\|\nabla f\|_{L^2} + \|f\|_{L^2}\right).$$

The general strategy of Chen and Wang is to establish a compactness theorem for segments of the flow over a fixed time interval, say $(T - 1 \leq t \leq T + 1)$. That is, they show that if $T_i$ is a sequence tending to infinity then, after passing to a subsequence, these segments of the flow converge (to a possibly singular limit). They show that the limit is "stationary", in that it is the solution of the Kähler-Ricci flow given by a Kähler-Ricci soliton, evolving by the action of holomorphic automorphisms. The proofs of Chen and Wang are based on a blow-up argument and the comparison with suitable "canonical neighbourhoods". For $\kappa > 0$, Chen and Wang define a class $\mathcal{KS}(n, \kappa)$ of non-compact length spaces. A space $W$ in $\mathcal{KS}(n, \kappa)$ is a smooth Ricci-flat Kähler $n$-manifold outside a closed singular set of codimension $> 3$, and with asymptotic volume ratio $\geq \kappa$, that is

$$\mathrm{liminf}_{r \to \infty} v_p(r) \geq \kappa.$$

In their application $\kappa = \kappa(C_3)$ is determined by the constant $C_3$ in the Sobolev bound. They prove from their definition that spaces in $\mathcal{KS}(n, \kappa)$ have many of the properties of the limit spaces treated by the Cheeger-Colding theory. In particular they adapt that theory to show the existence of metric tangent cones, They also establish a compactness property of $\mathcal{KS}(n, \kappa)$, with respect to based convergence over bounded sets. This compactness means that spaces in $\mathcal{KS}(n, \kappa)$ satisfy certain uniform estimates.

Chen and Wang's blow-up argument is governed by a canonical radius $cr(p) \in (0, \infty]$ which they define for any point $p$ in a Riemannian manifold $M$. This notion is in the same order of ideas as others in the literature such as the harmonic radius and curvature scale, but Chen and Wang's definition is tailored to the particular case at hand. The general idea is that $cr(p) \geq r$ if the $r$-ball centred at $p$, scaled to unit size, satisfies various definite estimates. The parameters in these estimates are chosen in line with the uniform estimates established in $\mathcal{KS}(n, \kappa)$, so that roughly speaking $cr(p) = \infty$ for a point in a space $W$ in $\mathcal{KS}(n, \kappa)$ (or more precisely $cr(p)$ is arbitrarily large for a point in a Riemannian manifold which is sufficiently close to some $W \in \mathcal{KS}(n, \kappa)$). Now Chen and Wang establish a lower bound $cr(p) \geq \epsilon > 0$ for any $p$ in a manifold $(X, \omega_t)$ along the Ricci flow. In outline, the argument is to suppose not, so there is a sequence of times $t_i$ and points $p_i \in (X, \omega_{t_i})$ such that $cr(p_i) \to 0$. Rescaling by $r_i^{-1}$ they arrive at a sequence of based manifolds $(p_i, M_i)$ with $cr$ bounded below by 1 and with $cr(p_i) = 1$. They show that these manifolds converge to a space in $\mathcal{KS}(n, \kappa)$ and derive a contradiction from the fact that $cr = \infty$ on $\mathcal{KS}(n, \kappa)$.

In this argument the bound $|R| \leq C_1$ on the scalar curvature enters in the following way. Under the Ricci flow the scalar curvature $R$ evolves by

$$\frac{\partial R}{\partial t} = \Delta R + |\mathrm{Ric}|^2 - n$$

After rescaling a portion of the Ricci flow, by a large factor $r_i^{-1}$ in the space direction and by $r_i^{-1/2}$ in the time direction, the scalar curvature $R'$ satisfies

$$\frac{\partial R'}{\partial t} = \Delta R' + |\mathrm{Ric}'|^2 - n r_i^2$$

and $|R'| \leq C_1 r_i^2$. Thus on any region where the rescaled flows, with a sequence of scalings $r_i^{-1} \to \infty$, converge in $C^\infty$ the limit must be a stationary Ricci-flat manifold. This is the fundamental mechanism which leads to the singular Ricci-flat spaces in $\mathcal{KS}(n, \kappa)$. The parameter $\kappa = \kappa(C_3)$ is determined by a standard relation between the Sobolev constant and volume ratio. (If a space has small asymptotic volume ratio one can write down a compactly supported function $f$ with $\|\nabla f\|_{L^2}$ small compared with $\|f\|_{L^{2n/n-1}}$. If the volume ratio is less than $\kappa(C_3)$, such a space cannot arise as a blow-up limit of manifolds with Sobolev bound $C_3$.)

The $L^2$ construction of holomorphic sections features in two ways in Chen and Wang's arguments. One is global, to produce a projective embedding of the limit space. The tangent cone information from the blow-up limit in $\mathcal{KS}(n, \kappa)$ is transferred to the manifolds in the limiting sequence and the techniques outlined in Section 3 apply. The other is local, to produce local holomorphic co-ordinates as ratios of suitable holomorphic sections. The bound on the potential $|h_t| \leq C_2$ is important here. First, since there is no lower bound on the Ricci curvature the argument based on the formula (7) does not immediately apply. Changing the metric on the line bundle by a factor $e^{h_t}$ introduces an extra term which precisely cancels the Ricci curvature contribution in (7) and the bound on the $h_t$ means that this change does not substantially effect the estimates. Second, the evolution equation (13) gives control of the change in the metric on the line bundle in time, and Chen and Wang are able to use this to obtain local holomorphic co-ordinates that are adapted to the metrics $\omega_t$ over definite time intervals. Alongside these complex geometry arguments they also use the Ricci flow techniques of Perelman.

The existence of a limit of the Kähler-Ricci flow, as established by Chen, Sun and Wang in [**17**] introduces further new ideas. The Chen and Wang result gives leaves open the possibility that different sequences of times $t_i \to \infty$ could lead to different limits. Let $\mathcal{C}$ be the set of all limits that arise.

By general principles this set if connected. One major step is to show that all $X_\infty$ in $\mathcal{C}$ can be embedded in projective space in such a way that the soliton vector fields are generated by the same fixed 1-parameter subgroup. This uses an algebro-geometric characterisation of the vector field of a Ricci soliton, via an generalisation of the Futaki invariant theory, which leads to a rigidity property.

**4.4. Proof by variational method.** The result proved by Berman, Boucksom and Jonsson in [**6**] involves a notion of "uniform K-stability". Let $\mathcal{X}$ be a test configuration for a polarised manifold $(X, L)$, so we have a $\mathbf{C}^*$-action on the central fibre $X_0$. Suppose first that $X_0$ is smooth and we fix an $S^1$-invariant Kähler metric in the class $c_1(\mathcal{L}_0)$ which yields a symplectic structure. The $S^1$-action is generated by a Hamiltonian function $H$ which we can normalise to have maximum value 0. Then we define $\|\mathcal{X}\|$ to be the $L^1$ norm of $H$. This is a quantity which is independent of the choice of metric in the cohomology class. The definition can be extended to any scheme, using the asymptotics of the trace of the action on sections of $\mathcal{L}^k$ as $k \to \infty$, similar to the definition of the Futaki invariant. Then $(X, L)$ is said to be *uniformly K-stable* if there is some $\epsilon > 0$ such that

$$\text{Fut}(\mathcal{X}) \geq \epsilon \|\mathcal{X}\| \tag{17}$$

for all non-trivial test configurations $\mathcal{X}$. The main result of [**6**] is that for a Fano manifold $X$, polarised by $K_X^{-1}$ and with finite automorphism group, the existence of a Kähler-Einstein metric is equivalent to uniform $K$-stability. Note that *a priori* uniform $K$-stability is a stronger condition than $K$-stability although *a posteriori* they are equivalent, for Fano manifolds with finite automorphism group. One can consider many other norms on test configurations and the general notion of uniform stability goes back to the thesis of Szekelyhidi. It is also related to another variant of $K$-stability developed by Szekelyhidi which considers filtrations of the co-ordinate ring $\bigoplus H^0(X, L^k)$ [**45**]. A definition of uniform stability which turns out to be equivalent to that in [**6**] was given by Dervan [**26**].

To give some indication of the proof in [**6**] we begin by considering an analogous situation in finite dimensions. Let $V$ be a complete (finite-dimensional) Riemannian manifold with the property that the each two points can be joined by a unique geodesic segment—for example a simply connected manifold of non-positive curvature. Let $v_0 \in V$ be a base point and let $F$ be function on $V$ which is convex along geodesics. If $\gamma : [0, \infty) \to V$ is a geodesic ray emanating from $v_0$, parametrised by arc length, the ratio $\gamma(t)/t$ is increasing with $t$ and we can define the *asymptotic slope* $S_\gamma \in [0, \infty]$ to be the limit as $t \to \infty$. If $S_\gamma > 0$ then for any $\delta < S_\gamma$ there is a $C_\gamma$ such that

$$F(\gamma(s)) \geq \delta s - C_\gamma. \tag{18}$$

An elementary argument, hinging on the compactness of the set of geodesics through $v_0$, shows that if there is some $\delta > 0$ such that if $S_\gamma \geq \delta$ for all such rays $\gamma$ then given some $\delta' < \delta$ we can find $C$ such

$$F(v) \geq \delta' d(v, v_0) - C \tag{19}$$

for all $v \in V$. It also follows easily that $F$ attains a minimum in $V$.

The relevance of this to the Kähler-Einstein problem on a Fano manifold $X$ is that, as we have discussed in Section 2, a Kähler-Einstein metric can be seen as a critical point of the Mabuchi functional $\mathcal{F}$ on the space of Kähler metrics $\mathcal{H}$. The pair $(\mathcal{H}, \mathcal{F})$ has many properties analogous to $(V, F)$ above. There is a Mabuchi metric which makes $\mathcal{H}$ formally a symmetric space of non-positive curvature and $\mathcal{F}$ is convex along geodesics. The programme roughly speaking, is to extend the arguments above to this infinite-dimensional setting and to relate the asymptotics of the Mabuchi functional, analogous to the asymptotic slope $S_\gamma$, to the condition (17) on test configurations.

A point $\omega$ in $\mathcal{H}$ defines a volume form on $X$ and the tangent space to $\mathcal{H}$ at $\omega$ can be identified with the functions $\delta\phi$ of integral 0. For each $p \geq 1$ the $L^p$ norm defines a Finsler structure on $\mathcal{H}$. The case $p = 2$ gives the infinite dimensional Riemannian structure first considered by Mabuchi [**38**] but the case $p = 1$ is also important, as shown by Darvas [**22**]. The completion, in the metric defined by this Finsler structure, is a space $\mathcal{E}^1$ of currents defined by "finite energy" potentials. Geodesics in $\mathcal{H}$ also have a good geometric meaning. Smooth geodesics segments parametrised by an interval $[a, b]$ correspond to $S^1$-invariant closed $(1,1)$-forms $\Omega$ on the product $X \times S^1 \times [a, b]$ which satisfy $\Omega^{n+1} = 0$ and which restrict to a metric in $\mathcal{H}$ on each copy of $X$ in the product. (The different Finsler structures share the same geodesics.) If we stay in the smooth category it is not true that any two points in $\mathcal{H}$ can be joined by a geodesic [**34**] but this is the case if one relaxes the definitions to allow singularities—in fact forms with $C^{1,1}$ potentials—as shown by Chen [**16**]. It is an elementary calculation that the Mabuchi functional is convex along smooth geodesics. The convexity in the general case is a deep recent result of Berman and Berndtsson [**4**].

Now fix a base point $\omega_0$ in $\mathcal{H}$—analogous to $v_0 \in V$. According to Darvas, the distance to $\omega_0$ in the $L^1$-Finsler structure is equivalent to functional $J$ which is well-known in the literature and which can be characterised by the property that $J(\omega_0) = 0$ and

$$\delta J = \int_X \delta\phi(\omega^n - \omega_0^n).$$

Thus the analogue of (19) is an inequality.

$$\mathcal{F}(\omega) \geq \delta' J(\omega) - C \tag{20}$$

for some fixed $\delta' > 0, C$ and $\omega$. This is sometimes referred to as the "properness" of the Mabuchi functional. Tian showed [**47**] that if this inequality (20) holds then a Kähler-Einstein metric exists. The Mabuchi functional is decreasing along the continuity path, as discussed in 4.2 above, and so $\mathcal{F}(\omega_s)$ controls $J(\omega_s)$ and Tian showed that this allows the continuity path to be continued to $s = 1$. Darvas and Rubinstein have given another proof of this result, and generalisations, recently [**23**].

There is a large circle of results relating geodesic rays in $\mathcal{H}$ to test configurations, see [**42**] for example. Using the conformal equivalence between $S^1 \times [0, \infty)$ and the punctured disc $\Delta^*$, geodesic rays correspond to $S^1$-invariant closed $(1,1)$-forms $\Omega$ on the product $X \times \Delta^*$ with $\Omega^{n+1} = 0$ and which are positive on each $X \times \{t\}$. For certain purposes one can work with *subgeodesics* which correspond to positive definite forms $\Omega$. In particular if $\mathcal{X}$ is a test configuration we can consider a "smooth" metric $\Omega$ on $\mathcal{X}$. For example we can embed $\mathcal{X}$ in some $\Delta \times \mathbf{CP}^N$ and take the restriction of a metric on the ambient manifold. The $\mathbf{C}^*$-action on $\mathcal{X}$ gives an open embedding $X \times \Delta^*$ and we get a subgeodesic ray $\omega_s$ in $\mathcal{H}$. Boucksom, Hisamoto and Jonsson [**10**] prove that (if the central fibre is reduced)

$$J(\omega_s) \sim \|\mathcal{X}\| s \quad ; \quad \mathcal{F}(\omega_s) \sim \mathrm{Fut}(\mathcal{X}) s,$$

as $s \to \infty$. (There are many earlier results of this kind in the literature, under various hypotheses.) Thus the uniform stability condition is equivalent to the statement that there is a $\delta > 0$ such that for any such subgeodesic ray, arising from a test configuration, we have

$$\mathcal{F}(\omega_s) \geq \delta J(\omega_s) - C, \tag{21}$$

where $C$ depends on the ray.

There are now two main aspects to the proof. (The exposition in [**6**] involves some sophisticated techniques which go well beyond this writers knowledge, and indeed [**6**] is described by the authors as an outline to be followed by a more detailed version. What we write below is extremely sketchy.)

- To pass from the subgeodesic rays arising from test configurations to general geodesic rays and establish an inequality (21) along any geodesic ray.

  A sub-geodesic ray comes with a family of Kähler potentials which can be viewed as a metric on the pull-back of $L$ to $X \times \Delta^*$ or as a singular metric on the pull-back to $X \times \Delta$. This metric defines a multiplier ideal sheaf: the local holomorphic sections which are in $L^2$ with respect to the metric. Fundamental results of Nadel show that this is a coherent sheaf and so one can construct corresponding blow-ups of $X \times \Delta$ along the powers of this ideal sheaf, which yield test configurations. Bermann, Boucksom and Jonsson use these to approximate the original ray by those arising from test configurations and eventually to pass from the algebro-geometric uniform K-stability hypothesis to the estimate (18) on general geodesic rays. For these purposes they also work with the Ding functional (which we mentioned briefly in Section 2). They also use ideas and results from non-Archimedean geometry.
- To carry the elementary arguments from the finite dimensional model over to the infinite dimensional situation. Results of Berman,

Boucksom, Eyssidieux, Guedj and Zeriahi [**5**] are used here to give the relevant compactness property in $\mathcal{E}^1$ for geodesics segments with bounded Mabuchi functional.

These variational techniques based on convex geometry in the space $\mathcal{H}$ of Kähler metrics have been used by Darvas and Rubinstein [**23**] and Berman, Darvas and Lu [**7**] to produce interesting results in the more general framework of constant scalar curvature metrics. The outstanding problem is to prove the regularity of weak solutions produced by minimising the Mabuchi functional.

## References


[1] M. Anderson *Convergence and rigidity of manifolds under Ricci curvature bounds* Invent. Math. 102 (1990) 429–445.

[2] S. Bando and T. Mabuchi *Uniqueness of Einstein Kähler metrics modulo connected group actions* Adv. Stud. Pure Math. 10 North Holland (1987) 11–40.

[3] R. Berman *K-polystability of Q-Fano varieties admitting Kähler-Einstein metrics* Inventiones Math. 203 (2016) 973–1025.

[4] R. Berman and B. Berndtsson *Convexity of the K-energy on the space of Käler metrics and uniqueness of extremal metrics.*

[5] R. Berman, S. Boucksom, P. Eyssidieux, V. Guedj and A. Zeriahi *Kähler-Einstein metrics and the Kähler-Ricci flow on log Fano varieties* arxiv:1111.7158.

[6] R. Berman, S. Boucksom and M. Jonsson *A variational approach to the Yau-Tian-Donaldson conjecture* arxiv:1509.04561.

[7] R. Berman, T. Darvas and C. Lu *Regularity of weak minimisers of the K-energy and applications to properness and K-stability* arxiv:1602.03114.

[8] R. Berman and H. Guenancia *Kähler-Einstein metrics on stable varieties and log canonical pairs* Geom. Funct. Analysis 24 (2014) 1683–1730.

[9] O. Biquard *Sur les fibrés paraboliques sur une surface complexe* J. London Math. Soc. 53 (1996) 302–316.

[10] S. Boucksom, T. Hisamoto and M. Jonsson *Uniform K-stability, Duistermaat-Heckman measures and singularities of pairs* arxiv:1504.06568.

[11] E. Calabi *On Kähler manifolds with vanishing canonical class.* In: Algebraic geometry and topology A symposium in honour of S. Lefschetz Princeton U.P. 1957 pp. 78–89.

[12] E. Calabi *The space of Kähler metrics.* In: Proc. Int Congress Math, 1954 E.P. Nordhoff 1956 pp. 206–207.

[13] J. Cheeger and T. Colding *On the structure of spaces with Ricci curvature bounded below, I* Jour. Diff Geom. 46 (1997) 406–480.

[14] J. Cheeger and A. Naber *Regularity of Einstein manifolds and the codimension 4 conjecture* Annals of Math. 182 (2015) 1093–1165.

[15] J. Cheeger, T. Colding and G. Tian *On the singularities of spaces with bounded Ricci curvature* Geom. Funct. Analysis 12 (2002) 873–914.

[16] X. Chen *Space of Kähler metrics* Jour. Differential Geometry 56 (2000) 189–234.

[17] X. Chen, S. Sun and B. Wang *Kähler-Ricci flow, Kähler-Einstein metrics and K-stability* arxiv:1508.04397.

[18] X. Chen, S. Donaldson and S. Sun *Kähler-Einstein mercs and stability* Int. Math. Res. Notices (2014) 2119–2125.

[19] X. Chen, S. Donaldson and S. Sun 2012 *Kähler-Einstein metrics on Fano manifolds I–III* Jour. Amer. Math. Soc 28 (2015) 183–278.

[20] X. Chen and B. Wang *Space of Ricci flows, II* arxiv:1405.6797.

[21] V. Datar and G. Székelyhidi *Kähler-Einstein metrics along the smooth continuity method* Geom. Funct. Analysis 26 (2016) 975–1010.

[22] T. Darvas *The Mabuchi geometry of finite-energy classes* Advances in Math. 285 (2015) 182–219.

[23] T. Darvas and Y. Rubinstein *Tian's properness conjecture and Finsler geometry of the space of Kähler metrics* arxiv:1506.07129.

[24] T. Delcroix *Kähler-Einstein metrics on group compactifications* arxiv:1510.07384.

[25] T. Delcroix *K-stability of spherical Fano varieties* arxiv:1608.01852.

[26] R. Dervan *Uniform stability of twisted constant scalar curvature Kähler metrics* arxiv:1412.0648.

[27] S. Donaldson *Scalar curvature and stability of toric varieties* J. Differential Geometry 62 (2002) 289–349.

[28] S. Donaldson *Kähler metrics with cone singularities along a divisor.* In: Essays in mathematics and its applications, Springer 2012 pp. 49–79.

[29] S. Donaldson *Stability of algebraic varieties and Kähler geometry.* To appear in Proceedings of the AMS Summer School in Algebraic Geometry, Salt Lake City 2015.

[30] S. Donaldson and S. Sun *Gromov-Hausdorff limits of Kähler manifolds and algebraic geometry* Acta Math. 213 (2014) 63–106.

[31] A. Futaki *An obstruction to the existence of Einstein Kähler metrics* Invent. Math. 73 (1983) 437–443.

[32] H.-J. Hein and S. Sun *Calabi-Yau manifolds with isolated conical singularities* arxiv:1607.02940.

[33] N. Ilten and H. Süss *K-stability for Fano manifolds with torus action of complexity one* arxiv:1507.04442.

[34] L. Lempert and L. Vivas *Geodesics in the space of Kähler metrics* Duke Math. Jour. 162 (2013) 1369–1381.

[35] C. Li *Greatest lower bounds on Ricci curvature for toric Fano manifolds* Adv. Math. 226 (2011) 4921–4932.

[36] C. Li, X. Wang and C. Xu *Quasiprojectvity of the moduli space of Kähler-Einstein Fano manifolds* arxiv:1502.06532.

[37] T. Mabuchi *K-energy maps integrating Futaki invariants* Tohoku Math. J. 38 (1986) 575–593.

[38] T. Mabuchi *Some symplectic geometry on compact Kähler manifolds, I* Osaka J. Math. 24 (1987) 227–252.

[39] Y. Matsushima *Sur las structure du groupsd'homéomorphismes d'une certaine varieté Kählerienne* Nagoya Math. J. 11 (1957) 145–150.

[40] Y. Odaka *The Calabi conjecture and K-stability* Int. Math. Res. Notices (2012) 2272–2288.

[41] Y. Odaka *Compact moduli spaces of Kähler-Einstein Fano varieties* Publ. Res. Int. Math. Sci. 51 (2015) 549–565.

[42] D. Phong and J. Sturm *Test configurations for K-stability and geodesic rays* J. Symplectic Geom. 5 (2007) 221–247.

[43] N. Sesum and G. Tian *Bounding scalar curvature and diameter along the Kähler-Ricci flow (after Perelman) and some applications* J. Inst. Math. Jussieu 7 (2008) 757–587.

[44] C. Spotti, S. Sun and C. Yao *Existence and deformation of Kähler-Einstein metrics on smoothable Q-Fano varieties* Duke Math. Jour. 165 (2016) 3043–3083.

[45] G. Székelyhidi *Filtrations and test configurations* With an appendix by S. Boucksom Math. Annalen 362 (2015) 451–484.

[46] G. Székelyhidi *The partial* $C^0$ *estimate along the smooth continuity method* J. Amer. Math. Soc. 29 (2016) 537–560.

[47] G. Tian *Kähler-Einstein metrics with positive scalar curvature* Inventiones Math. 130 (1997) 239–265.

[48] K. Uhlenbeck and S.-T. Yau *On the existence of Hermitian Yang-Mills connections in stable vector bundles* Comm. Pure Appl. Math. 39 (1986) 257–293.
[49] X.-J. Wang and X. Zhu *Kähler-Ricci solitons on toric manifolds with positive first Chern class* Advances in Math 188 (2004) 87–103.
[50] S.-T. Yau *On the Ricci curvature of a compact Kähler manifold and the complex Monge-Ampère equation, I* Commun. Pure Appl. Math. 31 (1978) 339–411.

Imperial College, London; and Simons Centre for Geometry and Physics, Stony Brook, New York
*E-mail address*: s.donaldson@imperial.ac.uk

Current Developments in Mathematics, 2015

# Order/disorder phase transitions: the example of the Potts model

Hugo Duminil-Copin

ABSTRACT. The critical behavior at an order/disorder phase transition has been a central object of interest in statistical physics. In the past century, techniques borrowed from many different fields of mathematics (Algebra, Combinatorics, Probability, Complex Analysis, Spectral Theory, etc) have contributed to a more and more elaborated description of the possible critical behaviors for a large variety of models (interacting particle systems, lattice spin models, spin glasses, percolation models). Through the classical examples of the Ising and Potts models, we survey a few recent advances regarding the rigorous understanding of such phase transitions for the specific case of lattice spin models. This review was written at the occasion of the Harvard/MIT conference Current Developments in Mathematics 2015.

CONTENTS



## 1. Introduction

Lattice models have been introduced as discrete models for real life experiments and were later on found useful to model a large variety of phenomena and objects, ranging from ferroelectric materials to lattice gas. They also provide discretizations of Euclidean and Quantum Field Theories and are as such important from the point of view of theoretical physics. While the original motivation came from physics, they appeared as extremely complex and rich mathematical objects, whose study required the developments of important new tools that found applications in many other domains of mathematics.

The zoo of lattice models is very diverse: it includes models of spin-glasses, quantum chains, random surfaces, spin systems, percolation models. Here, we focus on a smaller class of lattice models called spin systems. These systems are random collections of spin variables assigned to the vertices of a lattice. The archetypical example of such a model is provided by the Ising model, for which spins take value $\pm 1$. This model was introduced by Lenz in 1920 to model the temperature, called Curie's temperature, above which a magnet looses its ferromagnetic properties.

This review describes a few aspects of the behavior of lattice spin models near their critical point. Emphasis is put on some probabilistic techniques developed to study these models.

**1.1. Definition of the models.** Let $G$ be a finite subgraph of the lattice $\mathbb{Z}^d$. The set of vertices and edges of $G$ are denoted respectively by $V(G)$ and $E(G)$. Let $\Sigma$ be a subset of $\mathbb{R}^\nu$, where $\nu \in \mathbb{N}^*$. A *spin* variable $\sigma_x \in \Sigma$ is attributed to each vertex $x \in V(G)$. A *spin configuration* $\sigma = (\sigma_x : x \in V(G)) \in \Sigma^{V(G)}$ is given by the collection of all the spins.

For a family $(J_{xy})_{\{x,y\}\subset\mathbb{Z}^d}$ of *coupling constants*, introduce the Hamiltonian of a spin configuration $\sigma$ defined by

$$H_G^{\text{free}}(\sigma) := -\sum_{\{x,y\}\subset V(G)} J_{xy}\,\langle\sigma_x|\sigma_y\rangle,$$

where $\langle\cdot|\cdot\rangle$ denotes the standard scalar product on $\mathbb{R}^\nu$. The Gibbs measure on $G$ at inverse temperature $\beta$ with free boundary conditions is defined by the formula

$$\mu_{G,\beta}^{\text{free}}[f] := \frac{\displaystyle\int_{\Sigma^{V(G)}} f(\sigma)\exp\big[-\beta H_G^{\text{free}}(\sigma)\big]d\sigma}{\displaystyle\int_{\Sigma^{V(G)}} \exp\big[-\beta H_G^{\text{free}}(\sigma)\big]d\sigma} \tag{1.1}$$

for every $f : \Sigma^{V(G)} \longrightarrow \mathbb{R}$, where

$$d\sigma = \bigotimes_{x\in V(G)} d\sigma_x$$

is a product measure whose marginals $d\sigma_x$ are identical copies of a reference finite measure $d\sigma_0$ on $\Sigma$. Note that if $\beta = 0$, then spins are chosen independently according to the probability measure $d\sigma_0/\int_\Sigma d\sigma_0(\sigma)$.

A priori, $\Sigma$ and $d\sigma_0$ can be chosen to be arbitrary, thus leading to different examples of lattice spin models. The following (far from exhaustive list) of spin models already illustrates the vast variety of possibilities that such a formalism offers:

**Ising model:** $\Sigma = \{-1, 1\}$ and $d\sigma_0$ is the counting measure on $\Sigma$. This model was introduced by Lenz in 1920 [**Len20**] to model

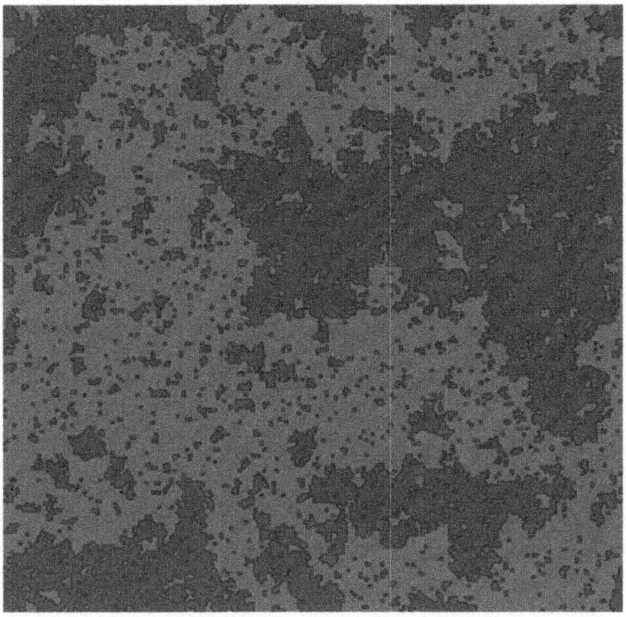

FIGURE 1. A simulation of the Ising model due to V. Beffara.

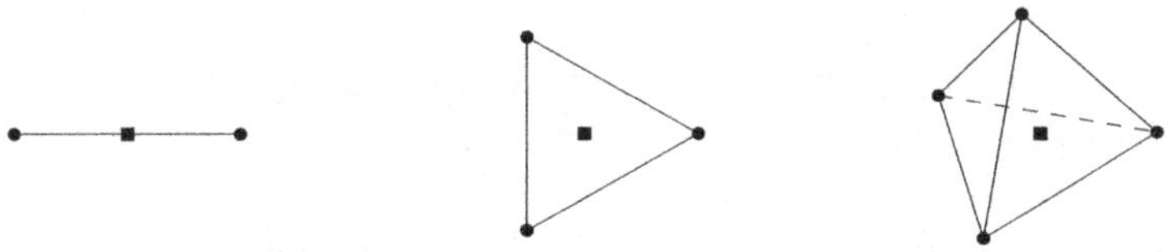

FIGURE 2. From left to right, $\mathbb{T}_2$, $\mathbb{T}_3$ and $\mathbb{T}_4$.

Curie's temperature for ferromagnets. It was studied in his PhD thesis by Ising [**Isi25**].

**Potts model:** $\Sigma = \mathbb{T}_q$ ($q \geq 2$ is an integer), where $\mathbb{T}_q$ is a simplex in $\mathbb{R}^{q-1}$ (see Fig. 2) containing $\vec{1} = (1, 0, \ldots, 0)$ such that

$$\langle \sigma_x | \sigma_y \rangle = \begin{cases} 1 & \text{if } \sigma_x = \sigma_y, \\ -\frac{1}{q-1} & \text{otherwise.} \end{cases}$$

and $d\sigma_0$ is the counting measure on $\Sigma$. This model was introduced by Potts [**Pot52**] following a suggestion of his adviser Domb. While the model received little attention early on, it became the object of great interest in the last forty years. Since then, mathematicians and physicists have been studying it intensively, and a lot is known on its rich behavior. It is now one of the foremost example of lattice models.

**Spin $O(n)$ model:** $\Sigma$ is the unit sphere in dimension $n$ and $d\sigma_0$ is the surface measure. This model was introduced by Stanley in [**Sta68**].

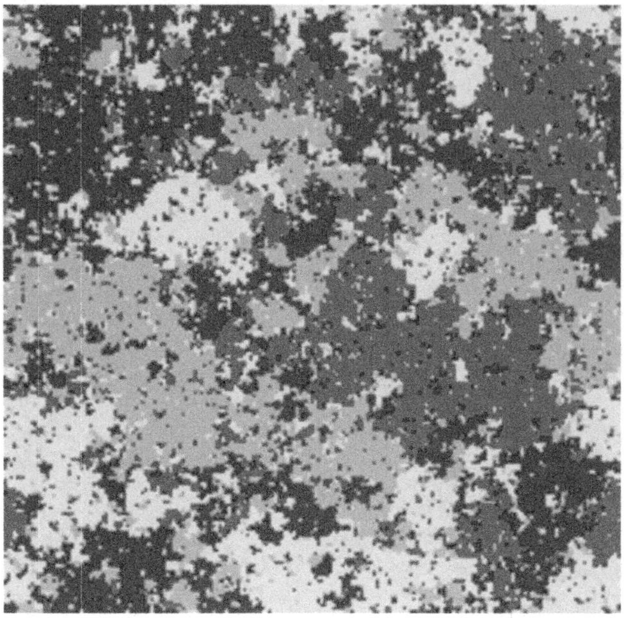

FIGURE 3. A simulation of the four-state Potts model due to V. Beffara.

This is yet another generalization of the Ising model (the case $n = 1$ corresponds to the Ising model). The $n = 2$ and $n = 3$ models were introduced slightly before the general case and are called the $XY$ and (classical) Heisenberg models respectively.

**Discrete Gaussian Free Field (GFF):** $\Sigma = \mathbb{R}$ and $d\sigma_0 = e^{-\sigma_0^2/2} d\lambda(\sigma_0)$, where $d\lambda$ is the Lebesgue measure on $\mathbb{R}$. The discrete GFF is a lattice version of the continuum GFF, sometimes called Euclidean bosonic massless GFF, which is the starting point of many physical theories (for instance Liouville Quantum Gravity, to mention an example which is studied here in Cambridge).

**The $\phi_d^4$ lattice model on $\mathbb{Z}^d$:** $\Sigma = \mathbb{R}$ and $d\sigma_0 = \exp(-a\sigma_0^2 - b\sigma_0^4) d\lambda(\sigma_0)$, where $a \in \mathbb{R}$ and $b \geq 0$. This model interpolates between the GFF corresponding to $a = 1/2$ and $b = 0$, and the Ising model corresponding to the limit as $b = -a/2$ tends to $+\infty$.

We will assume that the model has nearest neighbor ferromagnetic coupling constants, meaning that $J_{xy}$ is zero except if $x$ and $y$ are neighbors, in which case $J_{xy} = J \geq 0$ (notice that we may assume without loss of generality that $J = 1$, since multiplying $J$ by a constant corresponds to dividing $\beta$ by the same amount). The theories of non-ferromagnetic and non-nearest neighbors models (also called long-range models) are very interesting on their own right, nevertheless we prefer to reduce the scope of this lecture to the nearest neighbor ferromagnetic case to simplify the discussion. In the same spirit, and since the realm of continuous spin space is very different from the realm of discrete spin space, we choose to focus on the Potts model (which includes the Ising model as a special case).

**An important notation.** From now on, $\mu^{\text{free}}_{G,\beta,q}$ denotes the Gibbs measure for the nearest neighbor ferromagnetic $q$-state Potts model on $G$ with inverse temperature $\beta$ and free boundary conditions.

**1.2. Continuous/discontinuous transition between an ordered and a disordered phase.** Ferromagnetic models favor configurations in which spins point in the same direction. Furthermore, the parameter $\beta$ calibrates the strength of the interaction, or alternatively how much spins prefer to be aligned. For this reason, it does not come as a surprise that measures on very large graphs exhibit different behaviors at different inverse temperatures: for small $\beta$, the entropy should win on the energy, implying that spins should be roughly independent, while on the contrary for large values of $\beta$, the energy should be the most important factor, implying that spins should align in an ordered fashion.

Our goal is to study the appearance of this ordering, i.e. the appearance of a global alignment of the spins. In order to quantify this alignment, introduce the following random variable:

$$\mathscr{M}(G) := \frac{1}{|V(G)|} \sum_{x \in V(G)} \sigma_x.$$

In the case of the Potts model, the average of $\mathscr{M}(G)$ under the measure with free boundary conditions is always zero due to the symmetries of the spin space. In order to break the inherent symmetry, we introduce new *boundary conditions* as follows. Let

$$H^1_G(\sigma) := H^{\text{free}}_G(\sigma) - \sum_{x \in V(G), y \notin V(G)} J_{xy} \langle \sigma_x | \vec{1} \rangle$$

and $\mu^1_{G,\beta,q}$ defined as $\mu^{\text{free}}_{G,\beta,q}$ with $H^1_G$ replacing $H^{\text{free}}_G$ (recall that $\vec{1}$ is the vector in $\Sigma \subset \mathbb{R}^\nu$ with first coordinate 1 and other coordinates 0).

We may now speak of the average magnetization defined by

$$m(G, \beta, q) := \mu^1_{G,\beta,q}[\mathscr{M}(G)]$$

and of its limit

$$m^*(\beta, q) := \lim_{n \to \infty} m(\Lambda_n, \beta, q),$$

where $\Lambda_n := [\![-n, n]\!]^d$. At this point, the quantity on the right-hand side of the previous equation is not a priori well defined since $m(\Lambda_n, \beta, q)$ could a priori fail to converge as $n$ tends to infinity. We will see in the next section that a cute coupling argument shows that the limit indeed exists and that the following occurs:

THEOREM 1.1. *Let $d \geq 2$ and $q \geq 2$, there exists $\beta_c = \beta_c(d, q) \in (0, \infty)$ such that $m^*(\beta, q) = 0$ if $\beta < \beta_c$ and $m^*(\beta, q) \neq 0$ if $\beta > \beta_c$.*

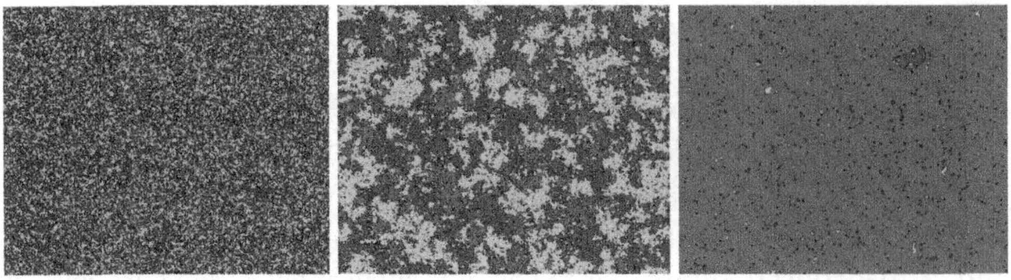

FIGURE 4. Simulations of three-state planar Potts model at subcritical, critical and supercritical temperatures.

The two ranges of $\beta$ for which $m^*(\beta, q) = 0$ and $m^*(\beta, q) \neq 0$ are respectively called the *disordered* and *ordered* phases. The value $\beta_c$ is called the *critical value*, and the model is said to undergo an *order/disorder phase transition*.

Note that the previous theorem leaves the behavior at $\beta_c$ open: $m^*(\beta_c, q)$ may a priori be zero or non-zero. While this may appear like a detail, let us highlight that many aspects of the phase transition are governed by the behavior of the model at $\beta = \beta_c$, and that for this reason this is maybe the most interesting value of $\beta$ to study from the physics point of view. We will say that the phase transition is *continuous* if $m^*(\beta_c, q) = 0$, and *discontinuous* otherwise.

Physicists have been interested in the behavior of the Potts model for a long time, and the following beautiful conjecture has emerged in the last fifty years.

CONJECTURE 1. *Let $d \geq 2$ and $q \geq 2$. The phase transition of the nearest neighbor ferromagnetic $q$-state Potts model is continuous if $q \leq q_c(d)$ and discontinuous if $q > q_c(d)$, where*

$$q_c(d) := \begin{cases} 4 & \text{if } d = 2 \\ 2 & \text{if } d \geq 3. \end{cases}$$

For $d = 2$, the conjecture is due to Baxter [**Bax78**]. For $d \geq 3$, the argument is based on considerations regarding the mean-field behavior of the model (see e.g. [**BC03**]). We postpone the discussion of known results to the beginning of each section (to maintain the suspense).

While the behavior of the model at the critical point of a discontinuous phase transition is fairly well understood (e.g. it features coexistence of different phases), the behavior at a continuous phase transition is much more mysterious and extraordinary rich. Indeed, continuous phase transitions are characterized by the absence of a *correlation length*. For this reason, the model at criticality may be studied via renormalization-type ideas and the large scale behavior of the model can be encoded via a limiting procedure (sometimes called *taking the scaling limit*) by Field Theories. The properties

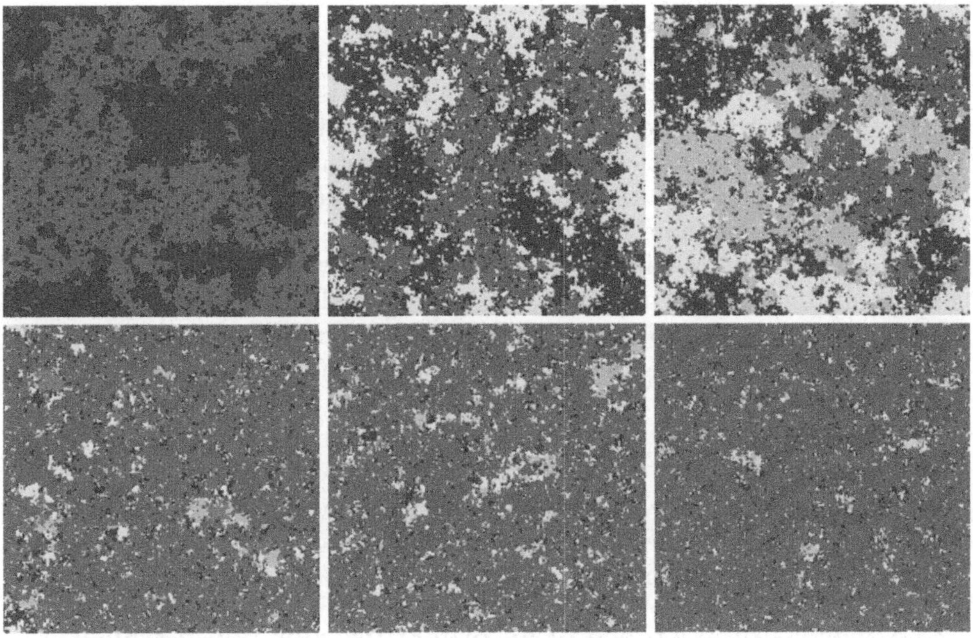

FIGURE 5. Simulations of the critical planar Potts model with $q$ equal to 2, 3, 4, 5, 6 and 9 respectively. The behavior for $q \leq 4$ is clearly different from the behavior for $q > 4$. In the first three pictures, each color (corresponding to each element of $\mathbb{T}_q$) seems to play the same role, while in the last three, one color wins over the other ones.

of these continuum versions of lattice models depend on the model and on the dimension.

If the model has a continuous phase transition in high enough dimension, then it exhibits its *mean-field behavior*, meaning that the critical exponents characterizing the phase transition are the same as those on the complete graph. For such dimensions, the Quantum Field Theory associated to the scaling limit of the critical models is trivial. For the Ising model, such a statement was proved by Aizenman and Fröhlich in dimension $d \geq 5$ [**Aiz82, Frö82**]. In dimension 4, critical exponents were proved to be equal to their mean-field value [**AF88**]. The question of the triviality of the limiting Field Theory (i.e. the fact that the renormalized coupling constant vanishes) is still open.

On the contrary, the scaling limit of *planar* models at their critical point is not trivial (once again we restrict this discussion to continuous phase transitions). It is conjectured to be described by conformally invariant objects encoded by Conformal Field Theory [**ISZ88**]. Let us highlight the extraordinary scope of such a prediction: while the lattice version of the model is clearly invariant under the automorphisms of the lattice, its scaling limit is invariant under a much larger group of transformations, namely the conformal maps. This large group of symmetries enables physicists and mathematicians to derive very delicate properties of the model at criticality, and it is fair to say that the understanding of critical behavior in dimension two has greatly progressed in the last few decades thanks to Conformal Field Theory and its mathematical counterpart (for instance the theory of discrete holomorphic observables, the SLE machinery, random surfaces, etc).

The two previous paragraphs leave the case of dimension $d = 3$, which is often the most interesting from the point of view of physics, open. There, new physical and mathematical ideas are required to construct the scaling limit of critical models, and this problem will probably occupy mathematicians and physicists for many years to come. Even at the physics level, the understanding is limited. Of course, conformal invariance may extend to arbitrary dimensions, yet the group of conformal transformations being reduced in dimension $d \geq 3$, the conclusions regarding the behavior of the scaling limit of the models are inevitably more limited (even though interesting developments using the so-called Conformal Bootstrap have shaken the physics community in the past few years, see e.g. [**SFD$^+$12**]).

## 2. A warm-up: existence of an order/disorder phase transition

As mentioned in the introduction, the existence of a phase transition itself is non-trivial. Indeed, several claims made in the previous section are not obvious:

- The quantity $m^*(\beta, q)$ is not clearly defined since the limit (as $n$ tends to infinity) of the quantity $m(\Lambda_n, \beta, q)$ may not exist.
- It is unclear whether $m^*(\beta, q)$ is equal to zero for small $\beta$, and is not equal to zero for large values of $\beta$. Interestingly, Ising wrongly predicted that the spontaneous magnetization would be equal to 0 for any values of $\beta$ independently of the dimension, based on the observation that it does so in dimension one (we will see later that he was completely wrong).
- While the intuition is convincing, it is a priori non trivial that there exists a critical value $\beta_c(q)$ separating an ordered from a disordered phase. It may very well be that the model alternates between ordered and disordered phases when $\beta$ increases. Let us illustrate the difficulty through the example of the Ising model. In order to rule out the possibility of alternating phases, we could prove that $\frac{d}{d\beta} m(G, \beta, 2)$ which is equal to

$$\frac{1}{|V(G)|} \sum_{x \in V(G)} \sum_{\{a,b\} \in E(G)} \mu^1_{G,\beta,2}[\sigma_x \sigma_a \sigma_b] - \mu^1_{G,\beta,2}[\sigma_x] \mu^1_{G,\beta,2}[\sigma_a \sigma_b],$$

must necessarily be positive. This would readily imply that the spontaneous magnetization would also be. The positivity would follow from

$$\mu^1_{G,\beta,2}[\sigma_x \sigma_a \sigma_b] - \mu^1_{G,\beta,2}[\sigma_x] \mu^1_{G,\beta,2}[\sigma_a \sigma_b] \geq 0$$

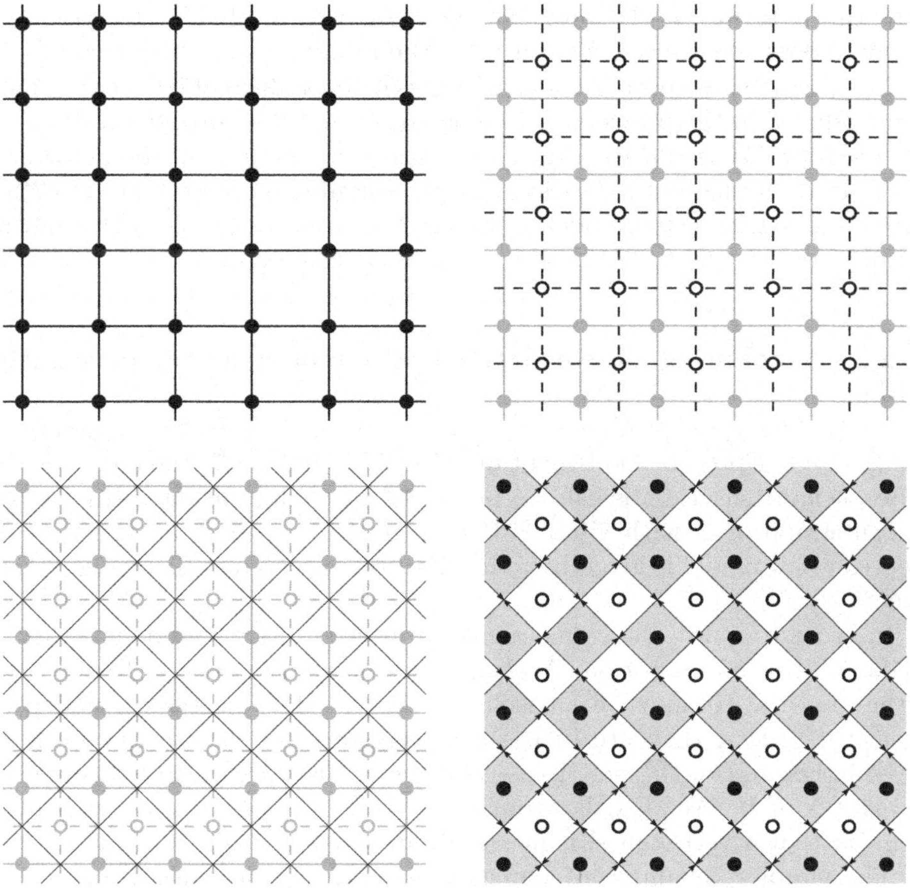

FIGURE 6. The square lattice (top left), its dual lattice (top right), its medial lattice (bottom left) and a natural orientation on the medial lattice (bottom right).

for any fixed $a$ and $b$, an inequality which is far from obvious (it is called the second Griffiths' inequality).

THEOREM 2.1 (Folklore). *Let $d \geq 2$ and $q \geq 2$. For any $\beta > 0$, there exists a probability measure $\mu^1_{\beta,q}$ on spin configurations on $\mathbb{Z}^d$ such that*

$$\lim_{G \nearrow \mathbb{Z}^d} \mu^1_{G,\beta,q} = \mu^1_{\beta,q},$$

*where the convergence is the weak convergence for measures.*

*Furthermore,*

$$\lim_{n\to\infty} m(\Lambda_n, \beta, q) = \lim_{n\to\infty} \mu^1_{\Lambda_n,\beta,q}(\sigma_0) = \mu^1_{\beta,q}(\sigma_0)$$

*and*

$$\beta_c(q) = \inf\{\beta > 0 : \mu^1_{\beta,q}(\sigma_0) \neq 0\} = \sup\{\beta > 0 : \mu^1_{\beta,q}(\sigma_0) = 0\}$$

*belongs to $(0, \infty)$.*

With the notation of the previous theorem, we see that $m^*(\beta, q)$ can be interpreted as the average spin at 0 for the infinite-volume measure $\mu^1_{\beta,q}$.

Proving all these properties using only the formalism of lattice spin models introduced in the previous section is not straightforward. It was done using *correlation inequalities* in the case of the Ising model, but remained open for the Potts model until the discovery by Fortuin and Kasteleyn [**FK72**] of a deep link between spin models and percolation-type models. The following sections are devoted to the description of this connection.

## 2.1. A coupling between the Potts model and a percolation model.

2.1.1. *The Fortuin-Kasteleyn percolation.* A *percolation configuration* $\omega = (\omega_e : e \in E(G))$ is an element of $\{0,1\}^{E(G)}$. If $\omega_e = 1$, the edge $e$ is said to be *open*, otherwise $e$ is said to be *closed.* A configuration $\omega$ can be seen as a subgraph of $G$ with $V(\omega) = V(G)$ and $E(\omega) = \{e \in E(G) : \omega_e = 1\}$. A percolation model is given by a distribution on percolation configurations on $G$.

In order to study the connectivity properties of the (random) graph $\omega$, we introduce some notation. A *cluster* is a maximal connected component of the graph $\omega$ (it may be an isolated vertex). Two vertices $x$ and $y$ are *connected in* $\omega$ if they are in the same cluster. We denote this event by $x \longleftrightarrow y$. For $A, B \subset \mathbb{Z}^d$, set $A \longleftrightarrow B$ if there exists a vertex of $A$ connected to a vertex of $B$. We also allow ourselves to consider $B = \infty$, in which case we mean that a vertex in $A$ is in an infinite cluster.

The simplest example of percolation model is provided by *Bernoulli percolation*: each edge is open with probability $p$, and closed with probability $1-p$, independently of the state of other edges. This model was introduced by Broadbent and Hammersley in [**BH57**] and has been one of the most studied probabilistic model. We refer to [**Gri99**] for a book on the subject.

Here, we will be interested in a slightly more complicated percolation model, named the *Fortuin-Kasteleyn percolation* (also called FK percolation or random-cluster model), which is a percolation model in which the states *open/closed* of edges depend on each others.

Let $G$ be a finite subgraph of $\mathbb{Z}^d$. Let $o(\omega)$ and $c(\omega)$ denote the number of open and closed edges of $\omega$. The *boundary conditions* $\xi$ are given by a partition $P_1 \sqcup \cdots \sqcup P_k$ of

$$\partial G := \{x \in V(G) : \exists y \notin V(G), \{x,y\} \in E(\mathbb{Z}^d)\},$$

and $k_\xi(\omega)$ denotes the number of clusters in the graph $\omega$ where clusters intersecting each $P_i$ are counted as one.

Definition 2.2. *The probability measure $\phi^\xi_{G,p,q}$ of the FK percolation on $G$ with* edge-weight $p \in [0,1]$, cluster-weight $q > 0$ *and boundary conditions*

*$\xi$ is defined by*

$$\phi^{\xi}_{G,p,q}[\{\omega\}] := \frac{p^{o(\omega)}(1-p)^{c(\omega)}q^{k_\xi(\omega)}}{Z^{\xi}_{G,p,q}} \tag{2.1}$$

*for every configuration $\omega \in \{0,1\}^{E(G)}$. The constant $Z^{\xi}_{G,p,q}$ is a normalizing constant, referred to as the* partition function, *defined in such a way that the sum over all configurations equals 1.*

For $q = 1$, FK percolation corresponds to Bernoulli percolation.

Let us provide two examples of boundary conditions. The free boundary conditions correspond to the partition composed of singletons only. Then $k_{\text{free}}(\omega)$ simply denotes the number of clusters in $\omega$. The *wired boundary conditions* correspond to the partition $\{\partial G\}$ and $k_{\text{wired}}(\omega)$ is the number of clusters obtained if all clusters touching the boundary are counted as 1.

2.1.2. *The coupling.* Consider an integer $q \geq 2$ and let $G$ be a finite graph. Assume that a configuration $\omega \in \{0,1\}^{E(G)}$ is given. One can deduce a spin configuration $\sigma \in \mathbb{T}_q^{V(G)}$ by assigning uniformly and independently to each cluster $\mathscr{C}$ of $\omega$ a spin $\sigma_{\mathscr{C}} \in \mathbb{T}_q$, except for the clusters intersecting the boundary $\partial G$ which are automatically associated to the spin $\vec{1}$. We then define $\sigma_x$ to be equal to $\sigma_{\mathscr{C}}$ for every $x \in \mathscr{C}$. Note that all the vertices in the same cluster automatically receive the same spin.

PROPOSITION 2.3. *Fix an integer $q \geq 2$. Let $p \in (0,1)$ and $G \subset \mathbb{Z}^d$ a finite graph. If the configuration $\omega$ is distributed according to $\phi^{\text{wired}}_{G,p,q}$, then the spin configuration $\sigma$ is distributed according to the $q$-state Potts measure $\mu^1_{G,\beta,q}$, where*

$$\beta = \beta(p,q) = -\tfrac{q-1}{q}\ln(1-p). \tag{2.2}$$

Since the proof is elementary, we include it here.

PROOF. Let $\Omega$ be the space of pairs $(\omega,\sigma)$ with $\omega \in \{0,1\}^{E(G)}$ and $\sigma \in \mathbb{T}_q^{V(G)}$, with the property that for any edge $e = \{x,y\}$,

$$\omega_e = 1 \quad \Rightarrow \quad \sigma_x = \sigma_y.$$

Consider a measure $\mathbf{P}$ on $\Omega$, where $\omega$ is a percolation configuration with wired boundary conditions and $\sigma$ is the corresponding spin configuration constructed as explained above. Then, for $(\omega,\sigma)$, we have:

$$\begin{aligned}\mathbf{P}[(\omega,\sigma)] &= \frac{1}{Z^{\text{wired}}_{G,p,q}}\, p^{o(\omega)}(1-p)^{c(\omega)}q^{k_{\text{wired}}(\omega)} \cdot q^{-k_{\text{wired}}(\omega)+1}\\ &= \frac{q}{Z^{\text{wired}}_{G,p,q}}\, p^{o(\omega)}(1-p)^{c(\omega)}.\end{aligned}$$

(The additive constant 1 is due to the fact that the spin of the clusters of $\omega$ touching the boundary is necessarily $\vec{1}$.)

Now, we construct another measure $\widetilde{\mathbf{P}}$ on $\Omega$ as follows. Let $\tilde{\sigma}$ be a spin-configuration distributed according to $\mu^1_{G,\beta,q}$, where $\exp[-\frac{q}{q-1}\beta] = 1-p$. We deduce $\tilde{\omega}$ from $\tilde{\sigma}$ by independently opening edges between neighboring vertices having the same spins with probability $p$. By definition, edges between vertices having different spins remain automatically closed. Then, for any $(\tilde{\omega},\tilde{\sigma})$,

$$\widetilde{\mathbf{P}}[(\tilde{\omega},\tilde{\sigma})] = \frac{\exp[-\frac{q}{q-1}\beta r(\tilde{\sigma})]\, p^{o(\tilde{\omega})}(1-p)^{c(\tilde{\omega})-r(\tilde{\sigma})}}{Z} = \frac{p^{o(\tilde{\omega})}(1-p)^{c(\tilde{\omega})}}{Z},$$

where $r(\tilde{\sigma})$ is the number of edges between vertices with different spins, and $Z$ is a normalizing constant.

In conclusion, $\mathbf{P}$ and $\widetilde{\mathbf{P}}$ are two probability measures on $\Omega$ assigning the same weights to configurations, they are therefore equal. Since the first marginal of $\mathbf{P}(=\widetilde{\mathbf{P}})$ is $\phi^{\text{wired}}_{G,p,q}$ and the second of $\widetilde{\mathbf{P}}(=\mathbf{P})$ is $\mu^1_{G,\beta,q}$, we obtain the result. $\square$

The coupling immediately extends to the FK percolation and the Potts model with free boundary conditions (we leave the question of what must be done with clusters touching the boundary as an exercise).

This coupling provides us with a dictionary between the properties of the FK percolation and the Potts model. In order to illustrate this fact, let us mention one consequence.

COROLLARY 2.4. *Let $d, q \geq 2$. Let $G$ be a finite subgraph of $\mathbb{Z}^d$. Let $\beta > 0$ and $p \in [0,1]$ be connected by* (2.2). *We have that for any $x \in V(G)$,*

$$\mu^1_{G,\beta,q}[\sigma_x] = \mathbf{E}\Big[\sigma_x \mathbf{1}_{\{x \overset{\omega}{\longleftrightarrow} \partial G\}}\Big] + \mathbf{E}\Big[\sigma_x \mathbf{1}_{\{x \overset{\omega}{\not\longleftrightarrow} \partial G\}}\Big] = \phi^{\text{wired}}_{G,p,q}[x \longleftrightarrow \partial G] \cdot \vec{1}.$$

PROOF. Simply use that $\sigma_x$ must be equal to $\vec{1}$ if $x$ is connected to $\partial G$, and $\sigma_x$ is uniformly chosen among spins in $\mathbb{T}_q$ otherwise. $\square$

This result will be used in the next sections to relate the phase transitions of the Potts model and of FK percolation. As a side remark, note that we just proved that $\langle \mu^1_{G,\beta,q}[\sigma_x] \,|\, \vec{1}\rangle$ is non-negative.

**2.2. Phase transition and critical point.** We now wish to prove that the Potts model undergoes a phase transition in dimension $d \geq 2$, meaning that $\beta_c(q,d)$ exists and belongs to $(0,\infty)$.

2.2.1. *Ordering in the FK percolation.* One of the advantages of percolation configurations compared to spin configurations is that $\{0,1\}^{E(G)}$ is naturally ordered (simply say that $\omega \leq \omega'$ if any edge $e \in E(G)$ satisfies $\omega_e \leq \omega'_e$). An event $A$ is *increasing* if any $\omega' \geq \omega$ with $\omega \in A$ satisfies that $\omega' \in A$. Let us mention briefly a few properties of FK percolation concerning increasing events. We refer to [**Gri06, Dum13**] for proofs.

Fix $q \geq 1$ and $G$ a finite graph. Let $p \leq p'$ and $\xi \leq \psi$ (meaning that any element of the partition $\xi$ is included in an element of the partition $\psi$). Also consider two increasing events $A$ and $B$, then

$$\begin{aligned} \phi^{\xi}_{G,p,q}[A\cap B] &\ge \phi^{\xi}_{G,p,q}[A]\phi^{\xi}_{G,p,q}[B] && \text{(FKG inequality)}\\ \phi^{\xi}_{G,p',q}[A] &\ge \phi^{\xi}_{G,p,q}[A] && \text{(monotonicity in } p)\\ \phi^{\psi}_{G,p,q}[A] &\ge \phi^{\xi}_{G,p,q}[A] && \text{(comparison of boundary conditions).}\end{aligned}$$

One should really emphasize the fact that the cluster weight $q$ must be larger than or equal to 1. These properties all fail for $q < 1$. The model is actually expected to be *negatively correlated* in this context, see [**Gri06**, Section 3.9] for a discussion.

2.2.2. *Phase transition in the FK percolation model.* Phase transitions are properties of infinite-volume systems. For FK percolation, the definition of a measure on $\mathbb{Z}^d$ is not direct since one cannot count the number of open or closed edges on $\mathbb{Z}^d$ (they could be and would be infinite). We thus define infinite-volume measures by taking a sequence of measures on larger and larger boxes $\Lambda_n$, where $n \ge 1$. The ordering between boundary conditions can be used to show that the sequences of measures $(\phi^{\text{wired}}_{p,q,\Lambda_n})_{n\ge0}$ and $(\phi^{\text{free}}_{p,q,\Lambda_n})_{n\ge0}$ converge weakly to measures $\phi^{\text{wired}}_{p,q}$ and $\phi^{\text{free}}_{p,q}$ on $\{0,1\}^{E(\mathbb{Z}^d)}$, called the infinite-volume measures with wired and free boundary conditions respectively.

One warning: while boundary conditions cannot be defined as a partition of the boundary in infinite volume, one still needs to keep track of the dependency on boundary conditions for finite-volume measures when constructing the measure. Therefore, the measures $\phi^{\text{wired}}_{p,q}$ and $\phi^{\text{free}}_{p,q}$ have no reason to be the same (we will see examples of values of $p$ and $q$ for which they are in fact different). In addition to this, one may imagine other infinite-volume measures obtained via limits of measures on finite graphs with arbitrary (and possibly random) boundary conditions. Nonetheless, the measures $\phi^{\text{wired}}_{p,q}$ and $\phi^{\text{free}}_{p,q}$ play very specific roles in the theory. First, they are invariant under translations and ergodic. Second, they are extremal infinite-volume measures, in the sense that any infinite-volume measure $\phi$ (see e.g. [**Dum13**, Definition 4.24] and references therein for a formal definition) with parameters $p$ and $q \ge 1$ satisfies

$$\phi^{\text{free}}_{p,q}[A] \le \phi[A] \le \phi^{\text{wired}}_{p,q}[A] \tag{2.3}$$

for any increasing events $A$. Third, an abstract theorem based on the convexity of the free energy (see [**Dum13**, Theorem 4.30]) shows that for a fixed $q \ge 1$, $\phi^{\text{free}}_{p,q} = \phi^{\text{wired}}_{p,q}$ (and therefore the infinite-volume measure is unique at this value of $p$) for all but possibly countably many values of $p$.

Let us make an additional remark. The properties of finite-volume measures (FKG inequality, monotonicity, ordering between boundary conditions) extend to infinite volume in a straightforward fashion.

We are now in a position to discuss the phase transition for the FK percolation.

THEOREM 2.5. *For $q, d \geq 1$, there exists a* critical point $p_c = p_c(q, d) \in [0, 1]$ *such that:*

- *For $p < p_c$, any infinite-volume measure has no infinite cluster almost surely.*
- *For $p > p_c$, any infinite-volume measure has an infinite cluster almost surely.*

PROOF. In order to prove this theorem, simply define

$$p_c := \inf\{p \in [0,1] : \phi^{\text{free}}_{p,q}[0 \longleftrightarrow \infty] > 0\} = \sup\{p \in [0,1] : \phi^{\text{free}}_{p,q}[0 \longleftrightarrow \infty] = 0\}.$$

(The second equality comes from the monotonicity of $\phi^{\text{free}}_{p,q}$ in $p$ and the fact that $0 \longleftrightarrow \infty$ is an increasing event.) For any $p > p_c$, the $\phi^{\text{free}}_{p,q}$-probability of having an infinite cluster is 1, since this event is invariant under translations and the measure $\phi^{\text{free}}_{p,q}$ is ergodic. Therefore, any infinite-volume measure has an infinite cluster almost surely by (2.3). For $p < p_c$, choose $p' \in (p, p_c)$ for which the infinite-volume measure is unique. In such case, we have that the $\phi^{\text{wired}}_{p',q}$-probability of having an infinite-cluster is 0 since $\phi^{\text{wired}}_{p',q} = \phi^{\text{free}}_{p',q}$. This is true for any $p < p'$ by monotonicity, and therefore for any infinite-volume measure since wired boundary conditions are the largest possible. □

One may easily check that $p_c(q, 1) = 1$ for any $q > 0$. Also, one may prove that $p_c(q, d) > 0$. Indeed, for any configuration $\omega$ on edges $f \neq e$, we find that

$$\phi^{\text{free}}_{p,q}[\omega_e = 1 | \omega_f : f \neq e] \leq p.$$

This implies by induction that for any set of disjoint edges $e_1, \dots, e_n$,

$$\phi^{\text{free}}_{p,q}[\omega_{e_i} = 1, \forall i \leq n] \leq p^n.$$

If 0 is connected to distance $n$, then there must exist a self-avoiding path of adjacent open edges (where adjacent means that for any $0 \leq i < n$, $e_i$ and $e_{i+1}$ share one endpoint). Therefore,

$$\phi^{\text{free}}_{p,q}[0 \longleftrightarrow \partial\Lambda_n] \leq \sum_{\text{path } e_1,\dots,e_n} \phi^{\text{free}}_{p,q}[\omega_{e_i} = 1, \forall i \leq n] \leq (2dp)^n,$$

where the sum is over self-avoiding paths of adjacent edges $e_1, \dots, e_n$. If $2dp < 1$, we obtain that this quantity tends to 0 as $n$ tends to infinity, thus proving that $p \leq p_c(q, d)$. In conclusion, $p_c(q, d) \geq 1/(2d)$.

2.2.3. *A special feature of dimension 2: planarity.* While trying to compute $p_c(q, d)$ appears very natural to do, a concrete formula is not really expected in general (for instance for $d \geq 3$, the value is probably not rational or algebraic). Nevertheless, dimension two enjoys a particularly nice property, called planar duality, enabling us to derive the critical value rigorously.

Consider the dual of $\mathbb{Z}^2$ to be the lattice $(\mathbb{Z}^2)^* := (\frac{1}{2}, \frac{1}{2}) + \mathbb{Z}^2$. By definition, each edge of $\mathbb{Z}^2$ crosses exactly one edge of $(\mathbb{Z}^2)^*$ which we now denote by $e^*$. Any configuration $\omega \in \{0,1\}^{E(\mathbb{Z}^2)}$ corresponds to a dual configuration $\omega^* \in \{0,1\}^{E((\mathbb{Z}^2)^*)}$ via the following relation

$$\omega^*_{e^*} = 1 - \omega_e, \quad \forall e \in E(\mathbb{Z}^2).$$

Planar duality refers to the fact that if $\omega$ is sampled according to $\phi^{\text{wired}}_{p,q}$, then $\omega^*$ is sampled according to $\phi^{\text{free}}_{p^*,q^*}$ (more precisely a translate by $(\frac{1}{2}, \frac{1}{2})$ of the measure, which is therefore defined on $(\mathbb{Z}^2)^*$), where

$$q^* = q \quad \text{and} \quad \frac{pp^*}{(1-p)(1-p^*)} = q.$$

Now, fix $q \geq 1$ and observe that $p_{\text{sd}} = p_{\text{sd}}(q) := \sqrt{q}/(1+\sqrt{q})$ is the only value of $p$ such that $p^* = p$.

Assume that $p_c < p_{\text{sd}}$, in such case there exists an infinite cluster of open edges at $p_{\text{sd}}$. But the dual model of $\phi^{\text{free}}_{p_{\text{sd}},q}$ is a translate by $(\frac{1}{2}, \frac{1}{2})$ of $\phi^{\text{wired}}_{p_{\text{sd}},q}$, and therefore there exists an infinite cluster of dual-open edges on $(\mathbb{Z}^2)^*$ as well. In other words, there is coexistence of an infinite cluster in $\omega$ and an infinite cluster in $\omega^*$. This fact can appear counter-intuitive physically, and can indeed be ruled out mathematically. This sketch of argument therefore suggests that $p_c \geq p_{\text{sd}}$.

Now assume that $p_c > p_{\text{sd}}$. Then there is no infinite cluster in $\omega$ or $\omega^*$ for any $p \in (p_{\text{sd}}, p_c)$. This fact can be proved to be occurring at one value of $p$ maximum (which turns out to be $p_{\text{sd}}$), and is the main object of the following theorem.

THEOREM 2.6 (Beffara, DC [**BD12a, DRT15**]). *For $q \geq 1$, the critical value $p_c = p_c(q,2)$ satisfies*

$$p_c = \frac{\sqrt{q}}{1+\sqrt{q}}.$$

One may easily check that $p_c(q,d)$ is decreasing in $d$. Therefore, $p_c(q,d) \leq \frac{\sqrt{q}}{1+\sqrt{q}}$ for $d \geq 2$, thus showing that $p_c(q,d) < 1$.

2.2.4. *Phase transition in the Potts model.* Corollary 2.4 implies that

$$\lim_{n\to\infty} \mu^1_{\Lambda_n,\beta,q}[\sigma_0] = \lim_{n\to\infty} \phi^{\text{wired}}_{\Lambda_n,p,q}[0 \leftrightarrow \infty] \cdot \vec{1} = \phi^{\text{wired}}_{p,q}[0 \leftrightarrow \infty] \cdot \vec{1},$$

where $\beta = -\frac{q-1}{q}\log[1-p]$. We deduce the following result:

THEOREM 2.7. *The quantity $\beta_c = \beta_c(q,d) := -\frac{q-1}{q}\log\left[1 - p_c(q,d)\right] \in (0,\infty)$ satisfies*

$$\lim_{n\to\infty} \mu^1_{\Lambda_n,\beta,q}[\sigma_0] \quad \begin{cases} = 0 & \textit{if } \beta < \beta_c \\ \neq 0 & \textit{if } \beta > \beta_c. \end{cases}$$

*Furthermore, we have that*

$$m^*(\beta,q) := \lim_{n\to\infty} m(\Lambda_n,\beta,q) = \lim_{n\to\infty} \mu^1_{\Lambda_n,\beta,q}[\sigma_0].$$

The first claim is due to the coupling between FK percolation and the Potts model. The second claim follows from the fact that the mean of the spins averaged on the whole box $\Lambda_n$ converges as $n$ tends to infinity to the mean of the spin at the origin. This can easily be checked using the ergodicity of $\phi_{p,q}^{\text{wired}}$, we omit this (fairly easy) proof here.

In words, the previous theorem states that the critical inverse temperature of the Potts model can be defined rigorously (it does not alternate between ordered and disordered phases as it a priori could), and this critical inverse-temperature can be expressed in terms of the critical value of the corresponding FK percolation. In two dimensions, we immediately obtain that $\beta_c(q,2) = \frac{q-1}{q}\log(1+\sqrt{q})$ thanks to Theorem 2.6.

Also note the icing on the cake: the phase transition of the Potts model is continuous if and only if $\phi_{p_c,q}^{\text{wired}}[0 \leftrightarrow \infty] = 0$. As a consequence, we can focus our attention on FK percolation in order to determine whether the phase transition is continuous or discontinuous. We do so in the next section.

## 3. Critical behavior on $\mathbb{Z}^2$

In this section, we focus on the two dimensional case. For now on, we work with FK percolation, and only briefly mention the consequences for the Potts model.

**3.1. An alternative between two possible critical behaviors.** Physicists possess several definitions of continuous phase transitions. For instance, it may refer to the divergence of the correlation length, the continuity of the order parameter (here the spontaneous magnetization), the uniqueness of the Gibbs states at criticality, the divergence of the susceptibility, the scale invariance at criticality, etc. From a mathematical point of view, these properties are not clearly equivalent (there are examples of models for which they are not), and they therefore refer to a priori different notions of continuous phase transition. The following result shows that all these properties are equivalent for the planar FK percolation (and therefore for the associated Potts model). As a consequence, the previous properties are alternative characterizations of a single notion, and we may think of *the* notion of a continuous phase transition.

THEOREM 3.1 (DC–Sidoravicius–Tassion [**DST15**]). *Let $q \geq 1$, the following assertions are equivalent at criticality:*

**P1** *(Absence of infinite cluster)* $\phi_{p_c,q}^{\text{wired}}[0 \longleftrightarrow \infty] = 0$.

**P2** *(Uniqueness of the infinite-volume measure)* $\phi_{p_c,q}^{\text{free}} = \phi_{p_c,q}^{\text{wired}}$.

**P3** *(Infinite susceptibility)* $\sum_{x\in\mathbb{Z}^2} \phi_{p_c,q}^{\text{free}}[0 \longleftrightarrow x] = \infty$.

**P4** *(Sub-exponential decay of correlations for free boundary conditions)*

$$\lim_{n\to\infty} \frac{1}{n} \log \phi_{p_c,q}^{\text{free}}[0 \longleftrightarrow \partial\Lambda_n] = 0.$$

**P5** *(RSW) Let $\alpha > 0$. There exists $c = c(\alpha) > 0$ such that for all $n \geq 1$ and any boundary conditions $\xi$,*

$$c \le \phi^{\xi}_{[-n,(\alpha+1)n]\times[-n,2n],p_c,q}\left[\begin{smallmatrix}[0,\alpha n]\times[0,n] \text{ is crossed from}\\ \text{left to right by an open path}\end{smallmatrix}\right] \le 1-c.$$

The previous theorem does not show that these conditions are all satisfied, only that they are equivalent. In fact, whether the conditions are satisfied or not depend on the value of $q$, as we will see later in this review.

Property **P5** is the strongest one. Note that we did not only require that the probability of being crossed by an open path remains bounded away from 0 uniformly in the size $n$ of the rectangle, but also *uniformly in boundary conditions*. This property is crucial for the applications of this theorem, in particular since it helps controlling the dependencies between edges in different parts of the graph. Among other results, **P5** implies

- the polynomial decay of correlations at criticality,
- the existence of sub-sequential scaling limits,
- the value for certain critical exponents called *universal* critical exponents (it has nothing to do with the universality for the model itself),
- the fractal nature of large clusters (with some explicit bounds on the Hausdorff dimension).

Furthermore, it represents an important tool for the following problems:

- understanding the conformal invariance of the model,
- understanding scaling relations between several critical exponents,
- proving the universal behavior at criticality.

This result was previously known in a few cases. For $q = 1$, the properties follow from a collection of results, among which the two fundamental papers of Russo [**Rus78**] and Seymour and Welsh [**SW78**]. For $q \gg 1$, the above properties fail (see [**KS82, LMMS$^+$91**]), while for $q = 2$ (which is coupled to the Ising model), all of these properties are proved to be true using specific properties of the Ising model; see [**Ons44, Sim80, DHN11**].

Before proceeding forward, let us discuss very briefly the ingredients in the proof for other values of $q \ge 1$. As mentioned before, **P5** is the strongest property. On the other hand, **P4** is the weakest. Therefore, it does not come as a surprise that **P5⇒P1⇒P2⇒P3⇒P4** is fairly easy to obtain. The main difficulty lies in the proof of **P4⇒P5**, or equivalently that

(3.1)
$$\inf\left\{\phi^{\text{free}}_{[-n,(\alpha+1)n]\times[-n,2n],p_c,q}\left[\begin{smallmatrix}[0,\alpha n]\times[0,n] \text{ is crossed from}\\ \text{left to right by an open path}\end{smallmatrix}\right] \quad , \quad n \ge 1\right\} = 0$$

implies that the probability of being connected to distance $n$ tends to 0 exponentially fast (i.e. **nonP4**). (Since the free boundary conditions are the smallest, and since the uniform upper bound on crossing probabilities can be obtained via duality and the uniform lower bound, (3.1) is equivalent to **nonP5**.)

In order to prove this implication, we developed a new geometric renormalization principle for crossing probabilities: crossing probabilities at scale $2n$ are expressed in terms of crossing probabilities at scale $n$. The renormalization scheme is built in such a way that as soon as the crossing probability passes below a certain threshold, crossing probabilities start decaying exponentially fast. As a consequence, either crossing probabilities remain bounded away from 0, or they decay to 0 exponentially fast.

The major difficulty arising in the implementation of this renormalization scheme is to handle long-range dependency and boundary conditions properly. The lack of independence renders the proof difficult, and requires the introduction of completely new tools. Let us finish by mentioning that historically, Kesten showed **P4⇒P5** for Bernoulli percolation but his proof cannot be extended to FK percolation with $q > 1$.

**3.2. Deciding between critical behaviors: the loop representation and Baxter's conjecture.** We should now decide whether, for a fixed $q \geq 1$, properties **P1–5** are satisfied or not. In order to do so, we introduce yet another representation of the Potts model. It is derived directly from the planar FK percolation. We start by defining the loop configuration $\overline{\omega}$ associated to a percolation configuration $\omega$. Let $(\mathbb{Z}^2)^\diamond$ be the lattice defined as follows. The set of vertices is given by the midpoints of edges of $\mathbb{Z}^2$. The edges are all the pairs of nearest vertices (i.e. vertices at distance $\sqrt{2}/2$ of each others). It is a rotated and rescaled version of the square lattice, see Fig. 6. For future reference, note that the edges of the medial lattice can be oriented in a counter-clockwise way around faces that are centered on a vertex of $\mathbb{Z}^2$.

Let $\Omega$ be a finite subgraph of $\mathbb{Z}^2$. Let $\Omega^*$ be the subgraph of $(\mathbb{Z}^2)^*$ induced by dual edges bordering faces of $(\mathbb{Z}^2)^*$ corresponding to vertices of $\Omega$. Let $\Omega^\diamond$ be the subgraph of $(\mathbb{Z}^2)^\diamond$ defined by the vertices corresponding to midpoints of edges in $E(\Omega)$, and edges between two vertices of $V(\Omega^\diamond)$ on the same face of $\Omega$.

Consider a configuration $\omega$ together with its dual configuration $\omega^*$ (recall that $\omega^*_{e^*} = 1 - \omega_e$). We draw $\omega^*$ in such a way that dual edges between vertices of $\partial\Omega^*$ are open in $\omega^*$ (we make such an arbitrary choice since these dual edges have no corresponding edges in $E(\Omega)$).

By definition, through every vertex of the medial graph $\Omega^\diamond$ of $\Omega$ passes either an edge of $\omega$ or an edge of $\omega^*$. Draw self-avoiding loops on $\Omega^\diamond$ as follows: a loop arriving at a vertex of the medial lattice always makes a $\pm\pi/2$ turn at vertices so as not to cross the edges of $\omega$ or $\omega^*$, see Fig. 7. The loop configuration is defined in an unequivocal way since:

- there is either an edge of $\omega$ or an edge of $\omega^*$ crossing non-boundary vertices in $\Omega^\diamond$, and therefore there is exactly one coherent way for the loop to turn at non-boundary vertices.
- there is only one possible $\pm\pi/2$ turn at boundary vertices keeping the loops in $\Omega^\diamond$.

From now on, the loop configuration associated to $\omega$ is denoted by $\overline{\omega}$.

We allow ourselves a slight abuse of notation: below, $\phi^0_{\Omega,p,q}$ denotes the measure on percolation configurations as well as its push-forward by the map

FIGURE 7. The configurations $\omega$ (in bold lines), $\omega^*$ (in dashed lines) and $\overline{\omega}$ (in plain lines).

$\omega \mapsto \overline{\omega}$. Therefore, the measure $\phi^0_{\Omega,p,q}$ will sometimes refer to a measure on loop configurations. Fix

$$x = x(p,q) := \frac{p}{\sqrt{q}(1-p)}.$$

PROPOSITION 3.2. *Let $\Omega$ be a connected finite subgraph of $\mathbb{Z}^2$ which complement in $\mathbb{Z}^2$ is connected. Let $p \in [0,1]$ and $q > 0$. For any configuration $\omega$,*

$$\phi^{\text{free}}_{\Omega,p,q}[\overline{\omega}] \quad = \quad \frac{x^{o(\omega)}\sqrt{q}^{\ell(\overline{\omega})}}{Z(\Omega,p,q)},$$

*where $\ell(\overline{\omega})$ is the number of loops in $\overline{\omega}$ and $Z(\Omega,p,q)$ is a normalizing constant.*

In particular, when $p = p_c(q)$, we obtain that $x = 1$ and the probability of a loop configuration is expressed in terms of the number of loops only.

PROOF. An induction on the number of open edges shows that

$$\ell(\overline{\omega}) = 2k(\omega) + o(\omega) - |V(\Omega)|. \tag{3.2}$$

Indeed, if there is no open edge, then $\ell(\overline{\omega}) = k(\omega) = |V(\Omega)|$. Now, adding an edge can either:

- join two connected components of $\omega$, thus decreasing the number of loops and the number of connected components by 1,
- close a cycle in $\omega$, thus increasing the number of loops by 1 and not changing the number of connected components.

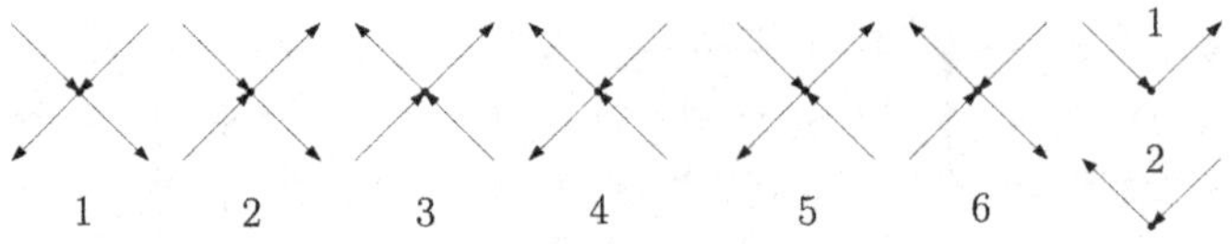

FIGURE 8. Different possible patterns for the six vertex model on the medial lattice. The two patterns on the right are boundary patterns.

Equation (3.2) implies that

$$\begin{aligned}
&p^{o(\omega)}(1-p)^{c(\omega)}q^{k_{\text{free}}(\omega)}\\
&= p^{o(\omega)}(1-p)^{|E(\Omega)|-o(\omega)}q^{k_{\text{free}}(\omega)}\\
&= (1-p)^{|E(\Omega)|}\sqrt{q}^{|V(\Omega)|}\left(\frac{p}{(1-p)\sqrt{q}}\right)^{o(\omega)}\sqrt{q}^{2k_{\text{free}}(\omega)+o(\omega)-|V(\Omega)|}\\
&= (1-p)^{|E(\Omega)|}\sqrt{q}^{|V(\Omega)|}x^{o(\omega)}\sqrt{q}^{\ell(\overline{\omega})}.
\end{aligned}$$

The proof follows by setting

$$\widehat{Z}(\Omega,p,q) := \frac{Z(\Omega,p,q)}{(1-p)^{|E(\Omega)|}\sqrt{q}^{|V(\Omega)|}}. \qquad \square$$

The loop representation of the FK percolation is a well-known representation. It allows to map the free energies of the Potts model and the FK percolation to the free energy of a solid-on-solid ice-type model. For completeness, we succinctly present the mapping here and we refer to [**Bax89**, Chapter 10] for more details on the subject. We focus on the case of a domain $\Omega$ with free boundary conditions.

From a loop configuration $\overline{\omega}$, one may deduce a configuration $\vec{\omega}$ of arrows on edges as follows. Orientate loops of $\overline{\omega}$ clockwise or counterclockwise. Then, forget the way loops turn at every vertex (simply remember which orientation the loops give to each edges of the medial graph). This procedure gives rise to a configuration $\vec{\omega}$ of arrows on edges of $\Omega^\diamond$ with the constraint that for every vertex of $\Omega^\diamond$, the number of incoming arrows is equal to the number of outgoing arrows. The set of possible configurations is denoted by $\mathrm{A}(\Omega)$.

Assume for a moment that we consider directly a model, called the 6 vertex model, on $\mathrm{A}(\Omega)$ for which the probability of $\vec{\omega}$ is given by

$$\mathrm{P}[\vec{\omega}] = \frac{x_1^{N_1(\vec{\omega})}x_2^{N_2(\vec{\omega})}x_3^{N_3(\vec{\omega})}x_4^{N_4(\vec{\omega})}x_5^{N_5(\vec{\omega})}x_6^{N_6(\vec{\omega})}y_1^{\widetilde{N}_1(\vec{\omega})}y_2^{\widetilde{N}_2(\vec{\omega})}}{Z_{6V}(x_1,x_2,x_3,x_4,x_5,x_6,y_1,y_2)},$$

where $x_1,\dots,x_6,y_1,y_2>0$, and $N_1,\dots,N_6,\widetilde{N}_1,\widetilde{N}_2$ correspond to the number of vertices with local arrangements of arrows of types 1–6 and boundary vertices with local arrangements of arrows of types 1–2; see Fig. 8.

We may consider the free energy of this model, defined by

$$f_{6V}(x_1,\ldots,x_6) := \lim_{n\to\infty} \frac{1}{|E(\Lambda_n)|} \log Z_{6V}\big(\Lambda_n, x_1,\ldots,x_6,y_1,y_2\big).$$

(We do not recall the dependency in $y_1$ and $y_2$ since the quantity can easily be seen to be independent of $y_1$ and $y_2$ in the limit $n \to \infty$.)

We already know from the coupling between the Potts model and FK percolation that the partitions functions of the two models are related. We also know that at criticality, the partition function of FK percolation is related to the partition function of the loop model. The transformation between the loop model and the 6 vertex model mentioned briefly above can be used to prove the following provided that the weight of a configuration of oriented loops obtained from $\overline{\omega}$ is proportional to the probability of $\overline{\omega}$ times $e^{2\pi i\tilde{\sigma}}$ to the number of loops oriented counterclockwise times $e^{-2\pi i\tilde{\sigma}}$ to the number of loops oriented clockwise, where $\tilde{\sigma}$ is chosen carefully.

Proposition 3.3 (Baxter [**Bax89**]). *Let $G$ be a finite graph, we have*

$$f_{\text{Potts}}(\beta_c, q) = \frac{2\beta_c}{q-1} - \log q + 2 f_{6V}\big(1,1,1,1,2\cos(\pi\tilde{\sigma}),2\cos(\pi\tilde{\sigma})\big)$$

*where* $\cos(2\pi\tilde{\sigma}) = \sqrt{q}/2$.

The main advantage of the 6 vertex model over the Potts model is that it is exactly solvable. In other words, one may compute the free energy of the model via transfer matrices and the so-called Bethe Ansatz (see [**Bax89**] and references therein). As a result, the free energy of the critical Potts model can be computed explicitly. Note that this provides little information on the critical behavior of the Potts model since thermodynamical quantities of the model are expressed in terms of *derivatives* of the free energy. Computing the free energy at one point only is not sufficient to access these derivatives. However, Baxter [**Bax71, Bax73, Bax78, Bax89**] used this correspondence together with additional unproved assumptions to state the following conjecture.

Conjecture 2. *Consider the Potts model on the square lattice. For $q \le 4$, the phase transition is continuous (and therefore* **P1–5** *are satisfied), while for $q > 4$ it is discontinuous (thus* **P1–5** *are not satisfied).*

In [**LMR86, LMMS⁺91, KS82**], the FK percolation was proved to undergo a discontinuous phase transition at criticality when $q \ge 25.72$ via a Pirogov-Sinai type argument, see also the proof in [**Dum15**]. In the next section, we focus on the regime of values of $q$ for which the phase transition is continuous, i.e. $q \in [1,4]$. This leaves the range of parameter $q \in (4, 25.72)$ open to analysis (in such case one should prove that the phase transition is discontinuous).

**3.3. Values of $q$ for which the phase transition is continuous.** In this section, we focus on the $q \leq 4$ case. We shall discuss the following theorem.

THEOREM 3.4 (Duminil-Copin [**Dum12**]). *Let $1 \leq q \leq 4$, then*

$$\lim_{n\to\infty} \frac{1}{n} \log \phi^0_{p_c,q}[0 \longleftrightarrow \partial\Lambda_n] = 0.$$

This theorem, combined with Theorem 3.1, implies the following corollary.

COROLLARY 3.5. *The phase transition is continuous for the FK percolation with $1 \leq q \leq 4$, and therefore for the Potts models with 2, 3 and 4 colors.*

The case of the Ising model (i.e. $q = 2$) was solved by Onsager [**Ons44**].

3.3.1. *Definition of parafermionic observables.* In order to prove this result, we introduce a new tool called *parafermionic observable.* This observable is easier to understand in the context of Dobrushin boundary conditions, which are defined as follows. The following paragraphs are difficult to read, and Fig. 9 may replace conveniently the tedious definitions below.

Let $\Omega$ be a connected graph with connected complement in $\mathbb{Z}^2$, and $a$ and $b$ two vertices on its boundary. The triplet $(\Omega, a, b)$ is called a Dobrushin domain. The set $\partial\Omega$ is divided into two boundary arcs denoted by $\partial_{ab}$ and $\partial_{ba}$ (the first one goes from $a$ to $b$ when going counterclockwise around $\partial\Omega$, while the second goes from $b$ to $a$). The Dobrushin boundary conditions are

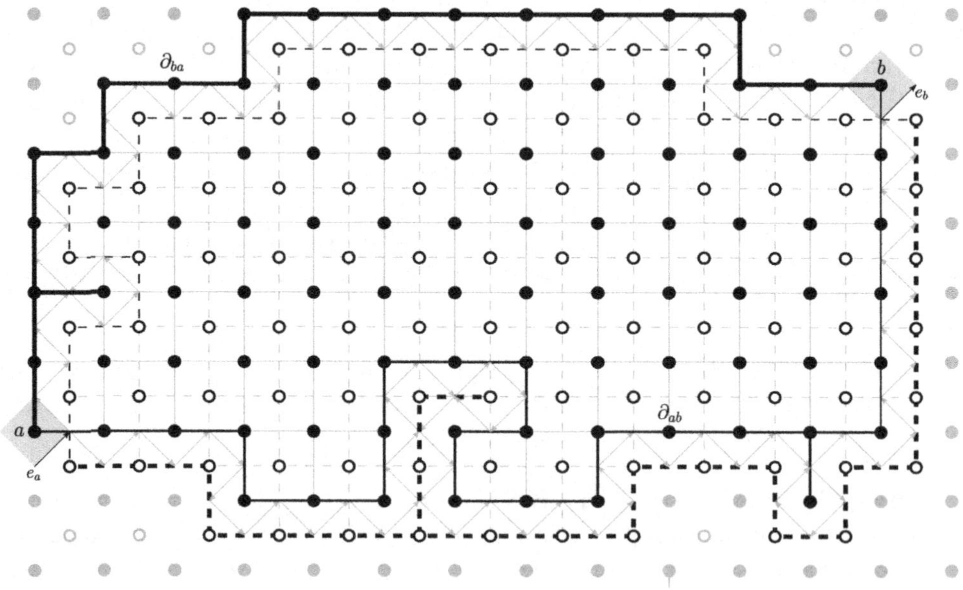


FIGURE 9. The primal and dual Dobrushin domains associated to a medial Dobrushin domain. Note the position of $a$ and $b$ and the definition of $\partial_{ba}$ and $\partial_{ab}$.

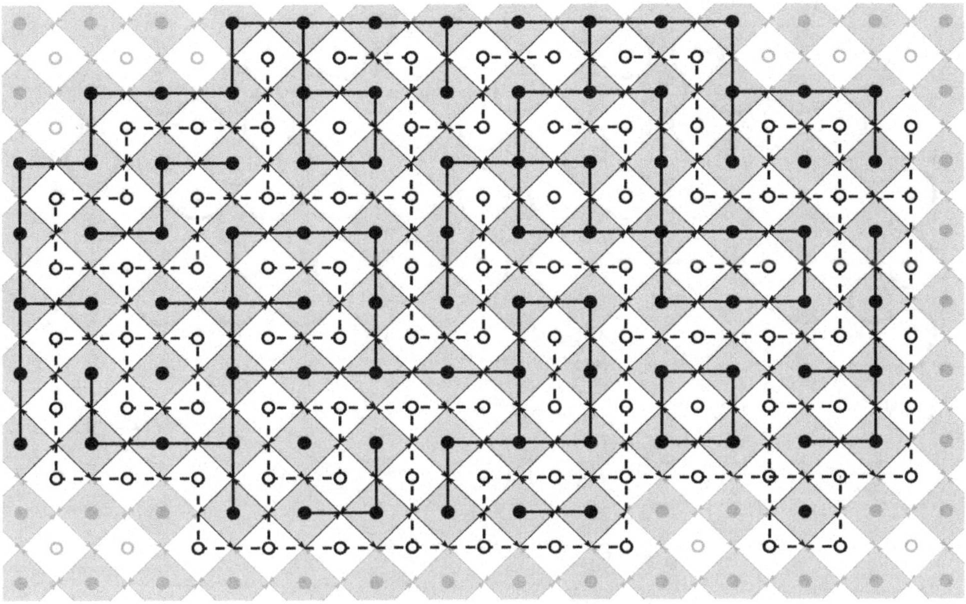

FIGURE 10. The configuration $\omega$ with its dual configuration $\omega^*$. Notice that the edges of $\omega$ are open on $\partial_{ba}$, and that those of $\omega^*$ are open on the part of the boundary of $\Omega^*$ bordering $\partial_{ab}$.

defined to be free on $\partial_{ab}$ and wired on $\partial_{ba}$. In other words, the partition is composed of $\partial_{ba}$ together with singletons.

Let $\Omega^*$ be the dual of the graph $\Omega$ whose set of edges is given by the edges of $(\mathbb{Z}^2)^*$ crossing an edge of $E(\Omega) \setminus \partial_{ba}$ and set of vertices given by the endpoints of these edges, see Fig. 9. Draw $\omega$ with the additional condition that edges of $\partial_{ba}$ are open, and the dual $\omega^*$ with the additional condition that edges on $\partial\Omega^*$ bordering $\partial_{ab}$ are open in $\omega^*$.

Define the medial graph $\Omega^\diamond$ of $\Omega$ as follows: the vertices are the vertices of $(\mathbb{Z}^2)^\diamond$ at the center of edges of $\Omega$ or $\Omega^*$. Let $e_a$ and $e_b$ be the two medial edges defined as on Fig. 9.

One may define a loop configuration $\overline{\omega}$ exactly as before, which this time contains loops together with a self-avoiding path going from $e_a$ to $e_b$, see Figures 9–11. This curve is called the *exploration path* and is denoted by $\gamma = \gamma(\omega)$. The loops correspond to the interfaces separating clusters from dual clusters, and the exploration path corresponds to the interface between the vertices connected in $\omega$ to $\partial_{ba}$ and the dual vertices connected in $\omega^*$ to $\partial^*_{ab}$.

The following definition will be instrumental in the reminder of this section.

DEFINITION 3.6. *The* winding $W_\Gamma(e, e')$ *of a curve* $\Gamma$ *(on the medial lattice) between two medial-edges* $e$ *and* $e'$ *of the medial graph is the total*

FIGURE 11. The loop configuration $\overline{\omega}$ associated to the primal and dual configurations $\omega$ and $\omega^*$ in the previous picture. The exploration path is drawn in bold. It starts at $e_a$ and finishes at $e_b$.

*signed rotation in radians that the (now oriented) curve makes from the mid-point of the edge e to that of the edge e'. By convention, if* $\Gamma$ *does not go through* $e'$, *we set* $W_\Gamma(e, e') = 0$.

The winding of $\Gamma$ can be computed in a very simple way: it corresponds to the number of $\frac{\pi}{2}$-turns on the left minus the number of $\frac{\pi}{2}$-turns on the right times $\pi/2$.

We are now in a position to define the parafermionic observable.

DEFINITION 3.7. *Consider a Dobrushin domain* $(\Omega, a, b)$. *The* parafermionic observable $F = F(\Omega, p, q, a, b)$ *is defined for any medial edge* $e \in E(\Omega^\diamond)$ *by*

$$F(e) := \phi_{\Omega,p,q}^{dobr}[e^{i\sigma W_\gamma(e,e_b)} \mathbf{1}_{e\in\gamma}],$$

*where* $\phi_{\Omega,p,q}^{dobr}$ *is the measure with Dobrushin boundary conditions on* $(\Omega, a, b)$ *and* $\gamma$ *is the exploration path and* $\sigma$ *is a solution of the equation*

$$\sin(\sigma\pi/2) = \sqrt{q}/2. \tag{3.3}$$

Note that $\sigma$ belongs to $\mathbb{R}$ for $q \le 4$ and to $1 + i\mathbb{R}$ for $q > 4$. This hints that the critical behavior of FK percolation is different for $q > 4$ and $q \le 4$. For $q \in [0, 4]$, $\sigma$ has the physical interpretation of a spin, which is fractional in general, hence the name parafermionic (fermions have half-integer spins while bosons have integer spins, there are no particles with fractional spin, but the use of such fractional spins at a theoretical level has been very fruitful in physics). For $q > 4$, $\sigma$ is not real anymore and does not have any physical interpretation.

These observables first appeared in the context of the Ising model (there they are called order-disorder operators) and dimer models. They were later on extended to FK percolation and the loop $O(n)$-model by Smirnov [**Smi06**] (see [**DS12a**] for more details). Since then, these observables have been at the heart of the study of these models. For an example of applications in a different context (for which the observable is simple to define and may appeal more to the reader), we refer to [**DS12b, BBMdG$^+$14, Gla13**] for a study of self-avoiding walks.

3.3.2. *Contour integrals of the parafermionic observable.* The parafermionic observable satisfies a spectacular property at criticality.

Let $(\Omega, a, b)$ be a Dobrushin domain. A *discrete contour* $\mathscr{C}$ is a finite sequence $z_0 \sim z_1 \sim \cdots \sim z_n = z_0$ in $V(\Omega) \cup V(\Omega^*)$ of neighboring points (meaning that they are at the center of adjacent faces of $\Omega^\diamond$) such that the path $(z_0, \ldots, z_n)$ is edge-avoiding. The discrete contour integral of the parafermionic observable $F$ along $\mathscr{C}$ is defined by

$$\oint_{\mathscr{C}} F(z)dz := \sum_{i=0}^{n-1} (z_{i+1} - z_i) F(\{z_i, z_{i+1}\}^*),$$

where the $z_i$ are considered as complex numbers and $\{z_i, z_{i+1}\}^*$ denotes the edge of $\Omega^\diamond$ intersecting $\{z_i, z_{i+1}\}$ in its center.

THEOREM 3.8 (Vanishing contour integrals). *Let $q > 0$, $p = p_c$, and a Dobrushin domain $(\Omega, a, b)$. For any discrete contour $\mathscr{C}$ of $(\Omega, a, b)$,*

$$\oint_{\mathscr{C}} F(z)dz = 0. \tag{3.4}$$

The property that discrete contour integrals vanish seems to correspond to a well-known property of holomorphic functions: their contour integrals are equal to 0. Nevertheless, one should be slightly careful when drawing such a parallel: the observable should rather be understood as the discretization of a form rather than a function. As a form, the fact that these discrete contour integrals vanish should be interpreted as the discretization of the property of being closed.

A troubling news is the fact that this property does not determine the function $F$ (for instance, even if boundary conditions are given, there is more than one function $F$ satisfying these boundary conditions and having zero contour integrals). Therefore, it is a priori not clear how one should be able to extract information from this property, yet the following section illustrates the fact that it can indeed be done.

3.3.3. *Sketch of the proof of Theorem 3.4.* Let us exploit the fact that the discrete integral along the boundary of a domain equals 0. Let $\mathbb{H} := \mathbb{N} \times \mathbb{Z}$. Taking the domain to be the rectangle $U_n := \Lambda_n \cap \mathbb{H}$, (3.4) applied to the discrete contour going around $U_n$ (i.e the unique contour crossing each edge of $U_n^\diamond$ incident to exactly one vertex of $\partial U_n^\diamond$) implies (after some work that we conveniently avoid presenting here) the following inequality: for $q \le 2$, there exists $c = c(q) > 0$ such that for any $n \ge 1$,

$$\sum_{x \in \partial\Lambda_n} \phi^{\text{free}}_{U_n,p_c,q}[0 \leftrightarrow x] \ge c. \tag{3.5}$$

Note that $\phi^{\text{free}}_{U_n,p_c,q}[0 \leftrightarrow x]$ is non zero for $x \in \partial\Lambda_n$ only if $x$ is also on $\partial U_n$. Therefore, the previous sum corresponds to the sum over vertices on the boundary of $U_n$ that are exactly at distance $n$ from the origin.

Using the comparison between boundary conditions, this implies that

$$\sum_{x \in \mathbb{Z}^2} \phi^{\text{free}}_{p_c,q}[0 \leftrightarrow x] \ge \sum_{n \ge 1} \sum_{x \in \partial\Lambda_n} \phi^{\text{free}}_{U_n,p_c,q}[0 \leftrightarrow x] = \infty. \tag{3.6}$$

The divergence of this series implies Theorem 3.4 for $q \le 2$.

The following discussion motivates why the case $q \in [2, 4]$ is more difficult. We do not provide much details but rather try to convey an important idea.

It is natural to predict that the following quantity decays like a power law:

$$\phi^{\text{free}}_{\mathbb{H},p_c,q}[0 \longleftrightarrow \partial\Lambda_{n/2}] = \frac{1}{n^{\alpha(q,\pi)+o(1)}},$$

where $\alpha(q, \pi)$ is a constant depending on $q$ only ($\pi$ refers to the "angle of the opening of $\mathbb{H}$" at 0), and $o(1)$ denotes a quantity tending to 0 as $n$ tends to infinity.

Moreover, one may argue (it is not straightforward) that the event that $x$ is connected to 0 in $U_n$ has a probability close to the probability that 0 and $x$ are connected to distance $n/2$ in $U_n$. For $x$ not too close to the corners of the rectangle, the boundary of $U_n$ looks like a straight line and it is therefore natural to predict that

$$\phi^{\text{free}}_{U_n,p_c,q}[0 \leftrightarrow x] = \frac{1}{n^{2\alpha(q,\pi)+o(1)}}.$$

Summing over all $x$ (for ease of exposition, let us ignore the problem of vertices at the corners), we deduce that

$$\sum_{x \in \partial\Lambda_n} \phi^{\text{free}}_{U_n,p_c,q}[0 \leftrightarrow x] = n^{1-2\alpha(q,\pi)+o(1)}. \tag{3.7}$$

Now, it is conjectured in physics that

$$\alpha(q, \pi) = 1 - 2\frac{\arccos(\sqrt{q}/2)}{\pi}.$$

Therefore for $q \in (2, 4]$, the quantity on the left-hand side of (3.7) is converging to 0 as $n \to \infty$ and it is therefore hopeless to get (3.5) for $q > 2$.

Nevertheless, we did not have to consider $\mathbb{H}$ in the first place. For instance, consider the graph $\mathbb{S}$ obtained by taking $\mathbb{Z}^2$ minus the half-line

$\{(-n,0), n \geq 0\}$. As before, one expects that

$$\phi^{\text{free}}_{\mathbb{S},p_c,q}[0 \longleftrightarrow \partial\Lambda_{n/2}] = \frac{1}{n^{\alpha(q,2\pi)+o(1)}},$$

where $\alpha(q, 2\pi)$ is a value which is a priori smaller than $\alpha(q, \pi)$ since $\mathbb{S}$ is larger ($2\pi$ refers this time to the "opening angle of $\mathbb{S}$ at" 0).

Therefore, if one applies the same reasoning as above to $\widetilde{U}_n := \Lambda_n \cap \mathbb{S}$, we may prove that

$$\sum_{x\in\partial\Lambda_n} \phi^{\text{free}}_{\widetilde{U}_n,p_c,q}[0 \leftrightarrow x] = n^{1-\alpha(q,\pi)-\alpha(q,2\pi)+o(1)}.$$

(We use that for $x \in \partial\Lambda_n$ not close to corners, the boundary looks straight, and that the boundary in $\widetilde{U}_n$ near 0 looks like in $\mathbb{S}$.) Since the map $z \mapsto z^2$ maps $\mathbb{R}^*_+ \times \mathbb{R}$ to $\mathbb{R}^2 \setminus -\mathbb{R}_+$, conformal invariance predicts that $\alpha(q, 2\pi) = \alpha(q, \pi)/2$. As a consequence,

$$\sum_{x\in\partial\Lambda_n} \phi^{\text{free}}_{\widetilde{U}_n,p_c,q}[0 \leftrightarrow x] = n^{1-\frac{3}{2}\alpha(q,\pi)+o(1)},$$

so that this quantity can indeed be larger or equal to 1 provided that $q \leq 3$.

The previous discussion remained at the level of predictions and does not seem to provide a good strategy for the proof since it would require to prove much more than we wish to get. A very good news is that the reasoning leading to (3.5) can indeed be applied to $\widetilde{U}_n$ instead of $U_n$ to give that for $q \leq 3$, there exists $c = c(q) > 0$ such that for any $n \geq 1$,

$$\sum_{x\in\partial\Lambda_n} \phi^{\text{free}}_{\widetilde{U}_n,p_c,q}[0 \leftrightarrow x] \geq c.$$

Since $\widetilde{U}_n$ is a subset of $\mathbb{Z}^2$, the comparison between boundary conditions implies that for any $q \leq 3$.

$$\sum_{x\in\mathbb{Z}^2} \phi^{\text{free}}_{p_c,q}[0 \leftrightarrow x] = \infty,$$

thus extending the result to every $q \leq 3$.

This reasoning does not directly extend to $q > 3$ since $\frac{3}{2}\alpha(q, \pi) > 1$ in this case. Nevertheless, one could consider a graph generalizing $\mathbb{H}$ and $\mathbb{S}$ with a "larger opening than $2\pi$" at 0. In fact, one may even consider a graph with "infinite opening" at 0 by considering subgraphs of the universal cover $\mathbb{U}$ of the plane minus a face of $\mathbb{Z}^2$, see Fig. 12. This is what was done in [**Dum12**]. The drawback of taking this set $\mathbb{U}$ is that it is not a subset of $\mathbb{Z}^2$ anymore. Thus, one has to translate the information obtained for the FK percolation on $\mathbb{U}$ into information for the FK percolation on $\mathbb{Z}^2$, which is a priori difficult since there is no easy comparison between the two graphs (for instance the comparison between boundary conditions is not sufficient).

To conclude this section, let us mention that the study of lattice models on discrete tori or more generally discrete Riemann surfaces has been the object of a lot of interest in recent years (for instance for dimers and the

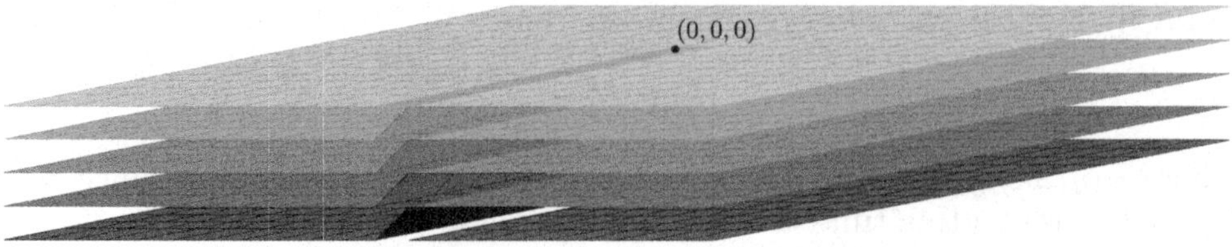


FIGURE 12. The graph $\mathbb{U}$.

Ising model). Some of the properties of these models are closely related to those on planar graphs, but new interesting features also emerge. The study of lattice models on $\mathbb{U}$ should also be very interesting, and the argument sketched above shows that $\mathbb{U}$ can be useful to understand the models on planar graphs.

**3.4. To infinity and beyond: conformal invariance at criticality.** We are interested in the rich behavior of FK percolations at the critical value of their continuous phase transition. From now on, we describe the large scale behavior of the macroscopic clusters by rescaling the lattice in such a way that the mesh (i.e. the length of edges) tends to zero. This procedure is called taking the *scaling limit*. In order to illustrate this procedure, let us consider a very explicit construction.

In this section, $(\Omega, a, b)$ always denotes a simply connected domain $\Omega$ together with two points $a$ and $b$ on its boundary. Also consider Dobrushin domains $(\Omega_\delta, a_\delta, b_\delta)$ of $\delta\mathbb{Z}^2$ converging to $(\Omega, a, b)$ in the Carathéodory sense (see [**Dum15**] and references therein for details). For smooth boundary for instance, one can for instance consider Dobrushin domains such that $\partial\Omega_\delta$ converges in the Hausdorff sense to the boundary of $\Omega$, and that $a_\delta$ and $b_\delta$ tend to $a$ and $b$ respectively.

In the Dobrushin domains $(\Omega_\delta, a_\delta, b_\delta)$ with Dobrushin boundary conditions, one can consider the exploration paths $\gamma_{(\Omega_\delta,a_\delta,b_\delta)}$. These exploration paths form a family of random variables indexed by $\delta > 0$ and we may study the convergence of this family as $\delta$ tends to 0.

Conformal Field Theory predicts that at criticality, $\gamma_{(\Omega_\delta,a_\delta,b_\delta)}$ converges as $\delta$ tends to 0 to a random, continuous, non-self-crossing curve $\gamma_{(\Omega,a,b)}$ from $a$ to $b$ staying in $\Omega$. Furthermore, the family of curves $(\gamma_{(\Omega,a,b)} : (\Omega, a, b))$ is expected to be conformally invariant in the following sense: for any $(\Omega, a, b)$ and any conformal (i.e. holomorphic and one-to-one) map $\psi : \Omega \to \mathbb{C}$,

$$\psi(\gamma_{(\Omega,a,b)}) \text{ has the same law as } \gamma_{(\psi(\Omega),\psi(a),\psi(b))}.$$

Notice that the conformal invariance of the limit implies the following fact. The random curve obtained by taking the scaling limit of the FK percolation in $(\psi(\Omega), \psi(a), \psi(b))$ has the same law as the image by $\psi$ of the random curve obtained by taking the scaling limit of the FK percolation in $(\Omega, a, b)$. This is clear for $\psi$ which corresponds to a symmetry of the lattice (for instance the rotation by $\frac{\pi}{2}k$ for some $k \in \mathbb{Z}$), but this claim implies that the property is true for any conformal transformation (therefore in particular for a rotation by any angle).

Fifteen years ago, Schramm [**Sch00**] proposed a natural candidate for the possible conformally invariant families of continuous non-self-crossing curves. He noticed that interfaces of models further satisfy the *domain Markov property* which, together with the assumption of conformal invariance, determines a one-parameter family of possible curves: the Stochastic Loewner Evolutions (SLE, which are now known as the Schramm–Loewner Evolutions). Let us briefly describe this object now (see [**Law05, Wer04, Wer05**] for comprehensive expositions). We start by recalling the definition of a Loewner chain.

Set $\mathbb{H}$ to be the upper half-plane $\mathbb{R} \times (0, \infty)$. Fix a simply connected subdomain $H$ of $\mathbb{H}$ such that $\mathbb{H} \setminus H$ is compact. Riemann's mapping theorem guarantees the existence of a conformal map from $H$ onto $\mathbb{H}$. Moreover, there are *a priori* three real degrees of freedom in the choice of the conformal map, so that it is possible to fix its asymptotic behavior as $z$ goes to $\infty$. Let $g_H$ be the *unique* conformal map from $H$ onto $\mathbb{H}$ such that

$$g_H(z) \ := \ z + \frac{C}{z} + O\left(\frac{1}{z^2}\right).$$

(The proof of the existence of this map is not completely obvious and requires Schwarz's reflection principle.)

Let $(W_t)_{t>0}$ be a continuous real-valued function (one usually requires a few things about this function, but let us omit these technical conditions here). Fix $z \in \mathbb{H}$ and consider the map $t \mapsto g_t(z)$ satisfying the following differential equation up to its explosion time:

$$\partial_t g_t(z) \ = \ \frac{2}{g_t(z) - W_t}. \tag{3.8}$$

For every fix $t$, let $H_t$ be the set of $z$ for which the explosion time of the differential equation above is strictly larger than $t$. One may verify that $H_t$ is a simply connected open set and that $\mathbb{H} \setminus H_t$ is compact. Furthermore, the map $z \mapsto g_t(z)$ is a conformal map from $H_t$ to $\mathbb{H}$.

If there exists a parametrized curve $(\Gamma_t)_{t>0}$ such that for any $t > 0$, $H_t$ is the connected component of $\mathbb{H} \setminus \Gamma[0, t]$ containing $\infty$, the curve $(\Gamma_t)_{t>0}$ is called *(the curve generating) the Loewner chain with driving process* $(W_t)_{t>0}$.

Then, the Loewner chain in $(\Omega, a, b)$ with driving function $(W_t)_{t>0}$ is simply the image of the Loewner chain in $(\mathbb{H}, 0, \infty)$ by a conformal from $(\mathbb{H}, 0, \infty)$ to $(\Omega, a, b)$.

DEFINITION 3.9. *For $\kappa > 0$ and $(\Omega, a, b)$, SLE($\kappa$) is the random Loewner evolution in $(\Omega, a, b)$ with driving process $\sqrt{\kappa} B_t$, where $(B_t)$ is a standard Brownian motion.*

By construction, the process is conformally invariant, random and fractal (see Fig. 13). In addition, it is possible to study quite precisely the behavior

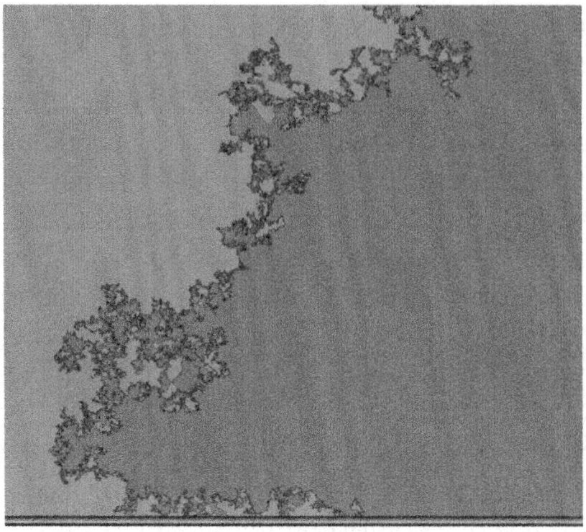

FIGURE 13. A simulation of SLE(6) due to V. Beffara.

of SLEs using stochastic calculus and to derive some of their path's properties (e.g. Hausdorff dimension, intersection exponents, etc). Because of the deep understanding of these random curves, the following conjecture is of first importance.

CONJECTURE 3 (Schramm). *Let* $(\Omega_\delta, a_\delta, b_\delta)$ *converging to* $(\Omega, a, b)$. *The exploration path* $\gamma_{(\Omega_\delta,a_\delta,b_\delta)}$ *of the* critical *FK percolation with parameters* $q \in [0,4]$ *and* $p = p_c(q)$ *converges weakly to SLE(*$\kappa$*) as* $\delta$ *tends to* 0, *where*

$$\kappa = \kappa(q) := \frac{4\pi}{\pi - \arccos(\sqrt{q}/2)}.$$

For completeness, let us mention the topology for the weak convergence mentioned in the conjecture. Let $X$ be the set of continuous parametrized curves and the distance $d(\gamma_1, \gamma_2)$ between $\gamma_1 : I \to \mathbb{C}$ and $\gamma_2 : J \to \mathbb{C}$ defined by

$$d(\gamma_1, \gamma_2) = \min_{\substack{\varphi_1:[0,1]\to I \\ \varphi_2:[0,1]\to J}} \sup_{t\in[0,1]} |\gamma_1(\varphi_1(t)) - \gamma_2(\varphi_2(t))|,$$

where the minimization is over *increasing bijective functions* $\varphi_1$ and $\varphi_2$. Note that $I$ and $J$ can be equal to $\mathbb{R}_+ \cup \{\infty\}$. The topology on $(X, d)$ gives rise to a notion of weak convergence for random curves on $X$.

Recently, the case $q = 2$ of Schramm's conjecture was settled by Chelkak, Duminil-Copin, Hongler, Kemppainen, Smirnov [**CDH**$^+$**14**]. The strategy of the proof is the following. First, prove that the family $(\gamma_{(\Omega_\delta,a_\delta,b_\delta)})_{\delta>0}$ is tight for the weak convergence, see e.g. [**KS12, CDH13, DS12a**]. Second, use the parafermionic observable with $q = 2$ in order to show that any sub-sequential limit of $(\gamma_{(\Omega_\delta,a_\delta,b_\delta)})_{\delta>0}$ is SLE(16/3). This second step is the crucial step of the proof, and it is based on the following result, which represents the main difficulty of the proof of conformal invariance.

Define the *normalized vertex fermionic observable (at criticality)* by

$$f_\delta(v) = f_\delta(\Omega_\delta, a_\delta, b_\delta, p_c, 2, v) := \frac{1}{\sqrt{2e_b}} \begin{cases} \frac{1}{2}\sum_{u\sim v} F_\delta(\{u,v\}) & \text{if } v \in \Omega_\delta^\diamond \setminus \partial\Omega_\delta^\diamond, \\ \frac{2}{2+\sqrt{2}}\sum_{u\sim v} F_\delta(\{u,v\}) & \text{if } v \in \partial\Omega_\delta^\diamond, \end{cases}$$

where $F_\delta$ is the edge fermionic observable defined in Definition 3.7, and $e_b$ is seen as a complex number (note that the modulus of $e_b$ is $\frac{\sqrt{2}}{2}\delta$).

THEOREM 3.10 (Smirnov [**Smi10**]). *Fix $q = 2$ and $p = p_c(2)$. Let $(\Omega, a, b)$ be a simply connected domain with two marked points on its boundary. Let $(\Omega_\delta, a_\delta, b_\delta)$ be a family of Dobrushin domains converging to $(\Omega, a, b)$ in the Carathéodory sense. Let $f_\delta$ be the* normalized vertex fermionic observable *in $(\Omega_\delta, a_\delta, b_\delta)$. We have*

$$f_\delta(\cdot) \longrightarrow \sqrt{\phi'(\cdot)} \quad \textit{when } \delta \to 0 \tag{3.9}$$

*uniformly on any compact subset of $\Omega$, where $\phi$ is any conformal map from $\Omega$ to the strip $\mathbb{R} \times (0,1)$ mapping $a$ to $-\infty$ and $b$ to $\infty$.*

Let us notice that the map $\phi$ is not unique since one could add any real constant to $\phi$, but this modification does not change its derivative.

The function $\sqrt{\phi'(\cdot)}$ is the holomorphic solution of a Riemann-Hilbert boundary value problem: its boundary values are orthogonal to the square root of the normal vector to $\partial\Omega$. It does not come as a surprise that the strategy of Smirnov's proof relies on this point. In fact, Smirnov proves that $f_\delta$ is a *discrete holomorphic* function satisfying a proper discretization of the boundary conditions mentioned above. Furthermore, he was able to show that any solution of this "discrete Riemann-Hilbert boundary value problem" must converge to the continuous solution of its continuum counterpart as $\delta$ tends to 0, thus implying that $f_\delta$ tends to $\sqrt{\phi'}$.

This result is heavily based on the theory of discrete holomorphic maps. Other than being interesting in themselves, discrete holomorphic functions have also found several applications in geometry, analysis, combinatorics, and probability. We refer the interested reader to the expositions by Lovász [**Lov04**], Stephenson [**Ste05**], Mercat [**Mer01**], Bobenko and Suris [**BS08**]. This beautiful tool has been applied to several statistical physics models, in particular in the work of Kenyon on dimers [**Ken00**].

Note that the FK percolation with $q = 2$ is coupled to the Ising model. The Ising model itself was proved to be conformally invariant in [**CS12**], and since then, conformal invariance of many other quantities describing the critical behavior have been derived (crossing probabilities [**BDH14**], interfaces with other boundary conditions [**HK13, Izy15**], energy and spin fields [**HS13, Hon10, CI13, CHI15**]), etc.

Let us mention that the parafermionic observable has even been used off criticality, where it does not satisfy discrete holomorphicity, to study the sub and supercritical phases; see [**BD12b, DGP14**].

We conclude this section by mentioning an important conjecture. Define $f_\delta(v)$ as before except that $\sqrt{2e_b}$ is replaced by $(2e_b)^\sigma$ (let us restrict ourselves to non-boundary vertices only).

CONJECTURE 4 (Smirnov [**Smi10**]). *Fix $q \in (0,4)$ and $p = p_c(q)$. Let $(\Omega, a, b)$ be a simply connected domain with two marked points on its boundary. Let $(\Omega_\delta, a_\delta, b_\delta)$ be a family of Dobrushin domains converging to $(\Omega, a, b)$ in the Carathéodory sense. Let $f_\delta$ be the* normalized vertex fermionic observable *in $(\Omega_\delta, a_\delta, b_\delta)$. We have*

$$f_\delta(\cdot) \longrightarrow \phi'(\cdot)^\sigma \quad \text{when } \delta \to 0 \tag{3.10}$$

*uniformly on any compact subset of $\Omega$, where $\phi$ is any conformal map from $\Omega$ to the strip $\mathbb{R} \times (0,1)$ mapping $a$ to $-\infty$ and $b$ to $\infty$.*

## 4. Critical behavior in dimension 3 and more

**4.1. Mean field prediction for the order of the phase transition.** Let $q \geq 2$. Consider the Potts model on the complete graph $G_n$ given by the vertex set $\{1,\dots,n\}$ and edge set composed of any pair of vertices in $\{1,\dots,n\}$. The Hamiltonian on $\mathbb{T}_q^{\{1,\dots,n\}}$ is given by

$$H^{\text{free}}_{G_n}[\sigma] := -\tfrac{1}{n}\sum_{x,y=1}^{n} \langle \sigma_x | \sigma_y \rangle$$

and the associated measure by $\mu^{\text{free}}_{G_n,\beta}$.

Introduce the following functionals: for $h \in \mathbb{R}^{q-1}$, define $G(h)$ to be the Laplace transform of $d\sigma_0$ at $h$, i.e.

$$G(h) := \log\Big(\int \exp(\langle h|\sigma_0\rangle) d\sigma_0\Big) = \log\Big(\sum_{v\in\mathbb{T}_q} e^{\langle h|v\rangle}\Big).$$

For $m \in \mathbb{R}^{q-1}$, define

$$\Phi_{\beta,q}(m) = -\tfrac{\beta}{2}\|m\|^2 - \inf_{h\in\mathbb{R}^{q-1}}\{G(h) - \langle m,h\rangle\}.$$

(Recall that $\|\cdot\|$ is the Euclidean norm on $\mathbb{R}^{q-1}$.)

The following fact justifies the introduction of these definitions: the average magnetization $\mathscr{M}(G_n)$ concentrates on values of $m$ minimizing $\Phi_{\beta,q}(m)$ in the following sense: let

$$M_\beta := \text{argmin}\{\Phi_{\beta,q}(m) : m \in \mathbb{R}^{q-1}\} \subset \mathbb{R}^{q-1},$$

then for any $\varepsilon > 0$,

$$\lim_{n\to\infty} \mu^{\text{free}}_{G_n,\beta}\big[d(\mathscr{M}(G_n), M_\beta) > \varepsilon\big] = 0. \tag{4.1}$$

(Above, $d(x,E)$ is the standard definition of distance from a point to a set.) In particular, if $M_\beta$ is a singleton (which is in fact equivalent in our case to $M_\beta = \{0\}$ due to the symmetries of $M_\beta$), then $\mathscr{M}(G_n)$ must converge to 0 in probability. This happens up to a certain inverse temperature $\beta_c(MF) \in$

$(0, \infty)$. On the other hand, for $\beta > \beta_c(MF)$, the set $M_\beta$ contains more than one element. In fact, the function $\Phi_{\beta,q}$ is easy to compute and study so that

- When $q = 2$,
$$M_\beta = \{-m^*(\beta, 2, \infty), m^*(\beta, 2, \infty)\} =: m^*(\beta, 2, \infty)\ \mathbb{T}_2$$
for $\beta \geq 0$, where $m^*(\beta, 2, \infty) \geq 0$ is continuous and equal to 0 if and only if $\beta \leq \beta_c(MF)$. In particular, $M_{\beta_c(MF)} = \{0\}$.
- When $q \geq 3$,
$$M_\beta = \begin{cases} m^*(\beta, q, \infty)\ \mathbb{T}_q & \text{if } \beta \neq \beta_c(MF) \\ \{0\} \cup m^*(\beta, q, \infty)\ \mathbb{T}_q & \text{if } \beta = \beta_c(MF) \end{cases},$$
where $m^*(\beta, q, \infty) \geq 0$ equals 0 if and only if $\beta < \beta_c(MF)$, is right continuous everywhere and left continuous except at $\beta_c(MF)$ where it has a discontinuity.

Because of (4.1), it is natural to interpret $m^*(\beta, q, \infty)$ as the spontaneous magnetization of some kind of mean-field infinite-volume Potts measure. The behavior for $q = 2$ and $q \geq 3$ is then very different, since the function $\beta \mapsto m^*(\beta, q, \infty)$ is continuous in the first case and discontinuous in the second.

**4.2. Discontinuity of the phase transition for $q \geq 3$.** This section will be devoted to the brief description of two results describing discontinuous phase transitions for the Potts model. They both deal with either $q \gg 1$ or $d \gg 1$.

Case of $q \geq 3$ and $d \gg 1$. Our goal is to compare the behavior of the model to the mean field behavior. We will use the following theorem. Let $\widetilde{\sigma}_x := \sigma_x - m^*(\beta, q, d)\vec{1}$ be the spin at $x$ centered around its mean (this definition is useful to speak of so-called truncated correlations).

THEOREM 4.1 (Biskup, Chayes [**BC03**]). *Let $d \geq 3$ and $q \geq 2$. Then,*

(4.2)
$$\Phi_{\beta,q}\big(m^*(\beta, q, d)\big) \leq \inf_{m \in \mathbb{R}^{q-1}} \Phi_{\beta,q}(m) + \tfrac{\beta}{2}\left(\left|\mu^1_{\beta,q}[\langle\sigma_0|\sigma_x\rangle]\right|^2 - \left|m^*(\beta, q, d)\right|^2\right)$$

(4.3)
$$\leq \inf_{m \in \mathbb{R}^{q-1}} \Phi_{\beta,q}(m) + \tfrac{\beta n}{2} I_d,$$

*where $x$ is a neighbor of the origin and $I_d$ is the so-called bubble diagram defined by*
$$I_d := \sum_{x \in \mathbb{Z}^d} \mu^1_{\beta,q}\big[\langle\widetilde{\sigma}_0|\widetilde{\sigma}_x\rangle\big]^2.$$

We will not use (4.2), but we included here to illustrate how (4.3) is obtained. The proof mainly uses convexity arguments involving the free energy of the model. We refer to [**BC03**] for details.

Equation (4.3) on the other hand is very useful. Assume for a moment that $I_d$ is tiny. Then, (4.3) yields that the spontaneous magnetization $m^*(\beta, q, d)$ is almost a minimizer of $\Phi_{\beta,q}$. This implies that $m^*(\beta, q, d)$ is close to an element of the set $M_\beta$. If one fixes $q \geq 3$, one observes that non-zero minimizers of $\Phi_{\beta,q}$ remain bounded away from 0 uniformly in $\beta$. If $I_d$ is sufficiently small, this excludes a whole range of possible values for the magnetization, and therefore forces it to be discontinuous provided that we show that $m^*(\beta, q, d)$ is not always close to 0 (note that we already know that for $\beta < \beta_c$, $m^*(\beta, q, d)$ is equal to 0). But the coupling with FK percolation implies easily that $m^*(\beta, q, d)$ converges to 1 as $\beta$ tends to $\infty$, and is therefore close to a non-zero element of $M_\beta$. This concludes the proof of the existence of a discontinuity for $m^*(\beta, q, d)$, once again provided that $I_d$ is tiny.

This last fact is not true in small dimensions. Nevertheless, $I_d$ tends to 0 as $d$ tends to infinity (see below), which implies that for any $q \geq 3$, there exists a dimension $d_c(q)$ such that for any $d \geq d_c(q)$, $m^*(\beta, q, d)$ has a discontinuity in $\beta$.

Note that the previous result does not prove that there is a jump in the spontaneous magnetization at $\beta_c$, but only at some $\beta$ (which must necessarily be larger or equal to $\beta_c$). It would be very interesting to prove that this value of $\beta$ is necessarily $\beta_c$. This should be the case, since it is predicted that $\beta \mapsto m^*(\beta, q)$ is continuous on $\mathbb{R} \setminus \{\beta_c\}$.

As mentioned above, we only need to prove that $I_d$ tends to zero as $d$ tends to infinity. This is a fairly direct and simple computation using an upper bound on spin-spin correlations given by the celebrated Infrared Bound, that we discuss now.

Let $G(x, y)$ be the Green function, i.e. the expected time spent at $y$ by a simple random walk starting at $x$. Note that the assumption that $d \geq 3$ guarantees that $G(x, y)$ is finite, since the random-walk is transient in dimension $d \geq 3$.

THEOREM 4.2 (Infrared Bound). *Consider the q-state Potts model on $\mathbb{Z}^d$ with $d \geq 3$. For any $\beta \in [0, \infty]$ and $v \in (\mathbb{R}^{q-1})^{\mathbb{Z}^d}$ with finite support,*

$$\sum_{x,y \in \mathbb{Z}^d} v_x v_y \mu_\beta^1[\langle \widetilde{\sigma}_x | \widetilde{\sigma}_y \rangle] \leq \frac{q-1}{2\beta} \sum_{x,y \in \mathbb{Z}^d} v_x v_y G(x, y).$$

Note that $G(x, y)$ is the spin-spin correlation for the discrete GFF. Since the $\phi_d^4$ lattice model interpolates between the Ising and the discrete GFF, it is not so surprising that spin-spin correlations of both models can be compared. The proof of this theorem is based on the so-called Reflection Positivity technique, see Fröhlich, Simon and Spencer [**FSS76**]. This technique has many applications in different fields of mathematical physics, we refer to [**Bis09**] and references therein for a more comprehensive study of this subject.

Case of $q \gg 1$ and $d \geq 2$. In this context, the Pirogov-Sinai theory can be harnessed to show that the phase transition is discontinuous. The original result uses Reflection Positivity [**KS82**] as well. Since then, the result was obtained without the help of Reflection Positivity by [**LMR86**] and can in fact be extended to FK percolation (also see [**DC**] for the 2D case).

**4.3. The case of the Ising model.** For many reasons, the Ising model is special among Potts models. One of these reasons is the $+/-$ gauge symmetry: flipping all the spins leaves the measure invariant (for free boundary conditions). In this section, we will explain how this property can be used to derive another geometric representation for the Ising model.

We do not restrict our attention to the nearest neighbor model anymore and treat general ferromagnetic models with interactions $(J_{xy} : \{x,y\} \subset \mathbb{Z}^d)$ satisfying $J_{xy} \geq 0$ for any $x, y$. In this section edges are simply pairs $\{x,y\} \subset \mathbb{Z}^d$. The definition of a percolation configuration is modified accordingly.

To adopt standard notation, we will write $\mu^+_{G,\beta}$ instead of $\mu^1_{G,\beta}$. Also note that the Ising model may be defined in infinite-volume thanks to the coupling with the FK percolation[1]: consider the infinite-volume FK measure with free boundary conditions and assign to each cluster a spin $+$ or $-$ to obtain a measure that we denote by $\mu^{\text{free}}_\beta$. Similarly, consider the infinite-volume FK measure with wired boundary conditions and assign to each finite cluster a spin $+$ or $-$, and to the infinite cluster (if it exists) a spin $+$, to obtain $\mu^+_\beta$. We have that $\mu^{\text{free}}_{G,\beta}$ and $\mu^+_{G,\beta}$ converge to $\mu^{\text{free}}_\beta$ and $\mu^+_\beta$ respectively as $G \nearrow \mathbb{Z}^d$.

4.3.1. *Harvesting the $+/-$ gauge symmetry: the random-current representation.* The following perspective on the Ising model's phase transition is driven by the observation that the onset of long range order coincides with a percolation transition in a dual system of currents. This point of view was developed in [**Aiz82**] and a number of subsequent works. Its advantage comes from the fact that we import the intuition and some tools from the theory of Bernoulli percolation that enable us to draw parallels between proofs in the contexts of Bernoulli percolation and the Ising model.

DEFINITION 4.3. *A* current $\mathbf{n}$ *on $G \subset \mathbb{Z}^d$ is a function from $\{\{x,y\} \subset V(G)\}$ to $\mathbb{N} := \{0,1,2,...\}$. A* source *of $\mathbf{n} = (\mathbf{n}_{xy} : \{x,y\} \subset V(G))$ is a vertex $x$ for which $\sum_{y\in V(G)} \mathbf{n}_{xy}$ is odd. The set of sources of $\mathbf{n}$ is denoted by $\partial\mathbf{n}$. The collection of currents on $G$ is denoted by $\Omega_G$. Also set*

$$w_\beta(\mathbf{n}) = \prod_{\{x,y\}\subset V(G)} \frac{(\beta J_{xy})^{\mathbf{n}_{xy}}}{\mathbf{n}_{xy}!}.$$

Let us describe the connection between currents and the Ising model. We start with the random current representation for free boundary conditions.

[1]In this case, the FK percolation is not a nearest neighbor model anymore; it can be interpreted as a FK percolation on the complete graph with vertex-set $\mathbb{Z}^d$. The edge-weights depend on the pairs $\{x,y\}$ through the same relation as for nearest neighbors, $p_{xy} = 1 - e^{-\beta J_{xy}}$.

For $\beta > 0$, $G$ a finite graph and $A \subset V(G)$, introduce the quantity

$$Z^{\text{free}}(G,\beta,A) = \sum_{\sigma\in\{-1,1\}^{V(G)}} \sigma_A \prod_{\{x,y\}\subset V(G)} \exp[\beta J_{xy}\sigma_x\sigma_y], \tag{4.4}$$

where $\sigma_A := \prod_{x\in A}\sigma_x$.

PROPOSITION 4.4. *Let $\beta > 0$, $G$ be a finite graph and $A \subset V(G)$, then*

$$Z^{\text{free}}(G,\beta,A) = 2^{|V(G)|} \sum_{\mathbf{n}\in\Omega_G:\,\partial\mathbf{n}=A} w_\beta(\mathbf{n}). \tag{4.5}$$

PROOF. Expanding $e^{\beta J_{xy}\sigma_x\sigma_y}$ for each $\{x,y\}$ to obtain

$$e^{\beta J_{xy}\sigma_x\sigma_y} = \sum_{\mathbf{n}_{xy}=0}^{\infty} \frac{(\beta J_{xy}\sigma_x\sigma_y)^{\mathbf{n}_{xy}}}{\mathbf{n}_{xy}!}$$

and substituting this relation in (4.4), one gets

$$Z^{\text{free}}(G,\beta,A) = \sum_{\mathbf{n}\in\Omega_G} w_\beta(\mathbf{n}) \sum_{\sigma\in\{-1,1\}^{V(G)}} \prod_{x\in V(G)} \sigma_x^{\mathbf{1}_{x\in A}+\sum_{y\in V(G)}\mathbf{n}_{xy}}.$$

Now, let $x \in V(G)$. Pairing configurations $\sigma$ with the configuration $\sigma^{(x)}$ which coincides with $\sigma$ except at $x$ where the spin is reversed, we see that $\mathbf{1}_{x\in A} + \sum_{y\in V(G)}\mathbf{n}_{xy}$ has to be even for the sum on configurations $\sigma$ not to be equal to 0. This implies that

$$\sum_{\sigma\in\{-1,1\}^{V(G)}} \prod_{x\in V(G)} \sigma_x^{\mathbf{1}_{x\in A}+\sum_{y\in V(G)}\mathbf{n}_{xy}} = \begin{cases} 0 & \text{if } \sum_{y\in V(G)}\mathbf{n}_{xy} \text{ is odd for some } x\notin A \text{ or even for } \textit{some } x\in A, \\ 2^{|V(G)|} & \text{otherwise.}\end{cases}$$

Thus, the definition of a current's source enables one to write

$$Z^{\text{free}}(G,\beta,A) = 2^{|V(G)|} \sum_{\mathbf{n}\in\Omega_G:\,\partial\mathbf{n}=A} w_\beta(\mathbf{n}) \qquad \square.$$

We deduce that for every $A \subset V(G)$,

$$\mu^{\text{free}}_{G,\beta}[\sigma_A] = \frac{Z^{\text{free}}(G,\beta,A)}{Z^{\text{free}}(G,\beta,\emptyset)} = \frac{\sum_{\mathbf{n}\in\Omega_G:\,\partial\mathbf{n}=A} w_\beta(\mathbf{n})}{\sum_{\mathbf{n}\in\Omega_G:\,\partial\mathbf{n}=\emptyset} w_\beta(\mathbf{n})}. \tag{4.6}$$

REMARK 4.5. *Equation* (4.6) *implies the first Griffiths' inequality: for any $A \subset V(G)$,*

$$\mu^{\text{free}}_{G,\beta}[\sigma_A] \ge 0.$$

Let us now turn to the random current representation for + boundary conditions. Introduce an additional vertex $\delta \notin \mathbb{Z}^d$ called the *ghost vertex*, and set the coupling $J_{x\delta} = J_{x\delta}(G)$ between it and vertices $x \in V(G)$ to be $\sum_{y\notin V(G)} J_{xy}$. We also set $J_{\delta\delta} = 0$.

Introduce, for $\beta > 0$, $G$ a finite graph and $A \subset V(G)$ the quantity

(4.7)
$$\begin{aligned}Z^+(G,\beta,A) &:= \sum_{\sigma\in\{-1,1\}^{V(G)}} \sigma_A \prod_{\{x,y\}\subset V(G)} \exp[\beta J_{xy}\sigma_x\sigma_y] \prod_{\substack{x\in V(G)\\ y\notin V(G)}} \exp[\beta J_{xy}\sigma_x]\\ &= \sum_{\sigma\in\{-1,1\}^{V(G)}} \sigma_A \prod_{\{x,y\}\subset V(G)} \exp[\beta J_{xy}\sigma_x\sigma_y] \prod_{x\in V(G)} \exp[\beta J_{x\delta}\sigma_x].\end{aligned}$$

A development similar to the above yields the following proposition.

PROPOSITION 4.6. *Let $\beta > 0$, $G$ be a finite graph and $A \subset V(G)$, then*

$$Z^+(G,\beta,A) = 2^{|V(G)|} \sum_{\mathbf{n}\in\Omega_{G\cup\{g\}}:\,\partial\mathbf{n}=A} w_\beta(\mathbf{n}). \tag{4.8}$$

Observe that (4.8) differs from (4.5) only through the fact that the summation is over all currents on $G \cup \{g\}$ instead of $G$. Also note that $J_{x\delta}$ depends on $G$.

We obtain a percolation configuration from a current by focusing on the trace $\widehat{\mathbf{n}} \in \{0,1\}^{E(G)}$ of the current $\mathbf{n}$ defined as follows:

$$\widehat{\mathbf{n}}_{xy} = \begin{cases} 1 & \text{if } \mathbf{n}_{xy} > 0,\\ 0 & \text{otherwise.}\end{cases}$$

As a consequence, we may speak of $x \overset{\widehat{\mathbf{n}}}{\longleftrightarrow} y$ for two vertices $x$ and $y$.

The percolation properties of the random currents are related to the correlations in the Ising model, very much like the percolation properties of FK percolation are related to correlations in the Potts model. In order to illustrate this fact, let us focus on the following proposition.

PROPOSITION 4.7. *Let $G$ be a finite graph, $\beta > 0$ and $x, y \in G$, then*

$$\mu^{\text{free}}_{G,\beta}[\sigma_x\sigma_y]^2 = \frac{\displaystyle\sum_{\partial\mathbf{n}_1=\partial\mathbf{n}_2=\emptyset} w_\beta(\mathbf{n}_1)w_\beta(\mathbf{n}_2)\mathbf{I}[x \overset{\widehat{\mathbf{n}_1+\mathbf{n}_2}}{\longleftrightarrow} y]}{\displaystyle\sum_{\partial\mathbf{n}_1=\partial\mathbf{n}_2=\emptyset} w_\beta(\mathbf{n}_1)w_\beta(\mathbf{n}_2)}.$$

This proposition follows from the following lemma. Below, $A\Delta B := (A\setminus B)\cup(B\setminus A)$ denotes the symmetric difference between $A$ and $B$.

LEMMA 4.8 (Switching lemma [**GHS70, Aiz82**]). *For any nested pair of finite graph $G$, any pair of vertices $x, y \in V(G)$, any $A \subset V(G)$, and any function $F : \Omega_G \to \mathbb{R}$:*

$$\sum_{\substack{\mathbf{n}_1\in\Omega_G:\,\partial\mathbf{n}_1=\{x,y\}\\ \mathbf{n}_2\in\Omega_G:\,\partial\mathbf{n}_2=A}} F(\mathbf{n}_1+\mathbf{n}_2)w_\beta(\mathbf{n}_1)w_\beta(\mathbf{n}_2)$$

$$= \sum_{\substack{\mathbf{n}_1\in\Omega_G:\,\partial\mathbf{n}_1=\emptyset\\ \mathbf{n}_2\in\Omega_G:\,\partial\mathbf{n}_2=A\Delta\{x,y\}}} F(\mathbf{n}_1+\mathbf{n}_2)w_\beta(\mathbf{n}_1)w_\beta(\mathbf{n}_2)\mathbf{I}[x \overset{\widehat{\mathbf{n}_1+\mathbf{n}_2}}{\longleftrightarrow} y].$$

This lemma is very useful when considering sums of two currents. It enables one to switch sources from one current $\mathbf{n}_1$ to the other one $\mathbf{n}_2$ at the cost of having $x$ connected to $y$ in $\widehat{\mathbf{n}_1+\mathbf{n}_2}$. This last condition looks necessary, since the sources at $x$ and $y$ in $\mathbf{n}_1$ force the existence of a path from $x$ to $y$ along which $\mathbf{n}_1$ (and therefore $\mathbf{n}_1+\mathbf{n}_2$) is positive – one may for instance look at the example of $F = 1-\mathbf{I}[x \overset{\widehat{\mathbf{n}_1+\mathbf{n}_2}}{\longleftrightarrow} y]$. Surprisingly, this is the only constraint one has to take into account. The switching lemma is at the base of almost every application of the random current representation.

Proof. The switching lemma applied to $F = 1$ and $A = \{x, y\}$ gives

$$\frac{\displaystyle\sum_{\partial\mathbf{n}_1=\partial\mathbf{n}_2=\emptyset} w_\beta(\mathbf{n}_1)w_\beta(\mathbf{n}_2)\mathbf{I}[x \overset{\widehat{\mathbf{n}_1+\mathbf{n}_2}}{\longleftrightarrow} y]}{\displaystyle\sum_{\partial\mathbf{n}_1=\partial\mathbf{n}_2=\emptyset} w_\beta(\mathbf{n}_1)w_\beta(\mathbf{n}_2)}$$

$$= \frac{\displaystyle\sum_{\partial\mathbf{n}_1=\partial\mathbf{n}_2=\{x,y\}} w_\beta(\mathbf{n}_1)w_\beta(\mathbf{n}_2)}{\displaystyle\sum_{\partial\mathbf{n}_1=\partial\mathbf{n}_2=\emptyset} w_\beta(\mathbf{n}_1)w_\beta(\mathbf{n}_2)} = \left(\frac{\displaystyle\sum_{\partial\mathbf{n}_1=\{x,y\}} w_\beta(\mathbf{n}_1)}{\displaystyle\sum_{\partial\mathbf{n}_1=\emptyset} w_\beta(\mathbf{n}_1)}\right)^2 = \mu^{\text{free}}_{G,\beta}[\sigma_x\sigma_y]^2. \square$$

The random current has been employed extensively in the study of the Ising model. For instance, it was archetypical in the derivation of Aizenman's celebrated result on the triviality of the Ising model for $d \geq 5$ [**Aiz82**], see below. It was also used in the derivations of several exponents, as well as in the derivation of the sharpness of the phase transition, see [**ABF87, DT15**]. In this section, we focus on a recent application of the random current representation, which illustrates perfectly the power of a geometric approach to spin-spin correlations.

4.3.2. *Continuity of the phase transition for Ising models on $\mathbb{Z}^d$, $d \geq 3$.* We are interested in the following result, which provides us with a criterion to show uniqueness of the infinite volume Ising measures.

Theorem 4.9 (Aizenman, DC, Sidoravicius [**ADS15**]). *Let $d \geq 1$ and consider a ferromagnetic Ising model on $\mathbb{Z}^d$ with coupling constants $(J_{xy})$. If*

$$\lim_{\|x-y\|\to\infty} \mu^{\text{free}}_\beta[\sigma_x\sigma_y] = 0, \tag{4.9}$$

*then there exists a unique Gibbs state at inverse temperature $\beta$, and in particular $\mu_\beta^{\text{free}} = \mu_\beta^+$.*

This fact is utterly wrong for Potts models with $q \geq 3$, as illustrated by the example of the *planar* Potts model with $q \geq 5$ colors: in such case the phase transition should be discontinuous, hence Theorem 3.1 implies that $\mu_{\beta_c}^1 \neq \mu_{\beta_c}^{\text{free}}$, since $\mu_{\beta_c}^1[\sigma_0] \neq 0$ (since $\phi_{p_c,q}^{\text{wired}}[0 \leftrightarrow \infty] > 0$) and there is exponential decay of the spin-spin correlations for $\mu_{\beta_c}^{\text{free}}$ (since $\phi_{p_c,q}^{\text{free}}[x \leftrightarrow y]$ decays exponentially fast in $\|x - y\|$).

It is tempting to say that, when applied to $\beta = \beta_c$, this theorem implies that the phase transition of the Ising model is continuous (observe that if $\mu_\beta^{\text{free}} = \mu_\beta^+$, then $\mu_\beta^+[\sigma_0] = 0$). This is not the case, since (4.9) has a priori no reason to be satisfied at $\beta_c$. In fact, the phase transition of the Ising model may be discontinuous, for instance for the model on $\mathbb{Z}$ with $J_{xy} = 1/|x-y|^2$, see [**ACCN88**]. Hence, the previous theorem provides us only with a sufficient condition to decide whether the phase transition is continuous or not.

Now, for the nearest neighbor Ising model on $\mathbb{Z}^d$ ($d \geq 3$), the infrared bound (4.10) can be improved as follows. In the case of the nearest-neighbor Ising model, one can easily deduce from the "averaged" bound given by Theorem 4.2 (which comes from the fact that the proof takes place in the Fourier space) the point-wise one for the free boundary conditions: for any $\beta < \beta_c$ and $x, y \in \mathbb{Z}^d$, $\tilde{\sigma}_x = \sigma_x$ and therefore

$$\mu_\beta^{\text{free}}[\sigma_x\sigma_y] \leq \tfrac{1}{2\beta} G(x,y) \tag{4.10}$$

which, by letting $\beta$ tend to $\beta_c$, implies by a classical argument of continuity from below of $\beta \mapsto \mu_\beta^{\text{free}}$ that

$$\mu_{\beta_c}^{\text{free}}[\sigma_x\sigma_y] \leq \tfrac{1}{2\beta_c} G(x,y).$$

Since the simple random walk is transient on $\mathbb{Z}^d$ for $d \geq 3$, $G(x,y)$ tends to 0 as $\|x-y\|$ tends to infinity. In particular, Theorem 4.9 implies (4.9) and therefore the following result.

COROLLARY 4.10 ([**ADS15**]). *For $d \geq 3$, the phase transition of the nearest neighbor ferromagnetic Ising model on $\mathbb{Z}^d$ is continuous.*

Recall that the phase transition is also continuous for $d = 2$ (see previous section), so that we conclude that the phase transition of the nearest-neighbor ferromagnetic Ising model is always continuous on $\mathbb{Z}^d$ for $d \geq 2$.

The proof of Theorem 4.9 is based on the study of the percolation properties of the infinite-volume limit of the random current representation. Let $\mathbf{P}_{G,\beta}^{\text{free}}$ be the law on sourceless currents on $G$ attributing a probability to $\mathbf{n}$ which is proportional to $w_\beta(\mathbf{n})$. Similarly, let $\mathbf{P}_{G,\beta}^+$ be the law on sourceless currents on $G \cup \{g\}$ attributing a probability to $\mathbf{n}$ which is proportional to $w_\beta(\mathbf{n})$. Define $\mathbb{P}_{G,\beta}$ to be the law of $\widehat{\mathbf{n}_1 + \mathbf{n}_2}$ restricted to edges of $G$, where

$\mathbf{n}_1$ and $\mathbf{n}_2$ are two independent currents with laws $\mathbf{P}^{\text{free}}_{G,\beta}$ and $\mathbf{P}^{+}_{G,\beta}$ respectively. This defines a family of percolation measures on finite graphs that gives rise to a percolation measure $\mathbb{P}_\beta$ on $\mathbb{Z}^d$ by taking the weak limit as $G \nearrow \mathbb{Z}^d$.

This measure can be proved to be invariant under translations and ergodic. Also, for any $\beta > 0$, the infinite cluster, if it exists, is unique almost surely. As a consequence of these properties,

$$(1) \quad \lim_{\|x-y\|\to\infty} \mathbb{P}_\beta[x \leftrightarrow y] = 0 \quad \Rightarrow \quad (2) \quad \mathbb{P}_\beta\left[0 \leftrightarrow \infty\right] = 0$$

(uniqueness of the infinite cluster is crucial: the existence of infinitely many infinite clusters would contradict this implication).

Now, the same proof as in Proposition 4.7 implies that for any $x, y \in \mathbb{Z}^d$,

$$\mathbb{P}_\beta[x \leftrightarrow y] = \mu^{\text{free}}_\beta[\sigma_x\sigma_y]\,\mu^{+}_\beta[\sigma_x\sigma_y] \le \mu^{\text{free}}_\beta[\sigma_x\sigma_y].$$

The hypothesis of the theorem implies that the right-hand side (1) is satisfied, and therefore that (2) also is. Yet, another application of the switching lemma (with some additional work) implies the existence, for every $x \in \mathbb{Z}^d$, of a constant $C_x > 0$ such that

$$\mu^{+}_\beta[\sigma_0\sigma_x] - \mu^{\text{free}}_\beta[\sigma_0\sigma_x] \le C_x\,\mathbb{P}_\beta\left[0 \leftrightarrow \infty\right].$$

Thus, (2) implies that $\mu^{+}_\beta[\sigma_0\sigma_x] = \mu^{\text{free}}_\beta[\sigma_0\sigma_x]$ for every $x \in \mathbb{Z}^d$. This property is classically known to imply that $\mu^{+}_\beta = \mu^{-}_\beta = \mu^{\text{free}}_\beta$. For instance, one may see that it implies the absence of infinite cluster in the FK percolation with wired boundary conditions, and therefore $\mu^{+}_\beta = \mu^{-}_\beta$ (the only difference between the two measures in the coupling would be that the infinite cluster of FK would be + or − respectively, but since there is none, the measures coincide). Thus, $\mu^{\text{free}}_\beta$ is sandwiched between $\mu^{+}_\beta$ and $\mu^{-}_\beta$, which are equal, so that $\mu^{\text{free}}_\beta = \mu^{+}_\beta$.

4.3.3. *The four-point function of the Ising model.* Let us discuss another application of the random current. The proof of the following proposition is fairly direct using the switching lemma (we leave it as a useful exercise for the reader).

PROPOSITION 4.11 (Aizenman [**Aiz82**]). *Let $\beta > 0$ and $G$ be a finite graph. Let $\langle\cdot\rangle$ denote $\mu^{\text{free}}_{G,\beta}$. Then, for any $x_1, x_2, x_3, x_4 \in V(G)$,*

$$\langle\sigma_{x_1}\sigma_{x_2}\sigma_{x_3}\sigma_{x_4}\rangle - \langle\sigma_{x_1}\sigma_{x_2}\rangle\langle\sigma_{x_3}\sigma_{x_4}\rangle - \langle\sigma_{x_1}\sigma_{x_3}\rangle\langle\sigma_{x_2}\sigma_{x_4}\rangle - \langle\sigma_{x_1}\sigma_{x_4}\rangle\langle\sigma_{x_2}\sigma_{x_3}\rangle$$

(4.11)

$$= -2\,\langle\sigma_{x_1}\sigma_{x_3}\rangle\langle\sigma_{x_2}\sigma_{x_4}\rangle \cdot \mathbf{P}^{\{1,3\}}_\beta \otimes \mathbf{P}^{\{2,4\}}_\beta\left[x_1 \overset{\widehat{\mathbf{n}_1+\mathbf{n}_2}}{\longleftrightarrow} x_2, x_3, x_4\right],$$

*where $\mathbf{P}^{\{a,b\}}_\beta$ denotes the law (on currents) attributing probability proportional to $w_\beta(\mathbf{n})$ to currents $\mathbf{n}$ with set of sources equal to $\{a,b\}$, and 0 to other currents.*

Above, $\otimes$ means the product of measures. Thus, we see that the connectivity properties of the sum of two independent currents (this time with sources) is once again involved in the estimation of useful truncated $n$-point functions (here the so-called Ursell's four-point function). Let us mention two applications of this result. The first one deals with the planar case.

COROLLARY 4.12. *Let $\beta > 0$. Let $G$ be a connected subgraph of $\mathbb{Z}^2$ with connected complement. As before, $\langle\cdot\rangle$ denotes $\mu^{\mathrm{free}}_{G,\beta}$. Let $x_1$, $x_2$, $x_3$ and $x_4$ be four vertices of $\partial G$ found in counter-clockwise order, then*

$$\langle\sigma_{x_1}\sigma_{x_2}\sigma_{x_3}\sigma_{x_4}\rangle = \langle\sigma_{x_1}\sigma_{x_2}\rangle\langle\sigma_{x_3}\sigma_{x_4}\rangle - \langle\sigma_{x_1}\sigma_{x_3}\rangle\langle\sigma_{x_2}\sigma_{x_4}\rangle + \langle\sigma_{x_1}\sigma_{x_4}\rangle\langle\sigma_{x_2}\sigma_{x_3}\rangle.$$

Note the minus sign in front of the second product on the right-hand side of the equality. The proof follows from the fact that the probability on the right-hand side of (4.11) is equal to 1. Indeed, a current in $G$ with sources at $x_1$ and $x_3$ contains a path from $x_1$ to $x_3$, which must intersect the path from $x_2$ to $x_4$ present in the current with sources at $x_2$ and $x_4$. Therefore, $x_1$, $x_2$, $x_3$ and $x_4$ must necessarily be all connected together in $\widehat{\mathbf{n}_1 + \mathbf{n}_2}$.

The previous result extends to any planar graph, and even to an arbitrary number of vertices $x_1, \ldots, x_{2n}$ (with some mild work). In such case, we obtain a *fermionic Wick rule* for the $2n$-point function

$$\langle\sigma_{x_1}\cdots\sigma_{x_{2n}}\rangle = \sum_{\pi\in\Pi_n} \varepsilon(\pi)\langle\sigma_{x_{\pi(1)}}\sigma_{x_{\pi(2)}}\rangle\cdots\langle\sigma_{x_{\pi(2n-1)}}\sigma_{x_{\pi(2n)}}\rangle,$$

where $\Pi_n$ is the set of pairings of $\{1,\ldots,2n\}$, i.e. the set of permutations $\pi$ such that $\pi(2j-1) < \pi(2j)$ for any $j \in \{1,\ldots,n\}$ and $\pi(2j-1) < \pi(2j+1)$ for any $j \in \{1,\ldots,n-1\}$. Above, $\varepsilon(\pi)$ is the signature of $\pi$, which can be seen as $-1$ to the power the parity of the number of intersections of the graph obtained by drawing arcs in $G$ between $\pi(2j-1)$ and $\pi(2j)$ for any $j \in \{1,\ldots,n\}$ in such a way that there is a minimal number of intersections.

The second application focuses on the $d \geq 5$ case. There, the quantity on the right-hand side of (4.11) tends to 0 as $x_1$, $x_2$, $x_3$ and $x_4$ are sent apart. Intuitively, this convergence comes from the fact that the paths from $x_1$ to $x_3$ in $\mathbf{n}_1$ and from $x_2$ to $x_4$ in $\mathbf{n}_2$ behave somehow like random walks. In particular, they do not intersect in dimension $d \geq 5$. Aizenman exploits this intuition and the infrared bound to prove in [**Aiz82**] that the four-point functions satisfies an asymptotic Wick rule at $\beta_c$ when $d \geq 5$. This was also proved by Fröhlich [**Frö82**] using a random-walk representation [**BFS82**] of the Ising model *à la* Symanzik [**Sym66**].

THEOREM 4.13 (Aizenman, Fröhlich [**Aiz82, Frö82**]). *Let $d \geq 5$ and $\beta = \beta_c$ and let $\langle\cdot\rangle = \mu^{\mathrm{free}}_{\beta_c}$. Then,*

$$\langle\sigma_{x_1}\sigma_{x_2}\sigma_{x_3}\sigma_{x_4}\rangle \sim \langle\sigma_{x_1}\sigma_{x_2}\rangle\langle\sigma_{x_3}\sigma_{x_4}\rangle + \langle\sigma_{x_1}\sigma_{x_3}\rangle\langle\sigma_{x_2}\sigma_{x_4}\rangle + \langle\sigma_{x_1}\sigma_{x_4}\rangle\langle\sigma_{x_2}\sigma_{x_3}\rangle.$$

*Above, $\sim$ means that the ratio of the left and right hand sides tends to 1 as the minimal distance between $x_1$, $x_2$, $x_3$ and $x_4$ tends to infinity.*

This theorem easily implies that the model has a trivial scaling limit.

**Acknowledgements.** This work was supported by a grant from the Swiss FNS, by the ERC AG COMPASP and the NCCR SwissMap also funded by the Swiss FNS.

## References

[ABF87] M. Aizenman, D. J. Barsky, and R. Fernández, *The phase transition in a general class of Ising-type models is sharp*, J. Statist. Phys. **47** (1987), no. 3–4, 343–374. MR 894398 (89f:82013)

[ACCN88] M. Aizenman, J. T. Chayes, L. Chayes, and C. M. Newman, *Discontinuity of the magnetization in one-dimensional $1/|x-y|^2$ Ising and Potts models*, J. Statist. Phys. **50** (1988), no. 1–2, 1–40. MR 939480 (89f:82072)

[ADS15] M. Aizenman, H. Duminil-Copin, and V. Sidoravicius, *Random currents and continuity of Ising model's spontaneous magnetization*, Communications in Mathematical Physics **334** (2015), 719–742.

[AF88] M. Aizenman and R. Fernández, *Critical exponents for long-range interactions*, Lett. Math. Phys. **16** (1988), no. 1, 39–49. MR 958462 (89h:82041)

[Aiz82] M. Aizenman, *Geometric analysis of $\varphi^4$ fields and Ising models. I, II*, Comm. Math. Phys. **86** (1982), no. 1, 1–48. MR 678000 (84f:81078)

[Bax71] R. J. Baxter, *Generalized ferroelectric model on a square lattice*, Studies in Appl. Math. **50** (1971), 51–69. MR 0307646 (46 #6766)

[Bax73] R. J. Baxter, *Potts model at the critical temperature*, Journal of Physics C: Solid State Physics **6** (1973), no. 23, L445.

[Bax78] R. J. Baxter, *Solvable eight-vertex model on an arbitrary planar lattice*, Philos. Trans. Roy. Soc. London Ser. A **289** (1978), no. 1359, 315–346. MR 0479213 (57 #18660)

[Bax89] R. J. Baxter, *Exactly solved models in statistical mechanics*, Academic Press Inc. [Harcourt Brace Jovanovich Publishers], London, 1989, Reprint of the 1982 original. MR 998375 (90b:82001)

[BBMdG$^+$14] N. R. Beaton, M. Bousquet-Mélou, J. de Gier, H. Duminil-Copin, and A. J. Guttmann, *The critical fugacity for surface adsorption of self-avoiding walks on the honeycomb lattice is $1+\sqrt{2}$*, Comm. Math. Phys. **326** (2014), no. 3, 727–754. MR 3173404

[BC03] M. Biskup and L. Chayes, *Rigorous analysis of discontinuous phase transitions via mean-field bounds*, Comm. Math. Phys. **238** (2003), no. 1–2, 53–93. MR 1989669 (2004f:82016)

[BD12a] V. Beffara and H. Duminil-Copin, *The self-dual point of the two-dimensional random-cluster model is critical for $q \geq 1$*, Probab. Theory Related Fields **153** (2012), no. 3–4, 511–542. MR 2948685

[BD12b] V. Beffara and H. Duminil-Copin, *Smirnov's fermionic observable away from criticality*, Ann. Probab. **40** (2012), no. 6, 2667–2689. MR 3050513

[BDH14] S. Benoist, H. Duminil-Copin, and C. Hongler, *Conformal invariance of crossing probabilities for the Ising model with free boundary conditions*, arXiv:1410.3715, 2014.

[BFS82] D. Brydges, J. Fröhlich, and T. Spencer, *The random walk representation of classical spin systems and correlation inequalities*, Comm. Math. Phys. **83** (1982), no. 1, 123–150. MR 648362

[BH57] S. R. Broadbent and J. M. Hammersley, *Percolation processes. I. Crystals and mazes*, Proc. Cambridge Philos. Soc. **53** (1957), 629–641. MR 0091567 (19,989e)

[Bis09] M. Biskup, *Reflection positivity and phase transitions in lattice spin models*, Methods of contemporary mathematical statistical physics, Lecture

Notes in Math., vol. 1970, Springer, Berlin, 2009, pp. 1–86. MR 2581604 (2011a:82016)

[BS08] A. I. Bobenko and Y. B. Suris, *Discrete differential geometry*, Graduate Studies in Mathematics, vol. 98, American Mathematical Society, Providence, RI, 2008, Integrable structure. MR 2467378 (2010f:37125)

[CDH13] D. Chelkak, H. Duminil-Copin, and C. Hongler, *Crossing probabilities in topological rectangles for the critical planar FK-Ising model*, arXiv:1312.7785, 2013.

[CDH+14] D. Chelkak, H. Duminil-Copin, C. Hongler, A. Kemppainen, and S. Smirnov, *Convergence of Ising interfaces to Schramm's SLE curves*, C. R. Acad. Sci. Paris Math. **352** (2014), no. 2, 157–161.

[CHI15] D. Chelkak, C. Hongler, and K. Izyurov, *Conformal invariance of spin correlations in the planar Ising model*, Ann. of Math. (2) **181** (2015), no. 3, 1087–1138. MR 3296821

[CI13] D. Chelkak and K. Izyurov, *Holomorphic spinor observables in the critical Ising model*, Comm. Math. Phys. **322** (2013), no. 2, 303–332. MR 3077917

[CS12] D. Chelkak and S. Smirnov, *Universality in the 2D Ising model and conformal invariance of fermionic observables*, Invent. Math. **189** (2012), no. 3, 515–580. MR 2957303

[DC] H. Duminil-Copin, *A new proof of first order phase transition a new proof of first order phase transition for the planar random-cluster and Potts models with $q \gg 1$*, preprint.

[DGP14] H. Duminil-Copin, C. Garban, and G. Pete, *The near-critical planar FK-Ising model*, Comm. Math. Phys. **326** (2014), no. 1, 1–35. MR 3162481

[DHN11] H. Duminil-Copin, C. Hongler, and P. Nolin, *Connection probabilities and RSW-type bounds for the two-dimensional FK Ising model*, Comm. Pure Appl. Math. **64** (2011), no. 9, 1165–1198. MR 2839298 (2012f:60332)

[DRT15] H. Duminil-Copin, A. Raoufi, and V. Tassion, *A new computation of the critical point for the planar random-cluster model with $q \geq 1$*, in preparation, 2015.

[DS12a] H. Duminil-Copin and S. Smirnov, *Conformal invariance of lattice models*, Probability and statistical physics in two and more dimensions, Clay Math. Proc., vol. 15, Amer. Math. Soc., Providence, RI, 2012, pp. 213–276. MR 3025392

[DS12b] H. Duminil-Copin and S. Smirnov, *The connective constant of the honeycomb lattice equals $\sqrt{2+\sqrt{2}}$*, Ann. of Math. (2) **175** (2012), no. 3, 1653–1665. MR 2912714

[DST15] H. Duminil-Copin, V. Sidoravicius, and V. Tassion, *Continuity of the phase transition for planar random-cluster and Potts models with $1 \leq q \leq 4$*, arXiv:1505.04159, 2015.

[DT15] H. Duminil-Copin and V. Tassion, *A new proof of the sharpness of the phase transition for Bernoulli percolation and the Ising model*, arXiv:1502.03050, 2015.

[Dum12] H. Duminil-Copin, *Divergence of the correlation length for critical planar FK percolation with $1 \leq q \leq 4$ via parafermionic observables*, Journal of Physics A: Mathematical and Theoretical **45** (2012), no. 49, 494013.

[Dum13] H. Duminil-Copin, *Parafermionic observables and their applications to planar statistical physics models*, Ensaios Matematicos, vol. 25, Brazilian Mathematical Society, 2013.

[Dum15] H. Duminil-Copin, *Geometric representations of lattice spin models*, book, Edition Spartacus, 2015.

[FK72] C. M. Fortuin and P. W. Kasteleyn, *On the random-cluster model. I. Introduction and relation to other models*, Physica **57** (1972), 536–564. MR 0359655 (50 #12107)

[Frö82] J. Fröhlich, *On the triviality of $\lambda\varphi_d^4$ theories and the approach to the critical point in $d \geq 4$ dimensions*, Nuclear Phys. B **200** (1982), no. 2, 281–296. MR 643591 (83h:81046)

[FSS76] J. Fröhlich, B. Simon, and T. Spencer, *Infrared bounds, phase transitions and continuous symmetry breaking*, Comm. Math. Phys. **50** (1976), no. 1, 79–95. MR 0421531 (54 #9530)

[GHS70] R. B. Griffiths, C. A. Hurst, and S. Sherman, *Concavity of magnetization of an Ising ferromagnet in a positive external field*, J. Mathematical Phys. **11** (1970), 790–795. MR 0266507 (42 #1412)

[Gla13] A. Glazman, *Connective constant for a weighted self-avoiding walk on $\mathbb{Z}^2$*, Preprint, 2013.

[Gri99] G. R. Grimmett, *Inequalities and entanglements for percolation and random-cluster models*, Perplexing problems in probability, Progr. Probab., vol. 44, Birkhäuser Boston, Boston, MA, 1999, pp. 91–105. MR 1703126 (2000i:60115)

[Gri06] G. Grimmett, *The random-cluster model*, Grundlehren der Mathematischen Wissenschaften [Fundamental Principles of Mathematical Sciences], vol. 333, Springer-Verlag, Berlin, 2006. MR 2243761 (2007m:60295)

[HK13] C. Hongler and K. Kytölä, *Ising interfaces and free boundary conditions*, J. Amer. Math. Soc. **26** (2013), no. 4, 1107–1189. MR 3073886

[Hon10] C. Hongler, *Conformal invariance of Ising model correlations*, Ph.D. thesis, université de Genève, 2010, p. 118.

[HS13] C. Hongler and S. Smirnov, *The energy density in the planar Ising model*, Acta Math. **211** (2013), no. 2, 191–225. MR 3143889

[Isi25] E. Ising, *Beitrag zur Theorie des Ferromagnetismus.*, Z. Phys. **31** (1925), 253–258.

[ISZ88] C. Itzykson, H. Saleur, and J. B. Zuber, *Conformal invariance and applications to statistical mechanics*, World Scientific, 1988.

[Izy15] K. Izyurov, *Smirnov's observable for free boundary conditions, interfaces and crossing probabilities*, Comm. Math. Phys. **337** (2015), no. 1, 225–252. MR 3324162

[Ken00] R. Kenyon, *Conformal invariance of domino tiling*, Ann. Probab. **28** (2000), no. 2, 759–795. MR 1782431 (2002e:52022)

[KS82] R. Kotecký and S. B. Shlosman, *First-order phase transitions in large entropy lattice models*, Comm. Math. Phys. **83** (1982), no. 4, 493–515. MR 649814 (83i:82033)

[KS12] A. Kemppainen and S. Smirnov, *Random curves, scaling limits and loewner evolutions*, arXiv:1212.6215, 2012.

[Law05] G. F. Lawler, *Conformally invariant processes in the plane*, Mathematical Surveys and Monographs, vol. 114, American Mathematical Society, Providence, RI, 2005. MR 2129588 (2006i:60003)

[Len20] W. Lenz, *Beitrag zum Verständnis der magnetischen Eigenschaften in festen Körpern.*, Phys. Zeitschr. **21** (1920), 613–615.

[LMMS+91] L. Laanait, A. Messager, S. Miracle-Solé, J. Ruiz, and S. Shlosman, *Interfaces in the Potts model. I. Pirogov-Sinai theory of the Fortuin-Kasteleyn representation*, Comm. Math. Phys. **140** (1991), no. 1, 81–91. MR 1124260 (93d:82021a)

[LMR86] L. Laanait, A. Messager, and J. Ruiz, *Phases coexistence and surface tensions for the Potts model*, Comm. Math. Phys. **105** (1986), no. 4, 527–545. MR 852089 (88i:82030)

[Lov04] L. Lovász, *Discrete analytic functions: an exposition*, Surveys in differential geometry. Vol. IX, Surv. Differ. Geom., IX, Int. Press, Somerville, MA, 2004, pp. 241–273. MR 2195410 (2007g:30067)

[Mer01] C. Mercat, *Discrete Riemann surfaces and the Ising model*, Comm. Math. Phys. **218** (2001), no. 1, 177–216. MR 1824204 (2002c:82019)

[Ons44] L. Onsager, *Crystal statistics. I. A two-dimensional model with an order-disorder transition*, Phys. Rev. (2) **65** (1944), 117–149. MR 0010315 (5,280d)

[Pot52] R. B. Potts, *Some generalized order-disorder transformations*, Proceedings of the Cambridge Philosophical Society, vol. 48, Cambridge University Press, 1952, pp. 106–109.

[Rus78] L. Russo, *A note on percolation*, Z. Wahrscheinlichkeitstheorie und Verw. Gebiete **43** (1978), no. 1, 39–48. MR 0488383 (58 #7931)

[Sch00] O. Schramm, *Scaling limits of loop-erased random walks and uniform spanning trees*, Israel J. Math. **118** (2000), 221–288. MR 1776084 (2001m:60227)

[SFD+12] S. El-Showk, M. F. Paulos, D. Poland, S. Rychkov, D. Simmons-Duffin, and A. Vichi, *Solving the 3d ising model with the conformal bootstrap*, Physical Review D **86** (2012), no. 2.

[Sim80] B. Simon, *Correlation inequalities and the decay of correlations in ferromagnets*, Comm. Math. Phys. **77** (1980), no. 2, 111–126. MR 589426 (82e:82004)

[Smi06] S. Smirnov, *Towards conformal invariance of 2D lattice models*, International Congress of Mathematicians. Vol. II, Eur. Math. Soc., Zürich, 2006, pp. 1421–1451. MR 2275653 (2008g:82026)

[Smi10] S. Smirnov, *Conformal invariance in random cluster models. I. Holomorphic fermions in the Ising model*, Ann. of Math. (2) **172** (2010), no. 2, 1435–1467. MR 2680496 (2011m:60302)

[Sta68] H. E. Stanley, *Dependence of critical properties on dimensionality of spins*, Physical Review Letters **20** (1968), no. 12, 589–592.

[Ste05] K. Stephenson, *Introduction to circle packing*, Cambridge University Press, Cambridge, 2005, The theory of discrete analytic functions. MR 2131318 (2006a:52022)

[SW78] P. D. Seymour and D. J. A. Welsh, *Percolation probabilities on the square lattice*, Ann. Discrete Math. **3** (1978), 227–245, Advances in graph theory (Cambridge Combinatorial Conf., Trinity College, Cambridge, 1977). MR 0494572 (58 #13410)

[Sym66] K. Symanzik, *Euclidean quantum field theory. I. Equations for a scalar model*, J. Mathematical Phys. **7** (1966), 510–525. MR 0187686

[Wer04] W. Werner, *Random planar curves and Schramm-Loewner evolutions*, Lectures on probability theory and statistics, Lecture Notes in Math., vol. 1840, Springer, Berlin, 2004, pp. 107–195. MR 2079672 (2005m:60020)

[Wer05] W. Werner, *SLE, conformal restriction, loops*, European Congress of Mathematics, Eur. Math. Soc., Zürich, 2005, pp. 515–528. MR 2185763 (2006j:82021)



Département de Mathématiques, Université de Genève, Genève Switzerland

*E-mail address*: hugo.duminil@unige.ch

Current Developments in Mathematics, 2015

# Hecke algebras and $p$-adic groups

Xuhua He

ABSTRACT. This survey article, is written as an extended note and supplement of my lectures in the current developments in mathematics conference in 2015. We discuss some recent developments on the conjugacy classes of affine Weyl groups and $p$-adic groups, and some applications to Shimura varieties and to representations of affine Hecke algebras.

CONTENTS

## Introduction

**0.1.** In [**Ti57**], Tits explains the analogy between the symmetric group $S_n$ and the general linear group over a finite field $\mathbb{F}_q$ and indicates that $S_n$ should be regarded as the general linear group over $\mathbb{F}_1$, the field of one element.

Following Tits' philosophy, we may informally regard affine Weyl groups as reductive groups over $\mathbb{Q}_1$, the 1-adic field. Although it might be premature to develop the theory of the 1-adic field at the current stage, we do have a fairly good understanding of conjugacy classes of affine Weyl groups, together with the length function on them, and such knowledge allows us to reveal a great part of the structure of conjugacy classes of $p$-adic groups.

In this note, we will focus on some questions related to conjugacy classes and we will look at the two sides of a coin:

- An affine Weyl group as a degeneration of a $p$-adic group;
- A $p$-adic group as a deformation of an affine Weyl group.

We will see how such considerations help us to understand several important problems and we will discuss some applications in arithmetic geometry (of Shimura varieties) and representation theory (of affine Hecke algebras). Note that both areas have been intensively studied. To enumerate the recent achievements in each area would lead to a more-than-100-page survey article and it is not my intention to do so here. I will just focus on a few problems to illustrate the mysterious connection between affine Weyl groups and $p$-adic groups.

**0.2.** A Coxeter group is generated by simple reflections and has a well-defined length function on it. It is obvious that in a given conjugacy class of the Coxeter group, there exists a sequence of conjugations by simple reflections, starting from any given element in the class and ending with a minimal-length element. For various reasons (both algebraic and geometric), it is desirable to have such a sequence, in which the lengths of the elements **weakly decrease**. If such a sequence exists, then for many questions (on

Hecke algebras, algebraic groups, Deligne-Lusztig varieties, etc.) one may reduce the questions on arbitrary elements to questions on minimal length elements.

Moreover, in a given conjugacy class, in general, there are more than one minimal length elements. For example, in type $E_8$, the conjugacy class $E_6(a_2) + A_2$ has 403200 elements and 16374 of them are of minimal length. Hence, it is also desirable to have a better understanding of the relations between these minimal length elements.

Minimal length elements were first studied by Geck and Pfeiffer [**GP93**]. They discover that the minimal length elements, in a finite Weyl group, have many remarkable properties.

Such properties have found important applications in the study of the "character table" of finite Hecke algebras (see e.g. [**GP93**], and [**GM97**]), in the study of Deligne-Lusztig theory on the representations of finite groups of Lie type (see, e.g. [**OR08**], [**BR08**], and [**HL12**]) and in the study of links between conjugacy classes in finite Weyl groups and unipotent conjugacy classes in reductive groups (see e.g. [**Lu11**]).

**0.3.** A natural question to ask is whether such properties hold for affine Weyl groups. In the study of this question, I find it enlightening to consider an affine Weyl group as a degeneration of a $p$-adic group. In fact, the theory of $p$-adic groups motivates us in the discovery of the following interesting features of affine Weyl groups:

(1) The arithmetic invariants of conjugacy classes of affine Weyl groups;
(2) The straight conjugacy classes and the reduction from non-straight conjugacy classes to straight conjugacy classes;
(3) The "special" partial conjugacy classes and their distinguished representatives;
(4) The partial order on the set of "special" partial conjugacy classes and on the set of straight conjugacy classes;
(5) The parameterization of conjugacy classes in terms of standard quadruples.

This is what we will discuss in Part I.

Items (1) and (2) are motivated by Kottwitz's classification of the Frobenius-twisted conjugacy classes of $G(\breve{\mathbb{Q}}_p)$, the reductive group $G$ over the completion of maximal unramified extension of $\mathbb{Q}_p$. They are key ingredients in the understanding of conjugacy classes of affine Weyl groups together with the length function on them.

Items (3) and (4) (for "special" partial conjugacy classes) were discovered in the process of understanding Lusztig's $G$-stable pieces and closure relations between these pieces.

Item (4) for straight conjugacy classes were discovered in the process of understanding the closure relations between Frobenius-twisted conjugacy classes of $G(\breve{F})$ and the closure relations between Newton strata of Shimura varieties.

Item (5) is motivated by the parametrization of Frobenius-twisted conjugacy classes of $G(\breve{F})$ in terms of basic Frobenius-twisted conjugacy classes of Levi subgroups. Such parametrization leads to the definition of the rigid cocenter of affine Hecke algebras, which play a crucial role in our discussion in Part III.

**0.4.** Now let us look at the other side of the coin. We will see how the knowledge of conjugacy classes of affine Weyl groups, together with the length function, helps us to study some problems related to conjugacy classes of $p$-adic groups.

Let $\sigma$ be the Frobenius morphism on $G(\breve{F})$. Let $\breve{\mathcal{I}}$ be a $\sigma$-stable Iwahori subgroup of $G(\breve{F})$. In Part II, we study the intersection of $\breve{\mathcal{I}}\dot{w}\breve{\mathcal{I}}$ with a $\sigma$-conjugacy class $[b]$ of $G(\breve{F})$. We focus on some basic questions:

- When the intersection is nonempty?
- If so, what is the dimension?

We also regard this intersection as a group-theoretic model to understand the relation between the Kottwitz-Rapoport strata (related to certain elements $w$) and the Newton strata (related to certain $\sigma$-conjugacy classes $[b]$) of Shimura varieties.

It is worth mentioning that there is another group-theoretic model to understand those strata of Shimura varieties, affine Deligne-Lusztig varieties

$$X_w(b) = \{g\breve{\mathcal{I}} \in G(\breve{F})/\breve{\mathcal{I}}; g^{-1}b\sigma(g) \in \breve{\mathcal{I}}\dot{w}\breve{\mathcal{I}}\}.$$

It is easy to see that $X_w(b) \neq \emptyset$ if and only if $\breve{\mathcal{I}}\dot{w}\breve{\mathcal{I}} \cap [b] \neq \emptyset$. However, there are a few differences between $X_w(b)$ and $\breve{\mathcal{I}}\dot{w}\breve{\mathcal{I}}$.

- We have the freedom to degenerate the $\sigma$-conjugacy class $[b]$ when considering $\breve{\mathcal{I}}\dot{w}\breve{\mathcal{I}} \cap [b]$. Thus $\breve{\mathcal{I}}\dot{w}\breve{\mathcal{I}} \cap [b]$ may be used to understand closure relations between Newton strata of Shimura varieties.
- Although $\breve{\mathcal{I}}\dot{w}\breve{\mathcal{I}} \cap [b]$ is infinite dimensional, there is a way to define the dimension of it (which is a finite number). This dimension is expected to be the dimension of the intersection of the corresponding Kottwitz-Rapoport stratum and corresponding Newton stratum in Shimura varieties.
- The affine Deligne-Lusztig variety $X_w(b)$, in general, may contain infinitely many irreducible components. The intersection $\breve{\mathcal{I}}\dot{w}\breve{\mathcal{I}} \cap [b]$, on the other hand, has only finitely many irreducible components.

Except these differences, the method we use to study affine Deligne-Lusztig varieties and the intersection $\breve{\mathcal{I}}\dot{w}\breve{\mathcal{I}} \cap [b]$ are similar. In **[He14]** we develop a reduction method to study affine Deligne-Lusztig varieties, and give a connection between affine Deligne-Lusztig varieties and class polynomials of the associated affine Hecke algebras. The method and the result remain valid, mutatis mutandis, for our new model $\breve{\mathcal{I}}\dot{w}\breve{\mathcal{I}} \cap [b]$ as well. The key idea is to do the reduction on $\breve{\mathcal{I}}\dot{w}\breve{\mathcal{I}}$ by $\sigma$-conjugation using a sequence of conjugations by simple reflections in affine Weyl group which weakly decrease the length.

After establishing the connection between the intersection $\breve{\mathcal{I}}\dot{w}\breve{\mathcal{I}} \cap [b]$ and class polynomials, we discuss several situations where the explicit non-emptiness pattern and/or dimension formula are established, including:

- The non-emptiness pattern and dimension formula of $\breve{\mathcal{K}}\epsilon^\mu\breve{\mathcal{K}} \cap [b]$ for any special maximal parahoric subgroup $\breve{\mathcal{K}}$.
- The non-emptiness pattern of $\breve{\mathcal{I}}\dot{w}\breve{\mathcal{I}} \cap [b]$ for basic $\sigma$-conjugacy class $[b]$.
- The dimension formula of $\breve{\mathcal{I}}\dot{w}\breve{\mathcal{I}} \cap [b]$ for basic $\sigma$-conjugacy class $[b]$ and an element $w$ in the Shrunken Weyl chamber.
- The non-emptiness pattern and dimension formula of $\breve{\mathcal{I}}\dot{w}\breve{\mathcal{I}} \cap [\epsilon^\mu]$ for a residually split group and an element $w$ in the very Shrunken Weyl chamber.
- The non-emptiness pattern of $\breve{\mathcal{K}}\,\mathrm{Adm}(\mu)_K\breve{\mathcal{K}} \cap [b]$ for any parahoric subgroup $\breve{\mathcal{K}}$.

**0.5.** Now we move to a different topic, the affine Hecke algebras.

A basic philosophy in representation theory is that *characters tell all.* This is the case for finite groups. What happens for affine Hecke algebras?

In Part III, we will discuss the case for affine Hecke algebras. The strategy is to first understand the cocenter of affine Hecke algebras, and then to investigate how the cocenter controls (both the ordinary and the modular) representations.

An affine Hecke algebra, is a deformation of the group algebra of an affine Weyl group. The cocenter of the group algebra, is very simple. The elements in the same conjugacy class have the same image in the cocenter and the cocenter has a standard basis indexed by conjugacy classes.

For affine Hecke algebras, the situation is more complicated. The elements in the same conjugacy class may not have the same image in the cocenter. In order to understand the cocenter of affine Hecke algebras, in addition to conjugacy classes, one also need to take into account the length of the elements inside the given conjugacy class. The upshot is that the cocenter still has a standard basis indexed by conjugacy classes, but they are the image of minimal length elements in the conjugacy class, not arbitrary element compared to the group algebra case.

Start from an element, not necessarily of minimal length in its conjugacy class, we have a reduction method to reach (possibly more than one) minimal length elements and write the image of this element in the cocenter as a linear combination of the standard basis, with coefficient the so-called class polynomials. In other words, class polynomials encode the information of the reduction method, which is used both in the study of cocenter of affine Hecke algebras and in the study of the intersection $\breve{\mathcal{I}}\dot{w}\breve{\mathcal{I}} \cap [b]$. This is the reason that the algebraic object class polynomial is a powerful tool to study the geometric object $\breve{\mathcal{I}}\dot{w}\breve{\mathcal{I}} \cap [b]$.

Back to representation theory, one difficulty is that there are infinitely many conjugacy classes of affine Weyl groups as well as infinitely many irreducible representations of affine Hecke algebras. Motivated by the basic $\sigma$-conjugacy classes of $G(\breve{F})$, we have a parametrization of conjugacy classes in terms of their Newton points, and there is a way to reduce any conjugacy class to a conjugacy class with a central Newton point. This leads to the definition of rigid cocenter and rigid quotient of the Grothendieck group of the representations of affine Hecke algebras. The main theorem in [**CHx**] is that the trace map induces a perfect pairing between the rigid cocenter and the rigid quotient for affine Hecke algebras with generic parameters. This result leads to the density Theorem and trace Paley-Wiener theorem on the relation between the cocenter and representations of affine Hecke algebras.

We expect that the rigid cocenter also plays a key role in the study of modular representations of affine Hecke algebras, and we propose two conjectures:

- The naive kernel conjecture, which predicts how the density theorem fails for affine Hecke algebras at roots of unity and in positive characteristic;
- The rigid determinant conjecture, which predicts the deformation of the representations with respect to the change of (generic) parameters.

**0.6.** Most of the results in this note are known by now (although scattered around the literature, with some mild assumption). However, we take this opportunity to present a few new materials, as well as remove some unnecessarily assumptions from the old results in the literature, including:

- Some results on the minimal length elements and partial order on the partial conjugacy classes for an affine Weyl group action (by conjugation) on a Coxeter group. See Theorem 1.14 and Proposition 1.27.
- A parametrization of conjugacy classes of extended affine Weyl groups in terms of standard quadruples. See Theorem 1.19.
- The use of $\breve{\mathcal{I}}\dot{w}\breve{\mathcal{I}} \cap [b]$ as a group-theoretic model to study some stratifications of Shimura varieties. See §2.12.
- The non-emptiness pattern and dimension formula for $\breve{\mathcal{K}}\epsilon^{\mu}\breve{\mathcal{K}} \cap [b]$ for special maximal parahoric subgroup $\breve{\mathcal{K}}$ (in the literature, it is only stated for hyperspecial maximal parahoric subgroups). See Theorem 2.24 and Theorem 2.29.
- The dimension formula for $\breve{\mathcal{I}}\dot{w}\breve{\mathcal{I}} \cap [b]$ for basic $[b]$ and $w$ in the Shrunken Weyl chamber (in the literature, it is only stated for residually split groups). See Theorem 2.31.
- The non-emptiness pattern and dimension formula for $\breve{\mathcal{I}}\dot{w}\breve{\mathcal{I}} \cap [\epsilon^{\mu}]$ for $G$ a residually split group and $w$ in the very Shrunken Weyl chamber. See Theorem 2.34.

- The naive kernel conjecture 3.13 and the rigid determinant conjecture 3.16 of affine Hecke algebras.

I am grateful to G. Lusztig for all he has taught me about the algebraic groups and Weyl groups over the years. Without his big influence this project would never be initiated. In more recent times, I also learned a lot from J. Adams, D. Ciubotaru, U. Görtz, T. Haines, J. Michel, S. Nie, X. Zhu, and especially M. Rapoport, who teaches me Shimura varieties. I am happy to express my gratitude to all of them. This survey article is written as an extended note of the talk at the Current Developments in Mathematics conference 2015 at Harvard. I thank the organizers for the invitation, and for the opportunity to write this survey. I also thank J. Adams, U. Görtz and T. Haines for their remarks on a preliminary version of this note. My research is partially supported by NSF grant DMS-1463852.

## 1. Part I. Combinatorics of affine Weyl groups

**1.1. Coxeter groups.** A *Coxeter system* is a pair $(W, \mathbb{S})$, where $\mathbb{S}$ is a finite set and $W$ is the group generated by the elements in $\mathbb{S}$, subject to certain relations. First, we have $s^2 = 1$ for all $s \in \mathbb{S}$. We call the elements in $\mathbb{S}$ the *simple reflections* of $W$. Next, we have relations between two elements $s \neq t$ in $\mathbb{S}$. Let $m_{st} \in \{2, 3, \cdots\} \cup \{\infty\}$ be the order of $st$. Since $s$ and $t$ have order 2, we have

$$sts \cdots = tst \cdots,$$

where both sides have exactly $m_{st}$ factors. If $m_{st} = 2$, then $st = ts$, i.e., the two simple reflections $s$ and $t$ commute. If $m_{st} = 3$, then $s$ and $t$ satisfy the Artin's braid relation $sts = tst$. If $m_{st} = \infty$, then there is no relation between $s$ and $t$.

In additional to the group structure, the length function $\ell : W \to \mathbb{N}$ plays a crucial role in this note. Here for any $w \in W$, $\ell(w)$ is the minimal integer $n$ such that $w = s_1 \cdots s_n$, where $s_i \in \mathbb{S}$.

Weyl groups (both finite and affine) are important examples of Coxeter groups. They play a crucial role in the study of Lie groups and $p$-adic groups. Now let me give two examples.

EXAMPLE 1.1. Let $W = S_n$ be the symmetry group of $\{1, 2, \cdots, n\}$. It is the Weyl group of $GL_n(\mathbb{C})$. It is a Coxeter group with the generators $s_i = (i, i+1)$ for $1 \leqslant i \leqslant n-1$. We have a presentation of $W$ in terms of generators and relations:

$$W = \langle s_1, \cdots, s_{n-1} \mid s_i^2 = 1, s_i s_{i+1} s_i = s_{i+1} s_i s_{i+1}, s_i s_j = s_j s_i \text{ if } |i-j| \geqslant 2 \rangle.$$

The length function is given by

$$\ell(\sigma) = \sharp\{(i,j); 1 \leqslant i < j \leqslant n; \sigma(i) > \sigma(j)\}.$$

EXAMPLE 1.2. Let $\tilde{W} = \mathbb{Z}^n \rtimes S_n$, where $S_n$ acts on $\mathbb{Z}^n$ by permutation. This is the Iwahori-Weyl group of $GL_n(\mathbb{Q}_p)$. Following [**Lu03**, 1.12], we may realize $W$ as the group of periodic permutations on $\mathbb{Z}$, i.e., the permutations

$f : \mathbb{Z} \to \mathbb{Z}$ such that $f(z+n) = f(z) + n$ for all $z \in \mathbb{Z}$. Define $\chi : W \to \mathbb{Z}$ by $\chi(f) = \sum_{z=1}^{n}(f(z) - z)$ and $W_a = \ker(\chi)$. Then $W_a$ is a Coxeter group with the generators $s_i$ for $i = 0, 1, \cdots n-1$, where $s_i : \mathbb{Z} \to \mathbb{Z}$ is defined by

$$s_i(z) = \begin{cases} z+1, & \text{if } z \equiv i \mod n; \\ z-1, & \text{if } z \equiv i+1 \mod n; \\ z, & \text{otherwise.} \end{cases}$$

The order $m_{ij}$ of $s_i s_j$ is given as follows:

If $n = 2$, then $m_{01} = \infty$.

If $n \geqslant 3$, then $m_{ij} = \begin{cases} 3, & \text{if } i - j \equiv \pm 1 \mod n; \\ 2, & \text{otherwise.} \end{cases}$.

The group $\tilde{W}$ is not a Coxeter group. But it is almost as good as a Coxeter group. We may realize $\tilde{W}$ as

$$\tilde{W} = W_a \rtimes \Omega.$$

Here $\Omega \cong \mathbb{Z}$ is the group generated by $\tau \in \tilde{W}$, where $\tau(m) = m+1$ for all $m \in \mathbb{Z}$. There is a length function $\ell$ on $\tilde{W}$, defined by

$$\ell(\sigma) = \sharp(Y_\sigma / \tau_n),$$

where $Y_\sigma = \{(i,j) \in \mathbb{Z} \times \mathbb{Z}; i < j, \sigma(i) > \sigma(j)\}$ and $Y_\sigma/\tau_n$ is the (finite) set of $\tau_n$-orbits on $Y_\sigma$ for the $\tau_n$ action given by $(i,j) \mapsto (i+n, j+n)$.

The restriction of this length function to $W_a$ is the length function of the Coxeter group $W_a$. The subgroup $\Omega$ of $\tilde{W}$ is the subgroup of length-zero elements.

**1.2. Weakly length-decreasing sequences.** Let $(W, \mathbb{S})$ be a Coxeter system. For any $J \subset \mathbb{S}$, let $W_J$ be the subgroup of $W$ generated by $J$. Then $(W_J, J)$ is again a Coxeter system. The two extreme cases are $W$ (for $J = \mathbb{S}$) and $\{1\}$ (for $J = \emptyset$).

Let $\mathcal{O}$ be a $W_J$-orbit on $W$ for a given $W_J$-action. We are interested in the relations between the elements, together with their length, in this orbit. More specifically, we would like to have the following property:

**Red-Min:** For any $w \in W$, there exists a sequence of simple reflections $s_1, \cdots, s_k \in J$ such that

- For any $1 \leqslant i \leqslant k$, $\ell(s_i s_{i-1} \cdots s_1 \cdot w) \leqslant \ell(s_{i-1} \cdots s_1 \cdot w)$;
- The element $s_k s_{k-1} \cdots s_1 \cdot w$ is of minimal length among all the elements in $\mathcal{O}$.

If this property holds, then one may reduce the questions on any element $w$ in $\mathcal{O}$ to the questions on the minimal length elements in $\mathcal{O}$ via the inductive step:

$$w_1 \rightsquigarrow s \cdot w_1 \text{ for some } s \in \mathbb{S} \text{ with } \ell(s \cdot w_1) \leqslant \ell(w_1).$$

This reduction method plays an important role in the study of many problems related to Coxeter groups.

EXAMPLE 1.3. We consider the action of $W_J$ on $W$ by left multiplication. Let ${}^JW$ be the set of elements $w$ in $W$ of minimal length in the cosets $W_Jw$. Then each $W_J$-orbit contains a unique element $x$ in ${}^JW$. Moreover, $\ell(wx) = \ell(w) + \ell(x)$ for any $w \in W_J$. In particular, $x$ is the unique minimal length element in the $W_J$-orbit of $x$. Using the structure of the Coxeter group $W_J$, it is easy to see that the **Red-Min** property holds for this orbit.

**1.3. Conjugation by simple reflections.** We are especially interested in the conjugation action of $W$ and the partial conjugation action (of $W_J$) on $W$.

Conjugacy classes of $W$ play an important role in the study of representations of $W$ and representations of the corresponding Hecke algebra, which is a deformations of the group algebra of $W$. They also play an important role in the study of conjugacy classes of algebraic groups and $p$-adic groups.

The $W_J$-conjugacy classes in $W$, for finite and affine Weyl groups $W$ and their proper subgroups $W_J$, are closely related to the conjugation action of parabolic subgroups on algebraic groups and to the conjugacy action of parahoric subgroups on $p$-adic groups. They also have important application to arithmetic geometry and to representation theory.

We are going to discuss the applications in more details in Part II and in Part III.

We first investigate a few examples of the elements in a fixed conjugacy class of $W$ and discuss some nice properties.

EXAMPLE 1.4. Let $W = S_5$ and $C$ be the conjugacy class of 5-cycles. Then $\sharp C = 4! = 24$ and the lengths of elements in $C$ are 4, 6 or 8.

Let $w = (12345)$ and $w' = (12453)$ be elements in $C$ of minimal length. They are related via

$$w = (12345) \overset{s_1}{\approx} (21345) \overset{s_2}{\approx} (31245) = w'.$$

Here $\approx$ means conjugation by simple reflections which preserve the length.

Suppose we start with an element of $C$ not of minimal length, say $w'' = (14235)$. Then $w''$ is related to a minimal length element in $C$ via

$$w'' = (14235) \xrightarrow{s_3} (13245) \xrightarrow{s_2} (12345) = w.$$

Here $\rightarrow$ means conjugation by simple reflections which weakly decrease the length.

EXAMPLE 1.5. Let $W = S_5$ and $C$ be the conjugacy class of transpositions. The minimal length elements in $C$ are the simple reflections. They are not $\approx$-equivalent to each other. However, they are still $\sim$-equivalent. Here $\sim$-equivalent means that the corresponding elements in the associated Braid group are conjugate.

Now let us investigate two examples of partial-conjugacy classes in $W$. Although left multiplication by $W_J$ and conjugate by $W_J$ are very different, the elements in ${}^JW$ still play a crucial role in the study of the conjugation action of $W_J$ on $W$.

The following example is a $W_J$-conjugacy class that contains an element in ${}^JW$.

EXAMPLE 1.6. Let $W = S_4$ and $J = \{2, 3\}$.

Let $w = s_1s_2s_3 \in {}^JW$ and $w' = s_2s_3s_2s_1s_2$ be an element in the $W_J$-orbit of $w$. The element $w'$ has larger length than $w$ and $w'$ is related to $w$ via

$$w' = s_2s_3s_2s_1s_2 \xrightarrow{s_2} s_3s_2s_1 \xrightarrow{s_3} s_2s_3s_1 \xrightarrow{s_2} s_3s_1s_2 \xrightarrow{s_3} s_1s_2s_3 = w.$$

The next example is a $W_J$-conjugacy class that does not contain an element in ${}^JW$.

EXAMPLE 1.7. Let $W = S_5$ and $J = \{1, 3, 4\}$.

Let $w = s_4s_3s_2s_1s_3s_2s_4$. It is not of minimal length in its $W_J$-orbit. We have

$$w = s_4s_3s_2s_1s_3s_2s_4 \xrightarrow{s_4} s_3s_2s_1s_3s_2 \xrightarrow{s_3} s_1s_2s_1s_3s_2 \xrightarrow{s_1} s_3s_2s_1s_3s_2.$$

The elements $s_1s_2s_1s_3s_2$ and $s_3s_2s_1s_3s_2$ are of minimal length in the $W_J$-orbit of $w$. Notice that these two elements have the same ${}^JW$-part $x = s_2s_1s_3s_2 \in {}^JW$ and prefix $s_1$ and $s_3$ are elements in $W_J$ conjugate to each other by the element $x$.

**1.4. $\sim$-equivalence and $\approx$-equivalence.** Conjugacy classes of $W$ play an important role in the study of split groups. Twisted conjugacy classes, on the other hand, play a similar role in the study of non-split groups. In the sequel, we will not only consider the conjugation action, but will consider the twisted conjugation action as well.

Now let us introduce some notation.

Let $(W, \mathbb{S})$ be a Coxeter system and $J, J' \subset \mathbb{S}$. Let $\delta : W_J \to W_{J'}$ be a group automorphism such that $\delta(J) = J'$. We consider the $\delta$-twisted partial conjugation of $W_J$ on $W$ defined by

$$w \cdot_\delta w' = ww'\delta(w)^{-1}.$$

We denote by $(W_J)_\delta = \{(w, \delta(w)); w \in W_J\} \subset W \times W$ the $\delta$-graph of $W_J$. We denote by $W/(W_J)_\delta$ the set of $\delta$-twisted $W_J$-conjugacy classes of $W$. If $\delta$ is the identity map, we may write $(W_J)_\Delta$ for $(W_J)_\delta$.

Let $w, w' \in W$. We write $w \to_{J,\delta} w'$ if there exists a sequence of simple reflections $s_1, \cdots, s_k \in J$ such that

- $w' = s_ks_{k-1} \cdots s_1 \cdot_\delta w$;
- for any $1 \leqslant i \leqslant k$, $\ell(s_is_{i-1} \cdots s_1 \cdot_\delta w) \leqslant \ell(s_{i-1} \cdots s_1 \cdot_\delta w)$.

We write $w \approx_{J,\delta} w'$ if $w \to_{J,\delta} w'$ and $w' \to_{J,\delta} w$.

As we have seen in the above examples, a $\delta$-twisted $W_J$-conjugacy class $\mathcal{O}$ of $W$ may contain more than one minimal length elements. We denote by $\mathcal{O}_{\min}$ the set of minimal length elements in $\mathcal{O}$ and by $\ell(\mathcal{O})$ the length of

$w$ for any $w \in \mathcal{O}_{\min}$. Now we introduce some equivalence relations between the elements in $\mathcal{O}_{\min}$.

We write $w \sim_{J,\delta} w'$ if there exists a sequence of elements $x_1, \cdots, x_k \in W_J$ such that $w' = x_k x_{k-1} \cdots x_1 \cdot_\delta w$ and for any $1 \leqslant i \leqslant k$,

- $\ell(x_i x_{i-1} \cdots x_1 \cdot_\delta w) = \ell(w)$;
- $\ell(x_i x_{i-1} \cdots x_1 w \delta(x_{i-1} x_{i-2} \cdots x_1)^{-1})$ or $\ell(x_{i-1} x_{i-2} \cdots x_1 w \delta(x_i x_{i-1} \cdots x_1)^{-1})$ equals $\ell(w) + \ell(x_i)$.

Note that both $\sim_{J,\delta}$ and $\approx_{J,\delta}$ are equivalence relations and for $w, w' \in W$,

$$w \approx_{J,\delta} w' \Longrightarrow w \sim_{J,\delta} w' \Longrightarrow w' \in W_J \cdot_\delta w \text{ and } \ell(w) = \ell(w').$$

But the converse, in general, does not hold.

If $\delta$ is the identity map, then we omit $\delta$ in the under script. If $J = \mathbb{S}$, then we omit $J$ in the under script.

### 1.5. The discovery of Geck and Pfeiffer.

1.5.1. *Minimal length elements in a finite Coxeter group.* In the study of characters of finite Hecke algebras, Geck and Pfeiffer [**GP93**] discovers that the minimal length elements in a given conjugacy class of a finite Weyl group has remarkable properties.

THEOREM 1.8. *Let $W$ be a finite Coxeter group and $\delta : W \to W$ is a group automorphism with $\delta(\mathbb{S}) = \mathbb{S}$. Let $\mathcal{O} \in W/W_\delta$. Then*

(1) ***Red-Min*** *property holds for $\mathcal{O}$.*
(2) *$\mathcal{O}_{\min}$ is a single $\sim_\delta$-equivalence class.*
(3) *If moreover, $\mathcal{O}$ is elliptic, i.e. $\mathcal{O} \cap W_J = \emptyset$ for any $J = \delta(J) \subsetneqq \mathbb{S}$, then $\mathcal{O}_{\min}$ is a single $\approx_\delta$-equivalence class.*

Part (1) and (2) for the ordinary conjugacy classes of a finite Weyl group are first proved by Geck and Pfeiffer in [**GP93**, Theorem 1.1]. The ordinary conjugacy classes for any finite Coxeter groups (including the non-crystallographic types) are studied in [**GP00**], where Part (3) is also proved. Part (1) and (2) for the twisted conjugacy classes of a finite Coxeter group are obtained by Geck, Kim and Pfeiffer in [**GKP00**, Theorem 2.6]. Part (3) for the twisted conjugacy classes is established in [**He07b**, Theorem 7.5]. The proofs in these paper involve a case-by-case analysis. A conceptual proof is found recently in [**HN12**].

1.5.2. *Parametrization of the conjugacy classes of a finite Coxeter group.* Let $\mathcal{O}$ be a $\delta$-twisted conjugacy class of $W$. Let $J$ be a minimal $\delta$-stable subset of $\mathbb{S}$ with $\mathcal{O} \cap W_J \neq \emptyset$. Then by definition, $\mathcal{O} \cap W_J$ is a union of $\delta$-twisted elliptic conjugacy classes of $W_J$. A nice property is that the elliptic classes never fuse.

THEOREM 1.9. *Let $\mathcal{O}$ be a $\delta$-twisted conjugacy class of $W$. Let $J$ be a minimal $\delta$-stable subset of $\mathbb{S}$ with $\mathcal{O} \cap W_J \emptyset$. Then $\mathcal{O} \cap W_J$ is a single $\delta$-twisted conjugacy class of $W_J$.*

For ordinary conjugacy classes, it is due to Geck and Pfeiffer in [**GP00**, Theorem 3.2.11]. The general case is in [**CH16**, Theorem 2.3.4].

Now we give a parametrization of $W/W_\delta$ in terms of parabolic subgroups $W_J$ and their elliptic conjugacy classes. Set

$$\Gamma_\delta = \{(J, C); J \subset \mathbb{S} \text{ with } \delta(J) = J, C \text{ is an elliptic } \delta\text{-twisted conjugacy class of } W_J\}.$$

We say that the pairs $(J, C)$ and $(J', C')$ in $\Gamma_\delta$ are equivalent if there exits $x \in W^\delta$ such that $J' = xJx^{-1}$ and $C' = xCx^{-1}$. The following result is obtained in [**CH16**, Proposition 2.4.1].

PROPOSITION 1.10. *Let $W$ be a finite Coxeter group. The map*

$$f : \Gamma_\delta \to W/W_\delta, \qquad (J, C) \mapsto W \cdot_\delta C$$

*induces a bijection between the equivalence classes of $\Gamma_\delta$ and the set of $\delta$-twisted conjugacy classes of $W$.*

**1.6. Partial conjugation action.** Let $(W, \mathbb{S})$ be a Coxeter system and $J, J' \subset \mathbb{S}$. Let $\delta : W_J \to W_{J'}$ such that $\delta(J) = J'$. We call a $\delta$-twisted $W_J$-conjugacy class *distinguished* if it contains an element in ${}^J W$. As we have seen from Example 1.6 and example 1.7, one may relate any $\delta$-twisted $W_J$-conjugacy class to a distinguished one. To make it precise, let me first introduce some notation.

For $w \in {}^J W$, we write $\mathrm{Ad}(w)\delta(J) = J$ if for any simple reflection $s \in J$, there exists a simple reflection $s' \in J$ such that $w\delta(s)w^{-1} = s'$. In this case, $w \in {}^J W^{\delta(J)}$. It is easy to see that for any $J_1, J_2 \subset \mathbb{S}$, and $w \in {}^{J_1 \cup J_2}\tilde{W}$, if $\mathrm{Ad}(w)\delta(J_i) = J_i$ for $i = 1, 2$, then $\mathrm{Ad}(w)\delta(J_1 \cup J_2) = J_1 \cup J_2$. Thus for any $J \subset \mathbb{S}$ and $w \in {}^J W$, the set $\{J' \subset J \mid \mathrm{Ad}(w)\delta(J') = J'\}$ contains a unique maximal element. We set

$$I(J, w, \delta) = \max\{J' \subset J \mid \mathrm{Ad}(w)\delta(J') = J'\}.$$

EXAMPLE 1.11. Let $W = S_4$, $J = \{2, 3\}$ and $w = s_1 s_2 s_3$. Then $I(J, w, id) = \emptyset$.

Let $W = S_5$, $J = \{1, 3, 4\}$ and $w = s_2 s_1 s_3 s_2$. Then $I(J, w, id) = \{1, 3\}$.

The classification of $\delta$-twisted $W_J$-conjugacy classes of $W$ is obtained in [**He07b**, Section 2].

THEOREM 1.12. *We have*

*(1) $W = \sqcup_{w \in {}^J W} W_J \cdot_\delta (W_{I(J,w,\delta)} w)$.*

*(2) For any $w \in {}^J W$, the inclusion $W_{I(J,w,\delta)} \to W_J \cdot_\delta (W_{I(J,w,\delta)} w)$ induces a natural bijection between the $\mathrm{Ad}(w) \circ \delta$-twisted conjugacy classes $C$ on $W_{I(J,w,\delta)}$ and the $\delta$-twisted $W_J$-conjugacy classes $\mathcal{O}$ on $W_J \cdot_\delta (W_{I(J,w,\delta)} w)$.*

We also have the following relation between the minimal length elements of $C$ and of $\mathcal{O}$. This is proved in [**He07b**, §3].

THEOREM 1.13. *Let $w \in {}^J W$ and $C$ be a* $\mathrm{Ad}(w) \circ \delta$*-twisted conjugacy class in $W_{I(J,w,\delta)}$. Let $\mathcal{O} = W_J \cdot_\delta (Cw)$. Then*

*(1) For any $x \in \mathcal{O}$, there exists $u \in C$ such that $x \to_{J,\delta} uw$.*

*(2) If $C$ satisfies the **Red-Min** property, then so is $\mathcal{O}$.*

*(3) If $C_{\min}$ is a single $\sim_{\mathrm{Ad}(w)\circ\delta}$-equivalence class, then $\mathcal{O}_{\min}$ is a single $\sim_\delta$-equivalence class.*

*(4) If $C_{\min}$ is a single $\approx_{\mathrm{Ad}(w)\circ\delta}$-equivalence class, then $\mathcal{O}_{\min}$ is a single $\approx_\delta$-equivalence class.*

In particular, we have

THEOREM 1.14. *Suppose that $W_J$ is a finite Coxeter group or an affine Weyl group. Let $\mathcal{O}$ be a $\delta$-twisted $W_J$-conjugacy class of $W$. Then*

(1) ***Red-Min** property holds for $\mathcal{O}$.*
(2) *$\mathcal{O}_{\min}$ is a single $\sim_\delta$-equivalence class.*
(3) *If moreover, $\mathcal{O}$ is distinguished, then $\mathcal{O}_{\min}$ is a single $\approx_\delta$-equivalence class.*

The case where $W_J$ is a finite Coxeter group is obtained in [**He07b**, §3] by combining Theorem 1.13 with Theorem 1.8. A different proof is obtained by Nie in [**Ni13**]. The case where $W_J$ is an affine Weyl group can be obtained in the same way by using Theorem 1.17 we are going to discuss instead of Theorem 1.8.

### 1.7. Affine Weyl groups.

1.7.1. *Definition.* We first recall the definition of affine Weyl groups.

Let $\mathscr{R} = (X^*, X_*, R, R^\vee, \Pi)$ be a based reduced root datum, i.e., $X^*$ and $X_*$ are free abelian groups of finite rank together with a perfect pairing $\langle\, ,\, \rangle : X^* \times X_* \to \mathbb{Z}$, $R \subset X^*$ is the set of roots, $R^\vee \subset X_*$ is the set of coroots and $\Pi \subset R$ is the set of simple roots. Let $V = X_* \otimes \mathbb{R}$. For any $\alpha \in R$, we have a reflection $s_\alpha$ on $V$ defined by $s_\alpha(x) = x - \langle \alpha, x\rangle \alpha^\vee$. For any $\alpha \in R$ and $k \in \mathbb{Z}$, we have an affine root $\tilde{\alpha} = \alpha + k$ and an affine reflection $s_{\tilde{\alpha}}$ on $V$ defined by $s_{\tilde{\alpha}}(x) = x - (\langle \alpha, x\rangle + k)\alpha^\vee$.

The finite Weyl group $W_0$ is the subgroup of $GL(V)$ generated by $s_\alpha$ for $\alpha \in R$. The affine Weyl group and the extended affine Weyl group are defined by

$$W_a = \mathbb{Z}R^\vee \rtimes W_0, \qquad \tilde{W} = X_* \rtimes W_0.$$

Let $\mathrm{Aff}(V)$ be the group of affine transformations on $V$. We may identify $\tilde{W}$ with the subgroup of $\mathrm{Aff}(V)$ generated by $t^\lambda$ for $\lambda \in X_*$ and $W_0$, where $t^\lambda$ is the translation $v \mapsto v - \lambda$ on $V$. We may also identify $W_a$ with the subgroup of $\mathrm{Aff}(V)$ generated by $s_{\tilde{\alpha}}$, where $\tilde{\alpha}$ runs over all the affine roots.

The set $\Pi$ of simple roots determines the set $R_+$ of positive roots, the dominant chamber

$$\mathcal{C} = \{x \in V; \langle \alpha, x\rangle > 0 \text{ for all } \alpha \in \Pi\},$$

the anti-dominant chamber

$$\mathcal{C}^- = \{x \in V; \langle \alpha, x\rangle < 0 \text{ for all } \alpha \in \Pi\}$$

and the base alcove

$$\mathfrak{a} = \{x \in V; -1 < \langle \alpha, x \rangle < 0 \text{ for all } x \in R_+\} \subset \mathcal{C}^-.$$

Let $\mathbb{S}_0 = \{s_\alpha; \alpha \in \Pi\}$. Then $(W_0, \mathbb{S}_0)$ is a Coxeter system. Let $\tilde{\mathbb{S}}$ be the set of $s_{\tilde{\alpha}}$, where the hyperplane $H_{\tilde{\alpha}} = V^{\tilde{\alpha}}$ is a wall of $\mathfrak{a}$. Then $(W_a, \tilde{\mathbb{S}})$ is again a Coxeter system.

The length of an element in $\tilde{W}$ is the number of walls between the base alcove $\mathfrak{a}$ and the image of $\mathfrak{a}$ under the action of that element. The explicit formula for length function is given by Iwahori and Matsumoto in [**IM65**]: for $\lambda \in X_*$ and $w \in W_0$.

$$\ell(t^\lambda w) = \sum_{\alpha \in R_+, w^{-1}(\alpha) \in R_+} |\langle \alpha, \lambda \rangle| + \sum_{\alpha \in R_+, w^{-1}(\alpha) \in R_-} |\langle \alpha, \lambda \rangle - 1|.$$

This length function extends the length function of the Coxeter group $W_a$ and we have a semi-direct product

$$\tilde{W} = W_a \rtimes \Omega,$$

where $\Omega = \{w \in \tilde{W}; \ell(w) = 0\}$ is the stabilizer of the base alcove $\mathfrak{a}$.

The definitions in §1.4 still make sense for $\tilde{W}$. The only modification that we need is the definition of $\approx$-equivalence relation on $\tilde{W}$. We write $w \approx_\delta w'$ if there exists $\tau \in \Omega$ such that $w \to_\delta \tau w' \delta(\tau)^{-1}$ and $\tau w' \delta(\tau)^{-1} \to_\delta w$.

For any $J \subset \Pi$, set $\tilde{W}_J = X_* \rtimes W_J$. We call $\tilde{W}_J$ a *parabolic subgroup* of $\tilde{W}$. This is the extended affine Weyl group associated to the root datum $\mathscr{R}_J = (X^*, X_*, R_J, R_J^\vee, J)$, where $R_J \subset R$ is the sub root system spanned by the roots in $J$ and $R_J^\vee \subset R^\vee$ is the set of corresponding coroots.

For any $K \subsetneq \tilde{\mathbb{S}}$ with $W_K$ finite, we call $W_K$ a *parahoric subgroup* of $\tilde{W}$.

1.7.2. *Arithmetic invariants.* Let $\delta : \tilde{W} \to \tilde{W}$ be a group automorphism with $\delta(\tilde{\mathbb{S}}) = \tilde{\mathbb{S}}$. We first discuss the classification of $\delta$-twisted conjugacy classes of $\tilde{W}$. Motivated by Kottwitz's work [**Ko85**] and [**Ko97**], we introduce two arithmetic invariants.

Since $\delta$ preserve the length function on $\tilde{W}$, we have $\delta(\Omega) = \Omega$. Let $\Omega_\delta$ be the set of $\delta$-coinvariants on $\Omega$. The Kottwitz map $\kappa = \kappa_{\tilde{W},\delta} : \tilde{W} \to \Omega_\delta$ is obtained by composing the natural projection map $\tilde{W} \to \tilde{W}/W_a \cong \Omega$ with the projection map $\Omega \to \Omega_\delta$. It is constant on each $\delta$-twisted conjugacy class of $\tilde{W}$. This gives one invariant.

Another invariant is given by the Newton map. For any $w \in \tilde{W}$, we regard the element $w\delta$ as an element in $\mathrm{Aff}(V)$. There exists $n \in \mathbb{N}$ such that $(w\delta)^n = t^\xi$ for some $\xi \in X_*$. Let $\nu_{w,\delta} = \xi/n$ and $\bar{\nu}_{w,\delta} \in \mathcal{C}$ be the unique dominant element in the $W_0$-orbit of $\nu_{w,\delta}$. We call $\bar{\nu}_{w,\delta}$ the *Newton point* of $w$ (with respect to the $\delta$-conjugation action). It is known that $\nu_{w,\delta}$ is independent of the choice of $n$ and $\bar{\nu}_{w,\delta} = \bar{\nu}_{w',\delta}$ if $w$ and $w'$ are $\delta$-conjugate. Let $\mathcal{O}$ be a $\delta$-twisted conjugacy class. We write $\nu_{\mathcal{O}} = \bar{\nu}_{w,\delta}$ for any $w \in \mathcal{O}$ and we call $\nu_{\mathcal{O}}$ the Newton point of $\mathcal{O}$.

Define a map

$$f : \tilde{W}/(\tilde{W})_\delta \to \Omega_\delta \times \mathcal{C}, \quad w \mapsto (\kappa(\mathcal{O}), \nu_{\mathcal{O}}).$$

This map is not surjective. The Newton point of an element $w \in \tilde{W}$ is a rational dominant coweight satisfying an integrality conditions. The map is not injective, either. Any $\delta$-twisted conjugacy class that intersects a parahoric subgroup is mapped to $(1, 0)$. However, there is a nice lifting of $\mathrm{Im}(f)$ in $\tilde{W}/(\tilde{W})_\delta$. This is given in terms of straight $\delta$-twisted conjugacy classes. We will discuss it in the next subsection.

**1.8. Straight and non-straight conjugacy classes.**

1.8.1. *Definition.* Let $w \in \tilde{W}$. We say that $w$ is *$\delta$-straight* if $\ell(w\delta(w) \cdots \delta^{n-1}(w)) = n\ell(w)$ for all $n \in \mathbb{N}$. We may formulate it in a slightly different way. Regard $\delta$ as an element in the group $\tilde{W} \rtimes \langle\delta\rangle$ and extend the length function from $\tilde{W}$ to $\tilde{W} \rtimes \langle\delta\rangle$ by requiring $\ell(\delta) = 0$. Then $(w\delta)^n = w\delta(w) \cdots \delta^{n-1}(w)\delta^n \in \tilde{W} \rtimes \langle\delta\rangle$ and $w$ is $\delta$-straight if and only if $\ell((w\delta)^n) = n\ell(w)$ for all $n \in \mathbb{N}$. By [**He14**, §2.4], $w$ is $\delta$-straight if and only if $\ell(w) = \langle 2\rho, \bar{\nu}_{w,\delta}\rangle$, where $\rho$ is the half sum of positive roots in $R$. We say that a $\delta$-twisted conjugacy $\mathcal{O}$ of $\tilde{W}$ is *straight* if it contains a $\delta$-straight element. The following result is obtained in [**HN14**, Theorem 3.3].

THEOREM 1.15. *The map $f : \tilde{W}/\tilde{W}_\delta \to \Omega_\delta \times \mathcal{C}$ induces a bijection between the set of straight $\delta$-twisted conjugacy classes and $Im(f)$.*

Note that in a straight $\delta$-twisted conjugacy class $\mathcal{O}$, the set of minimal length elements are exactly the set of $\delta$-straight element in it. Moreover, if $\mathcal{O}'$ is another $\delta$-twisted conjugacy classes with $f(\mathcal{O}') = f(\mathcal{O})$, then $\ell(w') > \langle 2\rho, \nu_{\mathcal{O}}\rangle$ for any $w \in \mathcal{O}'$.

1.8.2. *Minimal length elements.* We have seen in §1.6 that the map from non-distinguished $\delta$-twisted $W_J$-conjugacy classes to distinguished $\delta$-twisted $W_J$-conjugacy classes is "compatible" with the length function. There is a similar relation between non-straight conjugacy classes and straight conjugacy classes of $\tilde{W}$. It is proved in [**HN14**, Proposition 2.7].

THEOREM 1.16. *Let $\mathcal{O} \in \tilde{W}/(\tilde{W})_\delta$ and $\mathcal{O}'$ be a straight $\delta$-twisted conjugacy class with $f(\mathcal{O}) = f(\mathcal{O}')$. Then for any $w \in \mathcal{O}$, there exists a triple $(x, K, u)$ with $w \to_\delta ux$, where $x$ is a $\delta$-straight element in $\mathcal{O}'$, $K$ is a subset of $\tilde{\mathbb{S}}$ such that $W_K$ finite, $x \in {}^K\tilde{W}$ and $\mathrm{Ad}(x)\delta(K) = K$, and $u$ is an element in $W_K$.*

The following result is obtained in [**HN14**, Theorem 2.9] by combining Theorem 1.16 with Theorem 1.8.

THEOREM 1.17. *Let $W$ be an affine group and $\delta : W \to W$ is a group automorphism with $\delta(\mathbb{S}) = \mathbb{S}$. Let $\mathcal{O} \in W/W_\delta$. Then*

(1) ***Red-Min*** *property holds for $\mathcal{O}$.*
(2) *$\mathcal{O}_{\min}$ is a single $\sim_\delta$-equivalence class.*

In each distinguished $\delta$-twisted $W_J$-conjugacy classes, we have a canonical minimal length element, i.e., the element in that conjugacy class that is also of minimal length with respect to the left multiplication of $W_J$. The straight $\delta$-twisted conjugacy classes, in general, does not have such a canonical choice of minimal length element. However, all the minimal length elements in a given straight $\delta$-twisted conjugacy class still form a single $\approx$-equivalence class. This is proved in [**HN14**, Theorem 3.8].

THEOREM 1.18. *Let $\mathcal{O}$ be a straight $\delta$-twisted conjugacy class of $\tilde{W}$. Then $\mathcal{O}_{\min}$ is a single $\approx_\delta$-equivalence class.*

1.8.3. *Standard quadruples.* In this subsection, we give standard representative for the straight $\delta$-twisted conjugacy classes and give a parametrization of the $\delta$-twisted conjugacy classes.

For $J \subset \Pi$, set

$$\mathfrak{a}_J = \{x \in V; -1 < \langle \alpha, x \rangle < 0 \text{ for all } \alpha \in R_{J,+}\}.$$

Let $\tilde{J}$ be the set of simple reflections of the $\tilde{W}_J$, in other words, $\tilde{J}$ consists of the reflections $s_\alpha$, where the hyperplanes $H_\alpha$ is a wall of $\mathfrak{a}_J$. We denote by $\ell_J$ the length function on the quasi-Coxeter group $\tilde{W}_J$ and by $\Omega_J \subset \tilde{W}_J$ the stabilizer of $\mathfrak{a}_J$. Note that $\ell_J$ is different from the restriction to $\tilde{W}_J$ of the length function $\ell$ on $\tilde{W}$.

We say that $(J, x, K, C)$ is a *standard quadruple* if

(1) $J \subset \Pi$ with $\delta_0(J) = J$, where $\delta_0 \in GL(V)$ is the linear part of $\delta \in \text{Aff}(V)$;
(2) $\text{Ad}(x) \circ \delta$ normalizes $\tilde{W}_J$ and induces a permutation on the set of affine simple reflections of $\tilde{W}_J$ and $\langle \alpha, \nu_x \rangle > 0$ for all $\alpha \in \Pi - J$;
(3) $K \subset \tilde{J}$ with $W_K$ finite and $\text{Ad}(x)\delta(K) = K$.
(4) $C$ is an elliptic $\text{Ad}(x) \circ \delta$-twisted conjugacy class of $W_K$.

Note that the condition (2) implies that $J$ is the subset of $\Pi$ consisting of simple roots $\alpha$ with $\langle \alpha, \nu_{x,\delta} \rangle = 0$. In other words, $J$ is determined by the element $x$. Similarly, $C$ is contained in the affine Weyl group $W_{J,a}$ of the root datum $\mathscr{R}_J$ and $K$ is determined by $C$ by taking the support of any element in $C$. It is also worth mentioning that $\nu_{x,\delta}$ is dominant and $\nu_{ux,\delta} = \nu_{x,\delta}$ for any $u \in C$.

Let $\tilde{\Gamma}_\delta$ be the set of standard quadruples. We say that the standard quadruples $(J, x, K, C)$ and $(J', x', K', C')$ are *equivalent* in $\tilde{W}$ if $J = J'$, there exists $\tau \in \Omega_J$ with $x' = \tau x \delta(\tau)^{-1}$ and there exists $w \in \tilde{W}_J$ with $x'\delta(w)(x')^{-1} = w$ and $C' = w\tau C(w\tau)^{-1}$.

THEOREM 1.19. *Let $\tilde{W}$ be an extended affine Weyl group. The map*

$$f : \tilde{\Gamma}_\delta \to \tilde{W}/\tilde{W}_\delta, \qquad (J, x, K, C) \mapsto \tilde{W} \cdot_\delta Cx$$

*induces a bijection between the equivalence classes of standard quadruples and the set of $\delta$-twisted conjugacy classes of $\tilde{W}$.*

*Moreover, a $\delta$-twisted conjugacy class of $\tilde{W}$ is straight if and only if the standard quadruples corresponding to it are of the form $(J, x, \emptyset, \{1\})$.*

REMARK 1.20. (1) Let $(J, x, K, C)$ be a standard quadruples. Note that for $w, w' \in C$, $wx$ and $w'x$ are $\delta$-conjugate by an element in $W_K \subset \tilde{W}$. Thus $\tilde{W} \cdot_\delta Cx$ is a single $\delta$-twisted conjugacy class of $\tilde{W}$.

(2) A different parametrization is given in [**HN15**, Proposition 5.8]. It is more convenient to use this parameterization in the study of cocenter and representations of affine Hecke algebras.

(3) For any standard quadruple $(J, x, K, C)$, $x$ is contained in a straight $\delta$-twisted conjugacy class of $\tilde{W}$, but $x$ may not be a $\delta$-straight element.

PROOF. We first show that every $\delta$-twisted conjugacy class $\mathcal{O}$ of $\tilde{W}$ comes from a standard quadruple. By Theorem 1.16, there exists a triple $(x, K, u)$ with $ux \in \mathcal{O}_{\min}$, where $x$ is a $\delta$-straight element, $K$ is a subset of $\tilde{\mathbb{S}}$ such that $W_K$ finite, $x \in {}^K\tilde{W}$ and $\mathrm{Ad}(x)\delta(K) = K$, and $u$ is an element in $W_K$. We may assume furthermore that $K$ is the smallest subset of $\tilde{\mathbb{S}}$ that is $\mathrm{Ad}(x) \circ \delta$-stable and contains $\mathrm{supp}(u)$.

Let $J = \{\alpha \in \Pi; \langle \alpha, \bar{\nu}_{x,\delta} \rangle = 0\}$ and $z \in {}^J W_0$ with $z(\nu_{x,\delta}) = \bar{\nu}_{x,\delta}$. Let $x' = zx\delta(z)^{-1}$, $K' = z(K)$ and $C$ be the $\mathrm{Ad}(x) \circ \delta$-twisted conjugacy class of $W_{K'}$ that contains $zuz^{-1}$. Since $x$ is a $\delta$-straight element, $x'$ is a length-zero element in $\tilde{W}_J$. As explained in [**HNx**, §3.3], $z(K) \subset \tilde{J}$. By the smallest assumption on $K$, $C$ is an elliptic $\mathrm{Ad}(x) \circ \delta$-twisted conjugacy class of $W_K$. Therefore $(J, x', K', C)$ is a standard quadruple associated to $\mathcal{O}$.

Now let $(J_1, x_1, K_1, C_1), (J_2, x_2, K_2, C_2)$ be standard quadruples associated to the same $\delta$-twisted conjugacy class $\mathcal{O}$ of $\tilde{W}$. Then $\nu_{x_1,\delta} = \nu_{x_2,\delta} = \nu_{\mathcal{O}}$ and $k(x_1) = k(x_1) = k(\mathcal{O})$. Thus $J_1 = J_2$ and by Theorem 1.15, $x_1$ and $x_2$ are in the same straight $\delta$-twisted conjugacy class.

Set $J = J_1$. Let $u_i \in C_i$. Let $w \in \tilde{W}$ such that $u_2 x_2 = w u_1 x_1 \delta(w)^{-1}$. Since $w(\nu_{x_1,\delta}) = \nu_{x_2,\delta} = \nu_{x_1,\delta}$, we have $w \in \tilde{W}_J$. We write $w$ as $w = w_1 \tau$, where $w_1 \in W_{J,a}$ and $\tau$ is a length-zero element of $\tilde{W}_J$. Then we have $x_2 = \tau x_1 \delta(\tau)^{-1}$ as an element in $\Omega_J$. Set $K_2' = \mathrm{Ad}(\tau)(K_1)$ and $C_2' = \mathrm{Ad}(\tau)(C_1)$. Then $(J, x_2, K_2', C_2')$ is a standard quadruple and $(J, x_2, K_2', C_2')$ is equivalent to $(J, x_1, K_1, C_1)$.

We have that $C_2 x_2$ and $C_2' x_2$ are in the same $\delta$-conjugacy class of $\tilde{W}_J$. In other words, $C_2$ and $C_2'$ are in the same $\delta'$-twisted conjugacy class of $\tilde{W}_J$, where $\delta' = \mathrm{Ad}(x_2) \circ \delta$. Set $V_J = R_J^\vee \otimes_{\mathbb{Z}} \mathbb{R}$. For any $u_2 \in C_2$, $V_J^{u_2 \delta'}$ is spanned by the coweights $\omega_j^\vee$ for $j \notin K_2$. For any $u_2' \in C_2'$, $V_J^{u_2' \delta'}$ is spanned by the coweights $\omega_j^\vee$ for $j \notin K_2'$. Since $C_2$ and $C_2'$ are in the same $\delta'$-conjugacy class, by [**CH16**, Proposition 2.4.1] there exists $w_2 \in \tilde{W}_J$ with $\delta'(w_2) = w_2$ and $\mathrm{Ad}(w_2)(K_2) = K_2'$. Then $C_2'$ and $w_2 C_2 w_2^{-1}$ are both elliptic $\delta'$-twisted conjugacy classes of $W_{K_2'}$ with the same characteristic polynomials and the same minimal length. By [**He07b**, Theorem 7.5], $C_2' = w_2 C_2 w_2^{-1}$.

The "moreover" part is proved in [**HN14**, Proposition 3.2]. □

TABLE 1. Conjugacy classes of extended affine Weyl group of $GL_3$

| Conditions on $(a_1, a_2, a_3)$ | | Number | Standard quadruples |
|---|---|---|---|
| $a_1 > a_2 > a_3$ | | 1 | $(\emptyset, t^{(a_1,a_2,a_3)}, \emptyset, \{1\})$ |
| $a_1 = a_2 > a_3$ | $a_1 \in \mathbb{Z}$ | 2 | $(\{1\}, t^{(a_1,a_2,a_3)}, \emptyset, \{1\})$ |
| | | | $(\{1\}, t^{(a_1,a_2,a_3)}, \{1\}, \{s_1\})$ |
| | $a_1 = n + \frac{1}{2}$ | 1 | $(\{1\}, t^{(n+1,n,a_3)} s_1, \emptyset, \{1\})$ |
| $a_1 > a_2 = a_3$ | $a_2 \in \mathbb{Z}$ | 2 | $(\{2\}, t^{(a_1,a_2,a_3)}, \emptyset, \{1\})$ |
| | | | $(\{2\}, t^{(a_1,a_2,a_3)}, \{2\}, \{s_2\})$ |
| | $a_2 = n + \frac{1}{2}$ | 1 | $(\{2\}, t^{(a_1,n+1,n)} s_2, \emptyset, \{1\})$ |
| $a_1 = a_2 = a_3$ | $a_1 \in \mathbb{Z}$ | 3 | $(\{1,2\}, t^{(a_1,a_1,a_1)}, \emptyset, \{1\})$ |
| | | | $(\{1,2\}, t^{(a_1,a_1,a_1)}, \{i\}, \{s_i\})$ for $0 \leqslant i \leqslant 2$ |
| | | | $(\{1,2\}, t^{(a_1,a_1,a_1)}, \{i,j\}, \{s_i s_j, s_j s_i\})$ for $0 \leqslant i < j \leqslant 2$ |
| | $a_1 = n + \frac{1}{3}$ | 1 | $(\{1,2\}, t^{(n+1,n,n)} s_1 s_2, \emptyset, \{1\})$ |
| | $a_1 = n + \frac{2}{3}$ | 1 | $(\{1,2\}, t^{(n+1,n+1,n)} s_2 s_1, \emptyset, \{1\})$ |

1.8.4. *Examples.* In this subsection, we give two examples, the extended affine Weyl groups associated to $GL_3$ and $SL_3$, and their conjugacy classes.

EXAMPLE 1.21. Let us consider the based root datum of $GL_3$. Then $X^* = \mathbb{Z}e_1 \oplus \mathbb{Z}e_2 \oplus \mathbb{Z}e_3 \cong \mathbb{Z}^3$ and $X^* = \mathbb{Z}e_1^\vee \oplus \mathbb{Z}e_2^\vee \oplus \mathbb{Z}e_3^\vee \cong \mathbb{Z}^3$, where $\langle e_i, e_j^\vee \rangle = \delta_{ij}$. The simple roots are $\alpha_1 = e_1 - e_2$ and $\alpha_2 = e_2 - e_3$.

Let $\tilde{W}$ be the extended affine Weyl group associated to this root datum and $\delta$ be the identity map on $\tilde{W}$. Let $\mathcal{O}$ be a conjugacy class of $\tilde{W}$. The Newton point of $\mathcal{O}$ is an element $(a_1, a_2, a_3) \in \mathbb{Q}^3$ with $a_1 \geqslant a_2 \geqslant a_3$ that satisfies the integrality condition:

$$\text{For any } c \in \mathbb{Q}, c\sharp\{i; a_i = c\} \in \mathbb{Z}.$$

The image of $\mathcal{O}$ under the Kottwitz map is determined by the Newton point

$$\kappa(\mathcal{O}) = a_1 + a_2 + a_3 \in \mathbb{Z} \cong \Omega.$$

In Table 1, we list the number of conjugacy classes with given Newton point $(a_1, a_2, a_3)$, and the associated standard quadruples.

EXAMPLE 1.22. Let us consider the based root datum of $SL_3$. Here

$$V = \{ae_1^\vee + be_2^\vee + ce_3^\vee; a + b + c = 0\}.$$

Let $W_a$ be the associated affine Weyl group and $\delta$ be the identity map. Let $\mathcal{O}$ be a conjugacy class of $W_a$. The Newton point of $\mathcal{O}$ is an element $(a_1, a_2, a_3) \in \mathbb{Q}^3$ with $a_1 \geqslant a_2 \geqslant a_3$ and $a_1 + a_2 + a_3 = 0$.

In Table 2, we list some information on the conjugacy classes of $\tilde{W}$. Here the difference from Table 1 comes from the fact that the element $s_0 s_1, s_0 s_2$ and $s_1 s_2$ are conjugate in $\tilde{W}$, but not in $W_a$.

TABLE 2. Conjugacy classes of the affine Weyl group of $SL_3$

| Conditions on $(a_1, a_2, a_3)$ | | Number | Standard quadruples |
|---|---|---|---|
| $a_1 > a_2 > a_3$ | | 1 | $(\emptyset, t^{(a_1,a_2,a_3)}, \emptyset, \{1\})$ |
| $a_1 = a_2 > a_3$ | $a_1 \in \mathbb{Z}$ | 2 | $(\{1\}, t^{(a_1,a_2,a_3)}, \emptyset, \{1\})$ |
| | | | $(\{1\}, t^{(a_1,a_2,a_3)}, \{1\}, \{s_1\})$ |
| | $a_1 = n + \frac{1}{2}$ | 1 | $(\{1\}, t^{(n+1,n,a_3)} s_1, \emptyset, \{1\})$ |
| $a_1 > a_2 = a_3$ | $a_2 \in \mathbb{Z}$ | 2 | $(\{2\}, t^{(a_1,a_2,a_3)}, \emptyset, \{1\})$ |
| | | | $(\{2\}, t^{(a_1,a_2,a_3)}, \{2\}, \{s_2\})$ |
| | $a_2 = n + \frac{1}{2}$ | 1 | $(\{2\}, t^{(a_1,n+1,n)} s_2, \emptyset, \{1\})$ |
| $a_1 = a_2 = a_3 = 0$ | | 5 | $(\{1,2\}, 1, \emptyset, \{1\})$ |
| | | | $(\{1,2\}, 1, \{i\}, \{s_i\})$ for $0 \leqslant i \leqslant 2$ |
| | | | $(\{1,2\}, 1, \{1,2\}, \{s_1 s_2, s_2 s_1\})$ |
| | | | $(\{1,2\}, 1, \{0,1\}, \{s_1 s_0, s_0 s_1\})$ |
| | | | $(\{1,2\}, 1, \{0,2\}, \{s_0 s_2, s_2 s_0\})$ |

**1.9. Weakly length-increasing sequences.** Instead of considering the weakly length-decreasing sequences as in §1.2, one may consider the weakly length-increasing sequences in a given twisted conjugacy class.

Let $W$ be a finite Coxeter group and $w_0$ be the longest element of $W$. Then the map

$$W \to W, \qquad w \mapsto w w_0$$

is an order-reversing map and it sends $\delta$-twisted conjugacy classes to $\mathrm{Ad}(w_0) \circ \delta$-twisted conjugacy classes. Thus as a variation of Theorem 1.8 (1), we have

THEOREM 1.23. *Let $W$ be a finite Coxeter group and $\mathcal{O}$ be a $\delta$-twisted conjugacy class of $W$. Let $\mathcal{O}_{\max}$ be the set of maximal length elements in $\mathcal{O}$. Then for any $w \in \mathcal{O}$, there exists $w' \in \mathcal{O}_{\max}$ with $w' \to_\delta w$.*

A similar statement holds for affine Weyl groups.

THEOREM 1.24. *Let $\tilde{W}$ be an extended affine group and $\mathcal{O}$ be a $\delta$-twisted conjugacy class of $\tilde{W}$. Let $w \in \mathcal{O}$. If $w$ is not of maximal length in $\mathcal{O}$, then there exists $w' \in \mathcal{O}$ with $\ell(w') > \ell(w)$ and $w' \to_\delta w$.*

For ordinary conjugacy classes, this is proved by Rostami in [**Rox**, Main Theorem 1.1]. The general case can be proved using the method in [**HN14**].

The maximal length elements and the weakly length-increasing sequences play a role in the study of the center of Hecke algebras.

**1.10. The quotient space $W /\!\!/ (W_J)_\delta$.**

1.10.1. *The quotient $W /\!\!/ (W_J)_\delta$.* As we have seen in §1.6 and §1.8, some $\delta$-twisted $W_J$-conjugacy classes of $W$ are special and any non-special conjugacy classes can be "reduced" to a special conjugacy class.

- For partial conjugation action, the special classes are the distinguished ones.

- For (non-twisted and twisted) conjugation action of an extended affine Weyl group, the special classes are the straight conjugacy classes.

In either case, we denote by $W /\!/ (W_J)_\delta$ the set of special $\delta$-twisted conjugacy classes in $W/(W_J)_\delta$. We then have a natural map

$$\pi : W/(W_J)_\delta \to W /\!/ (W_J)_\delta.$$

Here the notation $/\!/$ is borrowed from the geometric invariant theory.

In the rest of this subsection, we will introduce a partial order on $W /\!/ (W_J)_\delta$ so that the quotient space becomes a topological space. We will discuss in Part II some applications of these topological spaces.

1.10.2. *The Bruhat order.* We first recall the definition of Bruhat order. Let $(W, \mathbb{S})$ be a Coxeter system. The Bruhat order on $W$ is defined as follows.

Let $w \in W$ and $w = s_1 \cdots s_n$ with $n = \ell(w)$ and $s_i \in \mathbb{S}$ for all $i$. Such expression is called a *reduced expression* of $w$. Let $w'$ be another element in $W$. We say that $w' \leqslant w$ if there exists $1 \leqslant i_1 < i_2 < \cdots < i_k \leqslant n$ such that $w' = s_{i_1} s_{i_2} \cdots s_{i_k}$. The definition of Bruhat order is in fact independent of the choice of reduced expressions of $w$. In the case where $W$ is a finite (resp. an affine) Weyl group, the Bruhat order describes the closures relations between the Schubert cells in the finite (resp. affine) flag varieties.

Now we discuss some general facts about the Bruhat order.

LEMMA 1.25. *Let $x, w, w' \in W$ with $x \leqslant w$ and $w' \to_{J,\delta} w$. Then there exists $x' \in W$ such that $x \to_{J,\delta} x'$ and $x' \leqslant w'$.*

PROOF. It suffices to prove the case where $w' = sw\sigma(s)$ for some $s \in J$.

If $w' > w$, then we may take $x' = x$. Now we assume that $\ell(w) = \ell(w')$. Without loss of generalization, we may assume furthermore that $sw < w$ and $w\delta(s) > w$.

If $sx > x$, then by [**Lu03**, Corollary 2.5], $x \leqslant sw$ and hence $x < w'$.

If $sx < x$, then $\ell(sx\delta(s)) \leqslant \ell(x)$ and $x \to_{J,\delta} sx\delta(s)$. By [**Lu03**, Corollary 2.5], $sx \leqslant sw$ and $sx\delta(s) \leqslant sw\delta(s)$. □

PROPOSITION 1.26. *Let $\mathcal{O} \in W/(W_J)_\delta$. If* **Red-Min** *property holds for $\mathcal{O}$, then $\mathcal{O}_{\min}$ is the set of minimal element in $\mathcal{O}$ with respect to the Bruhat order.*

PROOF. Let $w, w' \in \mathcal{O}$. If $w' \leqslant w$, then $\ell(w') \leqslant \ell(w)$. Thus if $w \in \mathcal{O}_{\min}$, then $w$ is a minimal element in $\mathcal{O}$ with respect to the Bruhat order.

On the other hand, let $w \in \mathcal{O}$. Suppose that $w \notin \mathcal{O}_{\min}$, then by **Red-Min** property, there exists $w' \in \mathcal{O}_{\min}$ with $w \to_{J,\delta} w'$. By Lemma 1.25, there exists $w_1 \approx_{J,\delta} w'$ with $w_1 < w$. □

1.10.3. *The partial order $\preceq$.*

PROPOSITION 1.27. *Suppose that $W_J$ is a finite Coxeter group or is an affine Weyl group. Let $\mathcal{O}, \mathcal{O}' \in W /\!/ (W_J)_\delta$. Then the following conditions are equivalent:*

*(1) For some* $w' \in \mathcal{O}'_{\min}$*, there exists* $w \in \mathcal{O}_{\min}$ *with* $w \leqslant w'$.
*(2) For any* $w' \in \mathcal{O}'_{\min}$*, there exists* $w \in \mathcal{O}_{\min}$ *with* $w \leqslant w'$.

The case where $W_J$ is a finite Coxeter group is obtained in [**He07b**, Corollary 4.6], by combining Lemma 1.25 with Theorem 1.8. The case where $W_J$ is an affine Weyl group is obtained in the same way by using Theorem 1.17 instead of Theorem 1.8.

Now we introduce a partial order on $W /\!\!/ (W_J)_\delta$. Let $\mathcal{O}, \mathcal{O}' \in W /\!\!/ (W_J)_\delta$. We write $\mathcal{O} \preceq \mathcal{O}'$ if the conditions in Proposition 1.27 are satisfied. By the equivalence of the two conditions in Proposition 1.27,

$$\mathcal{O} \preceq \mathcal{O}' \text{ and } \mathcal{O}' \preceq \mathcal{O}'' \Longrightarrow \mathcal{O} \preceq \mathcal{O}''.$$

Thus $\preceq$ gives a partial order on $W /\!\!/ (W_J)_\delta$.

It is worth mentioning that although we state the results for twisted partial conjugacy classes for a Coxeter group, the definition of $W /\!\!/ (W_J)_\delta$ and the partial order still works if we replace $W$ by a quasi-Coxeter group. In particular, the definition and results here still holds for extended affine Weyl groups as well.

The partial order $\preceq$, for finite Weyl group $W$ and $J \subsetneqq \mathbb{S}$, is first introduced in [**He07a**] to describe the closure relations between Lusztig's $G$-stable pieces [**Lu04a**] and [**Lu04b**]. The partial order $\preceq$, for affine Weyl group $W$, is used to describe the closure relations of several interesting stratifications of loop groups, as we will discuss in Part II, §2.6.

1.10.4. *The partial order on* ${}^J W$. If $J \subsetneqq \mathbb{S}$, then there is a natural bijection between $W /\!\!/ (W_J)_\delta$ and ${}^J W$. This partial order on ${}^J W$ can be described in an explicit way as follows:

Let $w, w' \in {}^J W$, then $W_J \cdot_\delta w \preceq W_J \cdot_\delta w'$ if and only if there exists $u \in W_J$ such that $uw\delta(u)^{-1} \leqslant w'$.

In this case, we also write $w \preceq_{J,\delta} w'$.

Note that the partial order $\preceq_{J,\delta}$ on ${}^J W$ is different from the restriction to ${}^J W$ of the Bruhat order on $W$. The latter one implies the first one, but not vice verse in general. The difference can be seen in the following example.

EXAMPLE 1.28. Let $W = S_4$ and $J = \{3\}$. The simple reflections of $W$ are $s_1, s_2, s_3$. We simply write $s_{abc\cdots}$ instead of $s_a s_b s_c \cdots$. Let $\delta$ be the nontrivial diagram automorphism, i.e., $\delta(s_1) = s_3$, $\delta(s_2) = s_2$ and $\delta(s_3) = s_1$. In Figure 1, we draw the Hasse diagram of ${}^J W$, with respect to the usual Bruhat order, the partial order $\preceq_{J,id}$ (the extra relation is in dotted line) and the partial order $\preceq_{J,\delta}$ (the extra relation is in double dotted line).

1.10.5. *The partial order on the set of straight conjugacy classes.* Let $\tilde{W}$ be an extended affine Weyl group. Proposition 1.27 gives a partial order $\preceq$ on the set of straight $\delta$-twisted conjugacy classes of $\tilde{W}$. On the other hand, we have an injective map

$$f : \tilde{W}/(\tilde{W})_\delta \to \Omega_\delta \times \mathcal{C}, \quad w \mapsto (\kappa(\mathcal{O}), \nu_{\mathcal{O}}).$$

There is a natural partial order on $\Omega_\delta \times \mathcal{C}$ given as follows.

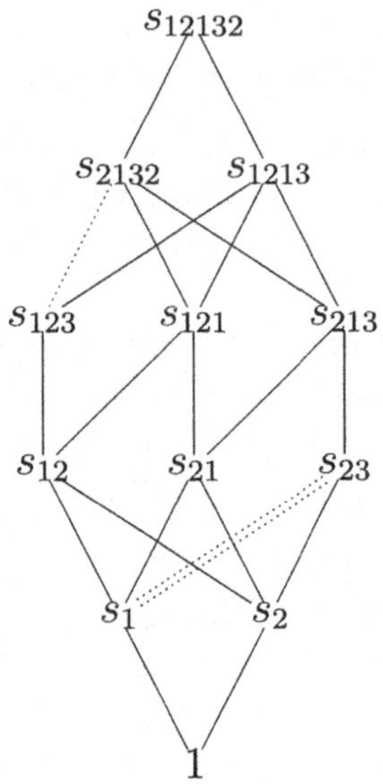


FIGURE 1. Hasse diagram for the partial orders on ${}^J W$.

For $(v,p),(v',p') \in \Omega_\delta \times \mathcal{C}$, we write $(v,p) \leqslant (v',p')$ if $v = v'$ and $p$ is less than or equal to $p'$ in the dominance order, i.e., $p' - p$ is a nonnegative linear combination of positive coroots. It is proved in [**Hex**, Theorem 3.1] that these two partial orders coincide.

THEOREM 1.29. *Let $\mathcal{O}, \mathcal{O}' \in \tilde{W}/(\tilde{W})_\delta$. Then $\mathcal{O} \preceq \mathcal{O}'$ if and only if $f(\mathcal{O}) \leqslant f(\mathcal{O}')$.*

## 2. Part II. Geometry of $G(\breve{F})$

**2.1. Bruhat decomposition on $G(\mathbf{k})$.** As a warm-up, we first recall the Bruhat decomposition of a complex reductive group.

Let $G$ be a connected reductive group over an algebraically closed field $\mathbf{k}$. Let $B$ be a Borel subgroup and $T \subset B$ be a maximal torus. Let $N$ be the normalizer of $T$. For any element $w$ in the Weyl group $W_0 = N(T)/T$, we choose a representative $\dot{w}$ in $N(T)$.

The group $G(\mathbf{k})$ is a union of Bruhat cells:

$$G(\mathbf{k}) = \sqcup_{w \in W_0} B\dot{w}B.$$

Moreover, the closure of a Bruhat cell $B\dot{w}B$ is a union of Bruhat cells and the closure relation is given by the Bruhat order of the Weyl group:

$$\overline{B\dot{w}B} = \sqcup_{w' \leqslant w} B\dot{w}'B.$$

The dimension of a Bruhat cell $B\dot{w}B$ is given in terms of length function of the Weyl group:

$$\dim(B\dot{w}B) - \dim B = \ell(w).$$

A variation is the decomposition of the flag variety $G/B$ into a union of Schubert cells

$$G/B = \sqcup_{w \in W_0} B\dot{w}B/B.$$

In fact, the subgroups $B$ and $N$ of $G(\mathbf{k})$ form a $(B, N)$-pair, i.e.,

- $G$ is generated by the subgroups $B$ and $N$.
- $B \cap N$ is a normal subgroup of $N$.
- The group $W = N/B \cap N$ is generated by a set of elements $s$ of order 2, for $s$ in some nonempty set $\mathbb{S}_0$.
- For any $s \in \mathbb{S}_0$ and $w \in W$, $\dot{s}Bw \subset B\dot{s}\dot{w}B \cup B\dot{w}B$.
- $\dot{s} \notin N_G(B)$ for all $s \in \mathbb{S}_0$.

An important consequence of the axioms of $(B, N)$-pairs is the following multiplication formula of Bruhat cells

$$B\dot{s}B\dot{w}B = \begin{cases} B\dot{s}\dot{w}B, & \text{if } sw > w; \\ B\dot{s}\dot{w}B \sqcup B\dot{w}B, & \text{if } sw < w. \end{cases}$$

$$B\dot{w}B\dot{s}B = \begin{cases} B\dot{w}\dot{s}B, & \text{if } ws > w; \\ B\dot{w}\dot{s}B \sqcup B\dot{w}B, & \text{if } ws < w. \end{cases}$$

**2.2. $G(\breve{F})$ and its Iwahori-Weyl groups.** Let $\mathbb{F}_q$ be a finite field with $q$ elements. Let $F$ be a non-archimedean local field with valuation ring $\mathcal{O}_F$ and residue field $\mathbb{F}_q$. We denote by $\epsilon$ its uniformizer. Let $\breve{F}$ be the completion of the maximal unramified extension of $F$ with valuation ring $\mathcal{O}_{\breve{F}}$ and residue field $\mathbf{k} = \bar{\mathbb{F}}_q$.

Let $G$ be a connected reductive group over $F$. Let $\sigma$ be the Frobenius automorphism of $\breve{F}/F$. We also denote the induced automorphism on $G(\breve{F})$ by $\sigma$.

Let $S \subset G$ be a maximal $\breve{F}$-split torus defined over $F$ and let $T$ be its centralizer. Since $G$ is quasi-split over $\breve{F}$, $T$ is a maximal torus of $G$. Let $\breve{\mathcal{I}}$ be the Iwahori subgroup fixing an alcove $\mathfrak{a}$ in the apartment $V$ attached to $S$. The *Iwahori-Weyl group* associated to $S$ is

$$\tilde{W} = N(\breve{F})/T(\breve{F}) \cap \breve{\mathcal{I}},$$

where $N$ denotes the normalizer of $S$ in $G$. The automorphism $\sigma$ on $G(\breve{F})$ induces an automorphism on the Iwahori-Weyl group $\tilde{W}$, which we still denote by $\sigma$. For $w \in \tilde{W}$, we choose a representative $\dot{w}$ in $N(\breve{F})$.

The relative Weyl group is $W_0 = N(\breve{F})/T(\breve{F})$. The Iwahori-Weyl group $\tilde{W}$ is a split extension of $W_0$ by the central subgroup $X_*(T)_\Gamma$, where $\Gamma = Gal(\bar{F}/\breve{F})$. The splitting depends on the choice of a special vertex of $\mathfrak{a}$.

In the split case, i.e., $S$ is a maximal torus of $G$, the Iwahori-Weyl group $\tilde{W}$ is the extended affine Weyl group associated to the root datum of $G$ and $\sigma : \tilde{W} \to \tilde{W}$ is the identity map. In the non-split case, the situation is more complicated.

EXAMPLE 2.1. Let $G$ be the special orthogonal group of a quadratic form in $2n+1$ variables. The local Dynkin diagram is of type $B_n$ if $G$ is split, and is of type ${}^2B_n$ if $G$ is non-split. The Iwahori-Weyl group of $G(\breve{F})$ for split and non-split quadratic forms are the same. However, the group automorphism on the Iwahori-Weyl group induced from the Frobenius automorphism $\sigma$ are different: it is the identity map in the split case and the unique nontrivial diagram automorphism in the non-split case.

TABLE 3. Local Dynkin diagrams for $SO(2n+1)$

| Name | Local Dynkin diagram |
|---|---|
| $B_n$ | 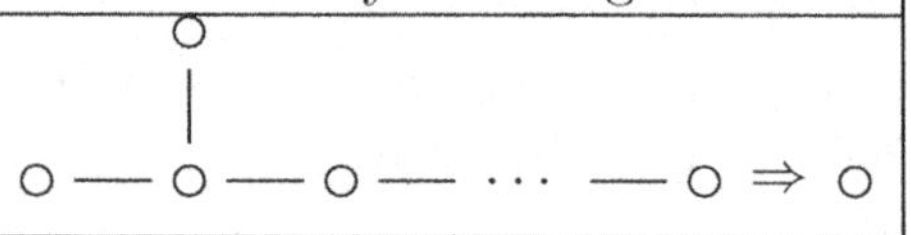 |
| ${}^2B_n$ | 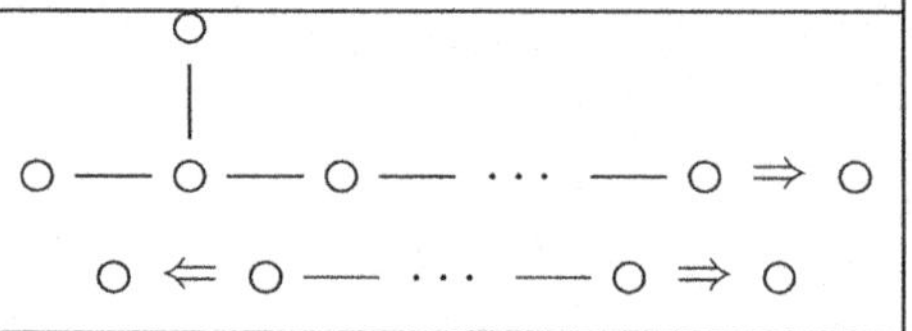 |

EXAMPLE 2.2. Let $G$ be the special unitary group of a hermitian form in $2n$ variables over a ramified quadratic extension of $F$. If the form $h$ is split, then the local Dynkin diagram is of type $B\text{-}C_n$. If the form $h$ is non-split, then the local Dynkin diagram is of type ${}^2B\text{-}C_n$. The Iwahori-Weyl group of $G(\breve{F})$ for split and non-split Hermitian forms are the same. However, the group automorphism on the Iwahori-Weyl group induced from the Frobenius automorphism $\sigma$ are different: it is the identity map if $h$ is split and the unique nontrivial diagram automorphism if $h$ is not split.

TABLE 4. Local Dynkin diagrams for ramified $SU(2n)$

| Name | Local Dynkin diagram |
|---|---|
| $B\text{-}C_n$ | 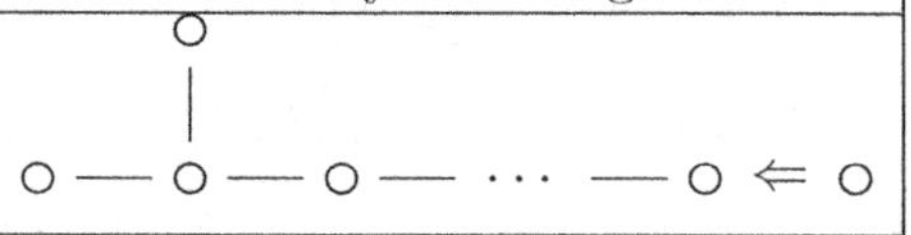 |
| ${}^2B\text{-}C_n$ | 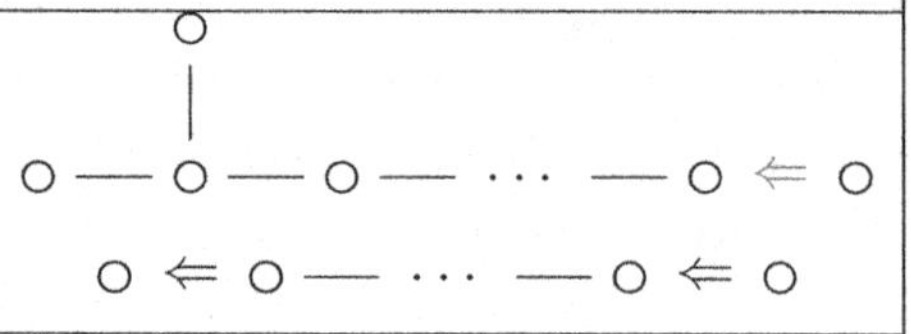 |

It is worth mentioning that although the local Dynkin diagram in Table 4 different from Table 3. The only difference is the orientation. As the change of orientation does not change the associated affine Weyl groups, the Iwahori-Weyl group $\tilde{W}$ together with the automorphism $\sigma : \tilde{W} \to \tilde{W}$ of ramified special unitary group $SU(2n)$ of a split (resp. non-split) hermitian form are the same as of split (resp. non-split) special orthogonal group $SO(2n+1)$.

In general, we may associate to $G \supset T$ a reduced root datum $\mathscr{R}$ such that the Iwahori-Weyl group of $G_{sc}$ can be identified with the affine Weyl group $W_a$ associated to $\mathscr{R}$. The set $\tilde{\mathbb{S}}$ of simple reflections of $W_a$ consists of the reflections along the walls of the alcove $\mathfrak{a}$ we fixed in the beginning. The group $\tilde{W}$ acts on the set of alcoves in the apartment attached to $S$, where $W_a$ acts transitively on the set of alcoves. We also have

$$\tilde{W} = W_a \rtimes \Omega,$$

where $\Omega$ is the isotropy subgroup of $\mathfrak{a}$.

The group $X_*(T)_\Gamma$ may have a nontrivial torsion part. However, as explained in [**HHx**, §8.1], the torsion part $X_*(T)_{\Gamma,tor}$ lies in the center of $\tilde{W}$ and thus may be ignored when considering the (non-twisted and twisted) conjugation action. In particular, the results we discussed in Part I on the minimal length elements, straight conjugacy classes, etc, apply to the Iwahori-Weyl groups as well.

The pair $(\tilde{W}/X_*(T)_{\Gamma,tor}, \sigma)$ for any group $G$ comes from the pair $(\tilde{W}', \sigma')$ for a suitable unramified group. This simple observation allows us to reduce the study of some questions on $G$ to the questions on an unramified group. This observation will be used in §2.9 and §2.10.

### 2.3. Iwahori-Bruhat decomposition on $G(\breve{F})$ and some variations.

2.3.1. *Double cosets of the Iwahori subgroup.* After the Bruhat decomposition was found in the 1950's, a rather unexpected extension was found by Iwahori and Matsumoto [**IM65**], for a split $p$-adic groups. Here in the decomposition, one use the (compact) Iwahori subgroup instead of the (non-compact) Borel subgroup and the (infinite) Iwahori-Weyl group instead of the (finite) Weyl group. This decomposition is then extended by Bruhat and Tits [**BT72**] to arbitrary $p$-adic groups.

THEOREM 2.3. *The pair $(\breve{\mathcal{I}}, N)$ is a $(B,N)$-pair. In particular,*

$$G(\breve{F}) = \sqcup_{w \in \tilde{W}} \breve{\mathcal{I}} \dot{w} \breve{\mathcal{I}}.$$

We also have an explicit formula on the multiplication of Bruhat cells

$$\breve{\mathcal{I}} \dot{s} \breve{\mathcal{I}} \dot{w} \breve{\mathcal{I}} = \begin{cases} \breve{\mathcal{I}} \dot{s} \dot{w} \breve{\mathcal{I}}, & \text{if } sw > w; \\ \breve{\mathcal{I}} \dot{s} \dot{w} \breve{\mathcal{I}} \sqcup \breve{\mathcal{I}} \dot{w} \breve{\mathcal{I}}, & \text{if } sw < w. \end{cases}$$

$$\breve{\mathcal{I}} \dot{w} \breve{\mathcal{I}} \dot{s} \breve{\mathcal{I}} = \begin{cases} \breve{\mathcal{I}} \dot{w} \dot{s} \breve{\mathcal{I}}, & \text{if } ws > w; \\ \breve{\mathcal{I}} \dot{w} \dot{s} \breve{\mathcal{I}} \sqcup \breve{\mathcal{I}} \dot{w} \breve{\mathcal{I}}, & \text{if } ws < w. \end{cases}$$

2.3.2. *Double cosets of a parahoric subgroup.* We have a similar decomposition for parahoric subgroups as well. Since all the Iwahori subgroups are conjugate, we will only consider the parabolic subgroups $\breve{\mathcal{K}}$ that contains $\breve{\mathcal{I}}$. For such $\breve{\mathcal{K}}$, we denote by $K \subset \tilde{\mathbb{S}}$ the corresponding set of simple reflections. The group $W_K$, generated by $K$, is the Weyl group of $\breve{\mathcal{K}}$. We denote by ${}^K\tilde{W}^K \subset \tilde{W}$ the set of minimal length representatives in their $W_K$-double cosets.

THEOREM 2.4. *Let $\breve{\mathcal{K}}$ be a parahoric subgroup of $G(\breve{F})$. Then*

$$G(\breve{F}) = \sqcup_{w \in {}^K\tilde{W}^K} \breve{\mathcal{K}}\dot{w}\breve{\mathcal{K}}.$$

Besides the Iwahori subgroup case, another case we are especially interested in is the special maximal parahoric subgroup case. Here we have a split extension of $\tilde{W} = X_*(T)_\Gamma \rtimes W_0$ with respect to the vertex corresponding to $\breve{\mathcal{K}}$ and $W_K \backslash \tilde{W}/W_K$ is in a natural bijection with the set of $W_0$-orbits of $X_*(T)_\Gamma$. We have the Cartan decomposition

$$G(\breve{F}) = \sqcup_{\{\mu\} \in X_*(T)_\Gamma/W_0} \breve{\mathcal{K}}\epsilon^\mu \breve{\mathcal{K}}.$$

Note that however, the multiplication formula for $\breve{\mathcal{K}}$-double cosets, in general, is more complicated than the multiplication formula for $\breve{\mathcal{I}}$-double cosets.

2.3.3. *Lusztig's finer decomposition.* Let $\breve{\mathcal{K}}$ be a parahoric subgroup of $G(\breve{F})$ and $\breve{\mathcal{I}} \subsetneqq \breve{\mathcal{K}}$. For $w \in {}^K\tilde{W}^K$, we decompose $\breve{\mathcal{K}}\dot{w}\breve{\mathcal{K}}$ further into finitely many subsets stable under the action of $\breve{K}_\sigma$, analogous to the $G$-stable piece decomposition (for reductive groups $G$ over algebraically closed fields) introduced by Lusztig in [**Lu04a**].

We have the following simple but very useful properties on the conjugation actions on the Bruhat cells:

- $\breve{\mathcal{K}} \cdot_\sigma \breve{\mathcal{I}}\dot{w}\breve{\mathcal{I}} = \breve{\mathcal{K}} \cdot_\sigma \breve{\mathcal{I}}\dot{w}'\breve{\mathcal{I}}$ if $w \approx_{K,\sigma} w'$;
- $\breve{\mathcal{K}} \cdot_\sigma \breve{\mathcal{I}}\dot{w}\breve{\mathcal{I}} = \breve{\mathcal{K}} \cdot_\sigma \breve{\mathcal{I}}\dot{s}\dot{w}\breve{\mathcal{I}} \cup \breve{\mathcal{K}} \cdot_\sigma \breve{\mathcal{I}}\dot{s}\dot{w}\sigma(\dot{s})\breve{\mathcal{I}}$ for $s \in K$ with $sw\sigma(s) < w$.
- If $w \in {}^K\tilde{W}$ and $u \in W_{I(K,w,\sigma)}$, then $\breve{\mathcal{K}} \cdot_\sigma \breve{\mathcal{I}}\dot{x}\dot{w}\breve{\mathcal{I}} = \breve{\mathcal{K}} \cdot_\sigma \breve{\mathcal{I}}\dot{w}\breve{\mathcal{I}}$.

The following decomposition is essentially contained in [**Lu10**, 1.4] and [**He11**, Proposition 2.5 & 2.6]. The general case is in [**GH15**, Theorem 3.2.1].

THEOREM 2.5. *For any $w \in {}^K\tilde{W}^K$,*

$$\breve{\mathcal{K}}\dot{w}\breve{\mathcal{K}} = \sqcup_{x \in W_K w W_K \cap {}^K\tilde{W}} \breve{\mathcal{K}} \cdot_\sigma \breve{\mathcal{I}}\dot{x}\breve{\mathcal{I}}.$$

Due to its importance, we give a sketch of the proof that

$$\breve{\mathcal{K}}\dot{w}\breve{\mathcal{K}} \subset \cup_{x \in W_K w W_K \cap {}^K\tilde{W}} \breve{\mathcal{K}} \cdot_\sigma \breve{\mathcal{I}}\dot{x}\breve{\mathcal{I}}.$$

Note that $\breve{\mathcal{K}}\dot{w}\breve{\mathcal{K}} = \sqcup_{w' \in W_K w W_K} \breve{\mathcal{I}}\dot{w}'\breve{\mathcal{I}}$. We argue by induction on $w'$ that

$$\breve{\mathcal{I}}\dot{w}'\breve{\mathcal{I}} \subset \cup_{x \in W_K w W_K \cap {}^K\tilde{W}} \breve{\mathcal{K}} \cdot_\sigma \breve{\mathcal{I}}\dot{x}\breve{\mathcal{I}}. \tag{2.1}$$

If $w'$ is a minimal length element in its $\sigma$-twisted $W_J$-conjugacy class, then by Theorem 1.16, $w' \approx_{K,\sigma} ux$ for some $x \in {}^K\tilde{W}$ and $u \in W_{I(K,x,\sigma)}$. In this case, $x \in W_J w' W_{\sigma(J)} = W_J w W_{\sigma(J)}$. The inclusion (2.1) holds for $w'$.

If $w'$ is not a minimal length element in its $\sigma$-twisted $W_J$-conjugacy class, then by Theorem 1.17, there exists $w'' \approx_{K,\sigma} w'$ and $s \in K$ with $sw''\sigma(s) < w''$. Then

$$\breve{\mathcal{K}} \cdot_\sigma \breve{\mathcal{I}}\dot{w}'\breve{\mathcal{I}} = \breve{\mathcal{K}} \cdot_\sigma \breve{\mathcal{I}}\dot{w}''\breve{\mathcal{I}} = \breve{\mathcal{K}} \cdot_\sigma \breve{\mathcal{I}}\dot{s}\dot{w}''\breve{\mathcal{I}} \cup \breve{\mathcal{K}} \cdot_\sigma \breve{\mathcal{I}}\dot{s}\dot{w}''\sigma(\dot{s})\breve{\mathcal{I}}.$$

The inclusion (2.1) for $w'$ follows from inductive hypothesis on $sw''$ and on $sw''\sigma(s)$.

**2.4. The set $B(G)$ of $\sigma$-conjugacy classes on $G(\breve{F})$.** For any $b \in G(\breve{F})$, we denote by $[b] = \{g^{-1}b\sigma(g); g \in G(\breve{F})\}$ its $\sigma$-conjugacy class. Let $B(G)$ be the set of $\sigma$-conjugacy classes of $G(\breve{F})$.[1] Kottwitz [**Ko85**] and [**Ko97**] has given a classification of the set $B(G)$. This classification generalizes the Dieudonné-Manin classification of isocrystals by their Newton polygons. For the purpose of this paper, it is convenient to view $B(G)$ as $\tilde{W} /\!\!/ \tilde{W}_\sigma$. Recall that $\Omega_\sigma$ is the set of $\sigma$-coinvariants on $\Omega$ and $\mathcal{C}$ is the dominant Weyl chamber of the reduced root datum $\mathscr{R}$ associated to $G$.

THEOREM 2.6. *The inclusion $N(\breve{F}) \to G(\breve{F})$ induces a map $\Psi : \tilde{W}/\tilde{W}_\sigma \to B(G)$. Moreover,*

*(1) the map $\Psi$ is surjective.*

*(2) We have the following commutative diagram*

$$\begin{array}{ccc} \tilde{W}/\tilde{W}_\sigma & \xrightarrow{\quad\Psi\quad} & B(G) \\ & {\scriptstyle f}\searrow \quad \swarrow{\scriptstyle f} & \\ & \Omega_\sigma \times \mathcal{C} & \end{array}$$

*(3) The map $f : B(G) \to \Omega_\sigma \times \mathcal{C}, b \mapsto (\kappa(b), \nu_b)$ is injective.*

Here part (3) is Kottwitz's classification of $B(G)$. The two arithmetic invariants on the set $\tilde{W}/\tilde{W}_\sigma$ of $\sigma$-twisted conjugacy classes of $\tilde{W}$ we introduced in §1.7.2 are the restriction of the invariants of $\sigma$-conjugacy classes of $G(\breve{F})$ to the $\sigma$-twisted conjugacy classes of $\tilde{W}$. The surjectivity of $\Psi$ is first proved in [**GHKR10**] for split groups. The general case is obtained in [**He14**, §3]. Another proof is given in [**GHN15**].

Note that we may identify $B(G)$ with $\mathrm{Im}(f)$. On the other hand, there is a natural bijection between $\mathrm{Im}(f)$ and $\tilde{W} /\!\!/ \tilde{W}_\sigma$. Therefore, we have the following commutative diagram

[1] To distinguish the conjugacy classes in $G(\breve{F})$ in $\tilde{W}$, we use **$\sigma$-conjugacy classes** for $G(\breve{F})$ and **$\sigma$-twisted conjugacy classes** for $\tilde{W}$.

$$\begin{array}{ccc} \tilde{W} /\!/ \tilde{W}_\sigma & \xleftrightarrow{\;\;1-1\;\;} & B(G)\,. \\ & {\scriptstyle f}\searrow \quad \swarrow{\scriptstyle f} & \\ & X_*(T)^+_{\mathbb{Q}} \times \pi_1(G)_{\Gamma_F} & \end{array}$$

For any $\sigma$-conjugacy class $[b]$ of $G(\breve{F})$, we denote by $\mathcal{O}_{[b]}$ (or just $\mathcal{O}_b$) the straight $\sigma$-twisted conjugacy class of $\tilde{W}$ that corresponds to $[b]$.

**2.5. Admissible subsets of $G(\breve{F})$.** We have discussed several interesting subsets of $G(\breve{F})$: $\breve{\mathcal{K}}\dot{w}\breve{\mathcal{K}}$ for $w \in {}^K\tilde{W}^K$, $\breve{\mathcal{K}} \cdot_\sigma \breve{\mathcal{I}}\dot{x}\breve{\mathcal{I}}$ for $w \in {}^K\tilde{W}$ and $[b] \in B(G)$. In the next few subsections, we will investigate some geometric properties (dimension, irreducible components, closure relations, etc.) of these subsets. In order to do so, let us assume furthermore that $F = \mathbb{F}((\epsilon))$. In this case, the affine flag variety and its deeper level generalization have a natural scheme structure. It is more difficult to do it over $p$-adic fields. We refer to the work of Zhu [**Zhx**] and the work of Bhatt and Scholze [**BSx**] in this direction.

For any $n \in \mathbb{N}$, let $\breve{\mathcal{I}}_n$ be the $n$-th congruence subgroup of $\breve{\mathcal{I}}$, i..e, the $n$-th Moy-Prasad subgroup associated to the barycenter of the base alcove. As in [**GHKR06**, §2.10], A subset $V$ of $G(\breve{F})$ is called *admissible* if for any $w \in \tilde{W}$, there exists $n \in \mathbb{N}$ such that $V \cap \breve{\mathcal{I}}\dot{w}\breve{\mathcal{I}}$ is stable under the right action of $\breve{\mathcal{I}}_n$. This is equivalent to say that for any $w \in \tilde{W}$, there exists $n' \in \mathbb{N}$ such that $V \cap \overline{\breve{\mathcal{I}}\dot{w}\breve{\mathcal{I}}}$ is stable under the right action of $\breve{\mathcal{I}}_{n'}$.

We say that an admissible subset $V$ of $G(\breve{F})$ is *locally closed* if for any $w \in \tilde{W}$, the underline reduced scheme of $(V \cap \overline{\breve{\mathcal{I}}\dot{w}\breve{\mathcal{I}}})/\breve{\mathcal{I}}_n$ is locally closed in $\overline{\breve{\mathcal{I}}\dot{w}\breve{\mathcal{I}}}/\breve{\mathcal{I}}_n$ for $n \gg 0$. In this case, we define the closure of $V$ to be

$$\overline{V} = \varinjlim_{w} V_w,$$

where $V_w$ be the inverse image under the projection $G(\breve{F}) \to G(\breve{F})/\breve{\mathcal{I}}_n$ of the closure of $(V \cap \overline{\breve{\mathcal{I}}\dot{w}\breve{\mathcal{I}}})/\breve{\mathcal{I}}_n$ in $G(\breve{F})/\breve{\mathcal{I}}_n$.

We say that $V$ is bounded if $V \cap \breve{\mathcal{I}}\dot{w}\breve{\mathcal{I}} = \emptyset$ for all but finitely many $w \in W$.

Let $V$ be a bounded, locally closed admissible subset of $G(\breve{F})$. By definition, there exists $n \in \mathbb{N}$ such that $V$ is stable under the right action of $\breve{\mathcal{I}}_n$. We say that $V_1$ is an irreducible component of $V$ if $V_1/\breve{\mathcal{I}}_n$ is an irreducible component of $V/\breve{\mathcal{I}}_n$ and we denote by $Irr(V)$ the set of irreducible components of $V$. We define

$$\dim_{\breve{\mathcal{I}}} V_1 = \dim(V_1/\breve{\mathcal{I}}_n) - \dim(\breve{\mathcal{I}}/\breve{\mathcal{I}}_n), \qquad \dim_{\breve{\mathcal{I}}} V = \max_{V_1 \in Irr(V)} \dim_{\breve{\mathcal{I}}} V_1.$$

We also denote by $Irr^{\max}(V)$ the set of irreducible components of $V$ of maximal possible dimension, i.e., of dimension equals $\dim_{\breve{\mathcal{I}}}(V)$.

By convenient, we set $\dim_{\breve{\mathcal{I}}}(\emptyset) = -\infty$ and $Irr^{\max}(\emptyset) = \emptyset$.

For any parahoric subgroup $\breve{\mathcal{K}}$, we set $\dim_{\breve{\mathcal{K}}}(V) = \dim_{\breve{\mathcal{I}}}(V) - \dim_{\breve{\mathcal{I}}}(\breve{\mathcal{K}})$.

In the rest of this subsection, we give some examples of admissible subsets. We start with a trivial example.

EXAMPLE 2.7. Let $w \in \tilde{W}$. Then $\breve{\mathcal{I}}\dot{w}\breve{\mathcal{I}}$ is a bounded, locally closed admissible subset of $G(\breve{F})$. The closure of $\breve{\mathcal{I}}w\breve{\mathcal{I}}$ is $\sqcup_{w' \leqslant w} \breve{\mathcal{I}}\dot{w}'\breve{\mathcal{I}}$ and $\dim_{\breve{\mathcal{I}}} \breve{\mathcal{I}}\dot{w}\breve{\mathcal{I}} = \ell(w)$.

The following result is easy to prove and we skip the details.

LEMMA 2.8. *Let $w \in \tilde{W}$ and $n \geqslant \ell(w)$. Then for any $g \in \breve{\mathcal{I}}w\breve{\mathcal{I}}$, $g\breve{\mathcal{I}}_n g^{-1} \subset \breve{\mathcal{I}}_{n-\ell(w)}$.*

EXAMPLE 2.9. Let $\breve{\mathcal{K}}$ be a parahoric subgroup of $G(\breve{F})$. By Lemma 2.8, there exists $n \in \mathbb{N}$ such that for any $g \in \breve{\mathcal{K}}$, $g\breve{\mathcal{I}}_n g^{-1} \subset \breve{\mathcal{I}}$. Hence for any $w \in \tilde{W}$, $\breve{\mathcal{K}} \cdot_\sigma \breve{\mathcal{I}}\dot{w}\breve{\mathcal{I}}$ is stable under the right action of $\breve{\mathcal{I}}_n$ and thus is an admissible subset of $G(\breve{F})$.

EXAMPLE 2.10. It is also known that each $\sigma$-conjugacy class of $G(\breve{F})$ is admissible. See [**Hex**, Theorem A.1].

**2.6. Closure relations.** Let $\breve{\mathcal{K}}$ be a parahoric subgroup of $G(\breve{F})$. We have the decomposition

$$G(\breve{F}) = \sqcup_{w \in {}^K\tilde{W}} \breve{\mathcal{K}} \cdot_\sigma \breve{\mathcal{I}}\dot{w}\breve{\mathcal{I}}.$$

Recall that the set ${}^K\tilde{W}$ is in natural bijection with $\tilde{W} /\!\!/ (W_K)_\sigma$ and there is a well-defined partial order on $\tilde{W} /\!\!/ (W_K)_\sigma \cong {}^K\tilde{W}$ (see §1.10). We show that this partial order describes the closure relations of the admissible subsets $\breve{\mathcal{K}} \cdot_\sigma \breve{\mathcal{I}}\dot{w}\breve{\mathcal{I}}$.

THEOREM 2.11. *Let $\breve{\mathcal{K}}$ be a parahoric subgroup and $x \in {}^K\tilde{W}$. Then*

$$\overline{\breve{\mathcal{K}} \cdot_\sigma \breve{\mathcal{I}}\dot{x}\breve{\mathcal{I}}} = \sqcup_{x' \in {}^K\tilde{W}, x' \leqslant_{K,\sigma} x} \breve{\mathcal{K}} \cdot_\sigma \breve{\mathcal{I}}\dot{x}'\breve{\mathcal{I}}.$$

It is proved in [**He11**, Proposition 2.5] in the case where $G$ is a split group and $\breve{\mathcal{K}}$ is the hyperspecial maximal subgroup. The general case is proved in the same way. We sketch the proof for completeness.

PROOF. The base idea is to prove by induction. However, in order to do so, we do not only need the elements in ${}^K\tilde{W}$, but have to use all the elements in $\tilde{W}$ as well. The more general statement we would like to prove is the following:

For any $w \in \tilde{W}$, $\overline{\breve{\mathcal{K}} \cdot_\sigma \breve{\mathcal{I}}\dot{w}\breve{\mathcal{I}}} = \sqcup_{x' \in {}^K\tilde{W}, x' \leqslant_{K,\sigma} w} \breve{\mathcal{K}} \cdot_\sigma \breve{\mathcal{I}}\dot{x}'\breve{\mathcal{I}}$. Here we say $x' \leqslant_{K,\sigma} w$ if there exists $x_1 \approx_{K,\sigma} x'$ with $x_1 \leqslant w$.

Since $\breve{\mathcal{K}}/\breve{\mathcal{I}}$ is proper,

$$\overline{\breve{\mathcal{K}} \cdot_\sigma \breve{\mathcal{I}}\dot{w}\breve{\mathcal{I}}} = \breve{\mathcal{K}} \cdot_\sigma \overline{\breve{\mathcal{I}}\dot{w}\breve{\mathcal{I}}} = \cup_{w' \in \tilde{W}, w' \leqslant w} \breve{\mathcal{K}} \cdot_\sigma \breve{\mathcal{I}}\dot{w}'\breve{\mathcal{I}}.$$

We argue by induction on $w'$ that $\breve{\mathcal{I}}\dot{w}'\breve{\mathcal{I}} \subset \sqcup_{x' \in {}^K\tilde{W}, x' \leqslant_{K,\sigma} w} \breve{\mathcal{K}} \cdot_\sigma \breve{\mathcal{I}}\dot{x}'\breve{\mathcal{I}}$ for any $w' \leqslant w$.

If $w'$ is a minimal length element in its $\sigma$-twisted $W_J$-conjugacy class, then by Theorem 1.16, $w' \approx_{K,\sigma} ux'$ for some $x' \in {}^K\tilde{W}$ and $u \in W_{I(K,x',\sigma)}$. So $\breve{\mathcal{K}} \cdot_\sigma \breve{\mathcal{I}}\dot{w}'\breve{\mathcal{I}} = \breve{\mathcal{K}} \cdot_\sigma \breve{\mathcal{I}}\dot{u}\dot{x}'\breve{\mathcal{I}} = \breve{\mathcal{K}} \cdot_\sigma \breve{\mathcal{I}}\dot{x}'\breve{\mathcal{I}}$. Since $x' \leqslant ux'$ and $w' \approx_{K,\sigma} ux'$, by Lemma 1.25 there exists $x'' \approx_{K,\sigma} x'$ with $x'' \leqslant w' \leqslant w$. By definition, $x' \preceq_{K,\sigma} w$.

If $w'$ is not a minimal length element in its $\sigma$-twisted $W_J$-conjugacy class, then there exists $w'' \approx_{K,\sigma} w'$ and $s \in K$ with $sw''\sigma(s) < w''$. Then

$$\breve{\mathcal{K}} \cdot_\sigma \breve{\mathcal{I}}\dot{w}'\breve{\mathcal{I}} = \breve{\mathcal{K}} \cdot_\sigma \breve{\mathcal{I}}\dot{w}''\breve{\mathcal{I}} = \breve{\mathcal{K}} \cdot_\sigma \breve{\mathcal{I}}\dot{s}\dot{w}''\breve{\mathcal{I}} \cup \breve{\mathcal{K}} \cdot_\sigma \breve{\mathcal{I}}\dot{s}\dot{w}''\sigma(\dot{s})\breve{\mathcal{I}}.$$

Note that $sw''\sigma(s) < sw'' < w''$. Thus for any $x' \in {}^K\tilde{W}$, if $x' \preceq_{K,\sigma} sw''\sigma(s)$ or $x' \preceq_{K,\sigma} sw''$, then $x' \preceq_{K,\sigma} w''$. By inductive hypothesis on $sw''$ and on $sw''\sigma(s)$, we have

$$\breve{\mathcal{K}} \cdot_\sigma \breve{\mathcal{I}}\dot{w}'\breve{\mathcal{I}} = \breve{\mathcal{K}} \cdot_\sigma \breve{\mathcal{I}}\dot{s}\dot{w}''\breve{\mathcal{I}} \cup \breve{\mathcal{K}} \cdot_\sigma \breve{\mathcal{I}}\dot{s}\dot{w}''\sigma(\dot{s})\breve{\mathcal{I}} \subset \sqcup_{x' \in {}^K\tilde{W}, x' \preceq_{K,\sigma} w''} \breve{\mathcal{K}} \cdot_\sigma \breve{\mathcal{I}}\dot{x}'\breve{\mathcal{I}}.$$

For any $x' \in {}^K\tilde{W}$ with $x' \preceq_{K,\sigma} w''$, by Lemma 1.25, there exists $x'' \approx_{K,\sigma} x'$ with $x'' \leqslant w' \leqslant w$. By definition, $x' \preceq_{K,\sigma} w$. □

Recall that there is a natural bijection between $\tilde{W} /\!/ \tilde{W}_\sigma$ and $B(G)$. On $\tilde{W} /\!/ \tilde{W}_\sigma$, there is a natural partial order $\preceq$ (see §1.10). It is proved in [**Hex**, Theorem B] that this partial order describes the closure relations between the $\sigma$-conjugacy classes of $G(\breve{F})$. The idea of the proof is similar to the proof of Theorem 2.11, but more technical.

THEOREM 2.12. *The natural bijection*

$$\tilde{W} /\!/ \tilde{W}_\sigma \to B(G), \qquad \mathcal{O}_{[b]} \to [b]$$

*is a bijection of partial order sets. Here the partial order on $\tilde{W} /\!/ \tilde{W}_\sigma$ is the combinatorial order $\preceq$ and the partial order on $B(G)$ is given by the closure relation.*

**2.7. Mazur's inequality.** In the next few subsections, we study the intersection of the subsets of $G(\breve{F})$ in §2.3 and §2.4.

The intersection of $\breve{\mathcal{K}}\epsilon^\mu\breve{\mathcal{K}} \cap [b]$, for $G = GL_n$ and $\breve{\mathcal{K}}$ the maximal hyperspecial subgroup, first appeared in Mazur's work [**Ma73**], in the study of the relation between the Hodge slope of a crystal and the Newton slope of the associated isocrystal.

By definition, an isocrystal is a pair $(N, g)$, where $N$ is a finite dimensional vector space over $\breve{F}$ and $g : N \to N$ is a semi-linear bijection. A crystal $M$ of an isocrystal $(N, g)$ is a $g$-stable $\mathcal{O}_{\breve{F}}$-lattice of $N$.

By Dieudonné-Manin theory, there is a natural bijection between the isomorphisms classes of isocrystals and the associated Newton slopes. On the other hand, for each crystal, one may associate a Hodge slope.

EXAMPLE 2.13. Let $N$ be a 3-dimensional vector space with standard basis $e_1, e_2, e_3$ and $g : N \to N$ is defined by $g(e_1) = e_2, g(e_2) = e_3, g(e_3) = \epsilon e_1$. Then all the eigenvalues of $g$ are the cubic roots of $\epsilon$ and Newton slope $\nu(N, g) = (\frac{1}{3}, \frac{1}{3}, \frac{1}{3})$.

Let $M = \oplus_{i=1}^{3} \mathcal{O}_{\breve{F}} e_i$ be a crystal of $(N, g)$. Then

$$\text{coker} g \mid_M \cong \mathcal{O}_{\breve{F}} e_1 / \epsilon \mathcal{O}_{\breve{F}} e_1 = \mathbf{k}^1 e_1 \oplus \mathbf{k}^0 e_2 \oplus \mathbf{k}^0 e_3.$$

The Hodge slope $\mu(M) = (1, 0, 0)$.

The Hodge polygon and the Newton polygon have the same end point and the Hodge polygon lies above the Newton polygon. In other words, $\mu(M) \geqslant \nu(N, g)$ with respect to the dominance order.

In general, we have the following result.

THEOREM 2.14. *(1) Let $(N, g)$ be an isocrystal and $M$ be a crystal of $(N, g)$. Then*

$$\mu(M) \geqslant \nu(N, g).$$

*(2) Let $(N, g)$ be an isocrystal of dimension $n$. Let $\mu = (a_1, \cdots, a_n) \in \mathbb{Z}^n$ with $a_1 \geqslant a_2 \geqslant \cdots \geqslant a_n$ and $\mu \geqslant \nu(N, g)$. Then there exists a crystal $M$ of $(N, g)$ with $\mu(M) = \mu$.*

Here part (1) is obtained by Mazur in [**Ma73**] and part (2) is obtained by Kottwitz and Rapoport in [**KR03**].

Now we reformulate the Mazur's inequality in a group-theoretic way.

An isocrystal of dimension $n$ corresponds to a $\sigma$-conjugacy class of $GL_n(\breve{F})$ and a crystal corresponds to a $\breve{\mathcal{K}}$-double coset in $GL_n(\breve{F})$, where $\breve{\mathcal{K}} = GL_n(\mathcal{O}_{\breve{F}})$. The above Theorem may be reformulated as follows:

Let $[b]$ be a $\sigma$-conjugacy class of $GL_n(\breve{F})$ and $\mu$ be a dominant coweight of $GL_n(\breve{F})$. Then $[b] \cap \breve{\mathcal{K}} \epsilon^\mu \breve{\mathcal{K}} \neq \emptyset$ if and only if $\mu \geqslant \nu_b$ and $\kappa([b]) = \kappa(\epsilon^\mu)$.

In the sequel, we will study the intersection $\breve{\mathcal{K}} \dot{w} \breve{\mathcal{K}} \cap [b]$ for any parahoric subgroup $\breve{\mathcal{K}}$ in a reductive group $G(\breve{F})$. Such intersection plays an important role in the study of reduction of Shimura varieties and we will discuss some applications in §2.12.

**2.8. The intersection of a Bruhat cell with a $\sigma$-conjugacy class.**

2.8.1. *Deligne-Lusztig reduction method.* Now we discuss a reduction method to study the intersection $\breve{\mathcal{I}} \dot{w} \breve{\mathcal{I}} \cap [b]$ after Deligne and Lusztig [**DL76**].

We first prove the following results.

PROPOSITION 2.15. *Let $X$ be an admissible subset of $G(\breve{F})$. Then the inverse image of $X$ under the multiplication map $m : G(\breve{F}) \times G(\breve{F}) \to G(\breve{F})$ is an admissible subset of $G(\breve{F}) \times G(\breve{F})$.*

PROOF. Let $w, w' \in \tilde{W}$. Then $\breve{\mathcal{I}} \dot{w} \breve{\mathcal{I}} \dot{w}' \breve{\mathcal{I}}$ is a bounded subset of $G(\breve{F})$. There exists $n \in \mathbb{N}$ such that $X \cap \breve{\mathcal{I}} \dot{w} \breve{\mathcal{I}} \dot{w}' \breve{\mathcal{I}}$ is stable under the right action of $\breve{\mathcal{I}}_n$. Suppose that $(g, g') \in \breve{\mathcal{I}} \dot{w} \breve{\mathcal{I}} \times \breve{\mathcal{I}} \dot{w}' \breve{\mathcal{I}}$ with $gg' \in X$. Then by Lemma 2.8,

$$g \breve{\mathcal{I}}_{n+\ell(w')} g' \breve{\mathcal{I}}_{n+\ell(w')} \subset gg' \breve{\mathcal{I}}_n \subset X \cap \breve{\mathcal{I}} \dot{w} \breve{\mathcal{I}} \dot{w}' \breve{\mathcal{I}}.$$

Hence $m^{-1}(X) \cap \breve{\mathcal{I}}\dot{w}\breve{\mathcal{I}} \times \breve{\mathcal{I}}\dot{w}'\breve{\mathcal{I}}$ is stable under the right action of $\breve{\mathcal{I}}_{n+\ell(w')} \times \breve{\mathcal{I}}_{n+\ell(w')}$ and $m^{-1}(X)$ is admissible. □

Let $w \in \tilde{W}$ and $s \in \tilde{\mathbb{S}}$ with $sw < w$. Then $\breve{\mathcal{I}}\dot{s}\breve{\mathcal{I}} \times \breve{\mathcal{I}}\dot{s}\dot{w}\breve{\mathcal{I}}$ is an admissible subset of $G(\breve{F}) \times G(\breve{F})$ and the multiplication map $G(\breve{F}) \times G(\breve{F}) \to G(\breve{F})$ induces

$$m : \breve{\mathcal{I}}\dot{s}\breve{\mathcal{I}} \times \breve{\mathcal{I}}\dot{s}\dot{w}\breve{\mathcal{I}} \to \breve{\mathcal{I}}\dot{w}\breve{\mathcal{I}}.$$

Notice that we do not have the notion of isomorphisms between the admissible subsets. However, we may still discuss the dimensions, irreducible components, connected components, etc. And for any admissible subset $X$ of $\breve{\mathcal{I}}w\breve{\mathcal{I}}$, there is a natural bijection between the irreducible components of $X$ and of $m^{-1}(X)$ and $\dim_{\breve{\mathcal{I}}}$ of an irreducible component of $X$ equals $\dim_{\breve{\mathcal{I}} \times \breve{\mathcal{I}}}$ of its inverse image in $\breve{\mathcal{I}}\dot{s}\breve{\mathcal{I}} \times \breve{\mathcal{I}}\dot{s}\dot{w}\breve{\mathcal{I}}$.

Let $p_{12} : G(\breve{F}) \times G(\breve{F}) \to G(\breve{F})$ be the map defined by $(g_1, g_2) \mapsto (g_2, \sigma(g_1))$. We consider the following diagram

(2.2)

$$\begin{array}{ccccccc}
\breve{\mathcal{I}}\dot{w}\breve{\mathcal{I}} \cap [b] & \xleftarrow{m} & m^{-1}(\breve{\mathcal{I}}\dot{w}\breve{\mathcal{I}} \cap [b]) & \xrightarrow{p_{12}} & Y & \xrightarrow{m} & Z \\
\downarrow & & \downarrow & & \downarrow & & \downarrow \\
\breve{\mathcal{I}}\dot{w}\breve{\mathcal{I}} & \xleftarrow{m} & \breve{\mathcal{I}}\dot{s}\breve{\mathcal{I}} \times \breve{\mathcal{I}}\dot{s}\dot{w}\breve{\mathcal{I}} & \xrightarrow{p_{12}} & \breve{\mathcal{I}}\dot{s}\dot{w}\breve{\mathcal{I}} \times \breve{\mathcal{I}}\sigma(\dot{s})\breve{\mathcal{I}} & \xrightarrow{m} & \breve{\mathcal{I}}\dot{s}\dot{w}\breve{\mathcal{I}}\sigma(\dot{s})\breve{\mathcal{I}}.
\end{array}$$

Here $Y = p_{12}m^{-1}(\breve{\mathcal{I}}\dot{w}\breve{\mathcal{I}} \cap [b])$ and $Z = m(Y)$.

Let $(g_1, g_2) \in m^{-1}(\breve{\mathcal{I}}\dot{w}\breve{\mathcal{I}} \cap [b])$. Then $g_1 g_2 \in [b]$. Hence $g_2\sigma(g_1) = m \circ p_{12}(g_1, g_2) \in [b]$. Thus $Z \subset \breve{\mathcal{I}}\dot{s}\dot{w}\breve{\mathcal{I}}\sigma(\dot{s})\breve{\mathcal{I}} \cap [b]$. On the other hand, for any $g \in \breve{\mathcal{I}}\dot{s}\dot{w}\breve{\mathcal{I}}\sigma(\dot{s})\breve{\mathcal{I}} \cap [b]$, there exists $g_2 \in \breve{\mathcal{I}}\dot{s}\dot{w}\breve{\mathcal{I}}$ and $g_1 \in \breve{\mathcal{I}}\dot{s}\breve{\mathcal{I}}$ such that $g_2\sigma(g_1) = g \in [b]$. So $g_1 g_2 \in [b]$ and $(g_1, g_2) \in m^{-1}(\breve{\mathcal{I}}\dot{w}\breve{\mathcal{I}} \cap [b])$. Therefore $Z = \breve{\mathcal{I}}\dot{s}\dot{w}\breve{\mathcal{I}}\sigma(\dot{s})\breve{\mathcal{I}} \cap [b]$.

THEOREM 2.16. *Let $w \in \tilde{W}$ and $s \in \tilde{\mathbb{S}}$ with $sw < w$. The diagram 2.2 induces a natural bijection*

$$Irr^{\max}(\breve{\mathcal{I}}\dot{w}\breve{\mathcal{I}} \cap [b]) \leftrightarrow Irr^{\max}(\breve{\mathcal{I}}\dot{s}\dot{w}\breve{\mathcal{I}}\sigma(\dot{s})\breve{\mathcal{I}} \cap [b]).$$

*We also have that*

$$\dim(\breve{\mathcal{I}}\dot{w}\breve{\mathcal{I}} \cap [b]) = \begin{cases} \dim(\breve{\mathcal{I}}\dot{s}\dot{w}\breve{\mathcal{I}}\sigma(\dot{s})\breve{\mathcal{I}} \cap [b]), & \text{if } sw\sigma(s) > sw; \\ \dim(\breve{\mathcal{I}}\dot{s}\dot{w}\breve{\mathcal{I}}\sigma(\dot{s})\breve{\mathcal{I}} \cap [b]) + 1, & \text{if } sw\sigma(s) < sw. \end{cases}$$

PROOF. If $sw\sigma(s) > sw$, then $\breve{\mathcal{I}}\dot{s}\dot{w}\breve{\mathcal{I}}\sigma(\dot{s})\breve{\mathcal{I}} = \breve{\mathcal{I}}\dot{s}\dot{w}\sigma(\dot{s})\breve{\mathcal{I}}$. The above diagram gives a natural bijection $C \leftrightarrow C'$ between the irreducible components of $\breve{\mathcal{I}}\dot{w}\breve{\mathcal{I}} \cap [b]$ and of $\breve{\mathcal{I}}\dot{s}\dot{w}\breve{\mathcal{I}}\sigma(\dot{s})\breve{\mathcal{I}} \cap [b]$. Moreover, we have $\dim_{\breve{\mathcal{I}}} C = \dim_{\breve{\mathcal{I}}} C'$.

If $sw\sigma(s) < sw$, then $\breve{\mathcal{I}}\dot{s}\dot{w}\breve{\mathcal{I}}\sigma(\dot{s})\breve{\mathcal{I}} = \breve{\mathcal{I}}\dot{s}\dot{w}\breve{\mathcal{I}} \sqcup \breve{\mathcal{I}}\dot{s}\dot{w}\sigma(\dot{s})\breve{\mathcal{I}}$ and $Z = Z_1 \sqcup Z_2$, where $Z_1 = \breve{\mathcal{I}}\dot{s}\dot{w}\breve{\mathcal{I}} \cap [b]$ and $Z_2 = \breve{\mathcal{I}}\dot{s}\dot{w}\sigma(\dot{s})\breve{\mathcal{I}} \cap [b]$. Here $Z_1$ is open in $Z$ and $Z_2$ is its closed complement. It is worth mentioning that in general $Z_1$ may not be dense in $Z$ (see Example 2.18). Also we do not have a nice correspondence between the irreducible components of $\breve{\mathcal{I}}\dot{w}\breve{\mathcal{I}} \cap [b]$ and of $Z$.

However, we still have a natural bijection $C \leftrightarrow D$ between the irreducible components of $\breve{\mathcal{I}}\dot{w}\breve{\mathcal{I}} \cap [b]$ and of $Y$. Under this bijection $\dim_{\breve{\mathcal{I}}} C = \dim_{\breve{\mathcal{I}}} D$.

On the other hand, $Y = Y_1 \sqcup Y_2$, where $Y_i = m_Y^{-1}(Z_i)$. For $i = 1, 2$, there is a natural bijection $D_i \leftrightarrow C'_i$ between the irreducible components of $Y_i$ and $Z_i$. Moreover, we have $\dim_{\breve{\mathcal{I}}} D_i = \dim_{\breve{\mathcal{I}}} C'_i + 1$. Note that if $C'$ is an irreducible component of $Z_2$, then $C'$ may not be an irreducible component of $Z$. However, if $\dim C' = \dim Z$, then it is an irreducible component of $Z$.

In particular, the above diagram gives a natural bijection $C \leftrightarrow C'$ between the irreducible components of maximal dimension of $\breve{\mathcal{I}}\dot{w}\breve{\mathcal{I}} \cap [b]$ and of $\breve{\mathcal{I}}\dot{s}\dot{w}\breve{\mathcal{I}}\sigma(\dot{s})\breve{\mathcal{I}} \cap [b]$. Moreover, we have $\dim_{\breve{\mathcal{I}}} C = \dim_{\breve{\mathcal{I}}} C' + 1$. □

Using Theorem 2.16, one may reduce the study of $\dim_{\breve{\mathcal{I}}}$ and $Irr^{\max}$ of $\breve{\mathcal{I}}\dot{w}\breve{\mathcal{I}} \cap [b]$ to the case where $w$ is of minimal length in its $\sigma$-twisted conjugacy class. The latter case is studied in [**He14**, Theorem 3.7].

THEOREM 2.17. *Let $w \in \tilde{W}$ be an element of minimal length in its $\sigma$-twisted conjugacy class. Then*

$$\breve{\mathcal{I}}\dot{w}\breve{\mathcal{I}} \subset [\dot{w}].$$

In other words, if $w \in \tilde{W}$ be an element of minimal length in its $\sigma$-twisted conjugacy class, then

$$\dim(\breve{\mathcal{I}}\dot{w}\breve{\mathcal{I}} \cap [b]) = \begin{cases} \ell(w), & \text{if } \dot{w} \in [b] \\ -\infty, & \text{if } \dot{w} \notin [b] \end{cases} \quad \text{and}$$

$$|Irr^{\max}(\breve{\mathcal{I}}\dot{w}\breve{\mathcal{I}} \cap [b])| = \begin{cases} 1, & \text{if } \dot{w} \in [b] \\ 0, & \text{if } \dot{w} \notin [b] \end{cases}.$$

Theorem 2.16, together with Theorem 2.17, provide a practical way to compute $\dim_{\breve{\mathcal{I}}}$ and $Irr^{\max}$ of $\breve{\mathcal{I}}\dot{w}\breve{\mathcal{I}} \cap [b]$ for arbitrary $w \in \tilde{W}$ and $[b] \in B(G)$. Let us look at an example.

EXAMPLE 2.18. This example comes from [**GH10**, §5]. Let $G = SL_4$. The simple reflections in $\tilde{W}$ are $s_0, s_1, s_2, s_3$. We simply write $s_{abc\cdots}$ instead of $s_a s_b s_c \cdots$. We take $w = s_{1201232101}$ and $b = 1$. In Figure 2, we illustrate how the reduction method is used to study $Y := \breve{\mathcal{I}}\dot{w}\breve{\mathcal{I}} \cap [b]$. Here we skip the step $w_1 \approx w_2$ and only record the step $w_1 \to s w_1 s$ with $s \in \mathbb{S}_0$ such that $s w_1 s < w_1$. The resulting figure is a binary tree. Here the left branch is the element $s w_1$ (which corresponds to an open part of the intersection) and the right branch is the element $s w_1 s$ (which corresponds to the closed complement of the intersection). Each box contains an element $w_1 \in \tilde{W}$ and a number, which equals to $\dim_{\breve{\mathcal{I}}}(\breve{\mathcal{I}}\dot{w}_1\breve{\mathcal{I}} \cap [b])$.

From the figure, we see that $Y = Y_1 \sqcup Y_2$, where $Y_1$ is open in $Y$ and $Y_2$ is its closed complement. Moreover, $\dim Y_1 = 7$ and $\dim Y_2 = 8$ and they are both irreducible. Thus both $Y_2$ and $\overline{Y_1}$ are the irreducible components of $Y$. In other words, $Y$ has two irreducible components, one is of dimension 8 and

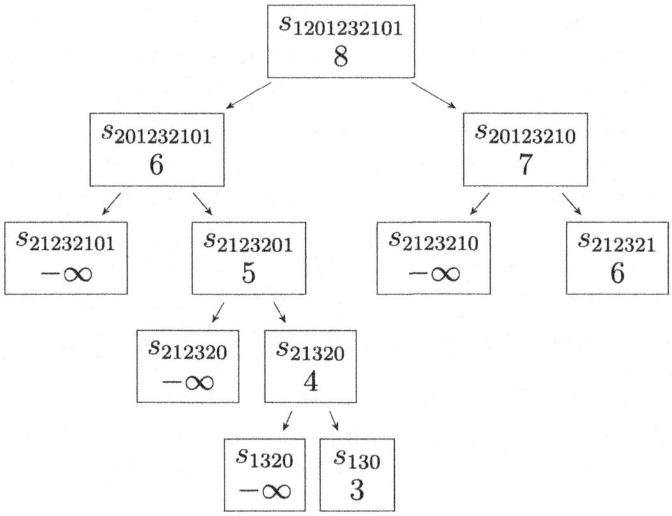


FIGURE 2. An example of Deligne-Lusztig reduction.

the other one is of dimension 7. This is an example where the intersection is not equidimensional.

2.8.2. *"Dimension=degree" theorem.* In general, such computation is quite difficult to run if the rank of the group is not very small. A better way to do this is to use the class polynomials of affine Hecke algebras, which encode the information of the reduction step.

We will give a systematic discussion of class polynomials in §3.6.3. In this section, we just give an algorithmic definition of the class polynomials.

To any $w \in \tilde{W}$ and a $\sigma$-twisted conjugacy class $\mathcal{O}$ of $\tilde{W}$, we associate the class polynomial $F_{w,\mathcal{O}} \in \mathbb{Z}[q]$. It is defined inductively on $w$.[2]

If $w$ is of minimal length in its $\sigma$-twisted conjugacy class, then

$$F_{w,\mathcal{O}} = \begin{cases} 1, & \text{if } w \in \mathcal{O}; \\ 0, & \text{otherwise.} \end{cases}$$

If $w$ is not of minimal length in its $\sigma$-twisted conjugacy class, then by Theorem 1.16, there exists $w' \in \tilde{W}$ and $s \in \tilde{\mathbb{S}}$ such that $w \approx_\sigma w'$ and $sw'\sigma(s) < w'$. We set

$$F_{w,\mathcal{O}} = (q-1) F_{sw',\mathcal{O}} + q F_{sw'\sigma(s),\mathcal{O}}.$$

In particular, if we regard $F_{w,\mathcal{O}}$ as a polynomial of $q-1$, then all the coefficients are nonnegative integers.

By convention, we define the degree of the polynomial 0 to be $-\infty$ and the leading coefficient of the polynomial 0 to be 1.

[2]The definition here differs from the definition of class polynomials $f_{w,\mathcal{O}}$ defined in [**He14**, §2] as the normalization of Hecke algebras we use here is different from [**He14**]. The relation is $F_{w,\mathcal{O}} = v^{\ell(w)-\ell(\mathcal{O})} f_{w,\mathcal{O}} \mid_{q=v^2}$.

THEOREM 2.19. *Let $w \in \tilde{W}$ and $[b] \in B(G)$. Let $F_{w,[b]} = \sum_{\mathcal{O} \in \tilde{W}/\tilde{W}_\sigma, f(\mathcal{O})=f([b])} q^{\ell(\mathcal{O})} F_{w,\mathcal{O}}$. Then*

*(1) $\dim_{\breve{\mathcal{I}}}(\breve{\mathcal{I}}\dot{w}\breve{\mathcal{I}} \cap [b]) = \deg F_{w,[b]}$.*

*(2) The cardinality of $Irr^{\max}(\breve{\mathcal{I}}\dot{w}\breve{\mathcal{I}} \cap [b])$ equals the leading coefficient of $F_{w,[b]}$.*

The proof is similar to the proof of [**He14**, Theorem 6.1] and is proved by induction on $w$.

As a consequence, we have

COROLLARY 2.20. *Let $w \in \tilde{W}$ and $[b] \in B(G)$. If $\ell(w) < \langle 2\rho, \nu_b \rangle$, then $\breve{\mathcal{I}}\dot{w}\breve{\mathcal{I}} \cap [b] = \emptyset$.*

The reason is that if the reduction method always decreases the length of the elements involved. In particular, if $\ell(w) < \langle 2\rho, \nu_b \rangle$, then after apply the reduction, one can only get $\sigma$-twisted conjugacy classes of $\tilde{W}$ different from $\mathcal{O}_{[b]}$.

We have similar "dimension=degree" theorem for parahoric subgroups.

THEOREM 2.21. *Let $\breve{\mathcal{K}}$ be a $\sigma$-stable parahoric subgroup of $G(\breve{F})$ and $w \in {}^K\tilde{W}^K$. Set $F_{w,K,[b]} = \sum_{w' \in W_K w W_K} F_{x,[b]}$. Then*

*(1) $\dim_{\breve{\mathcal{K}}}(\breve{\mathcal{K}}\dot{w}\breve{\mathcal{K}} \cap [b]) = \deg F_{w,K,[b]} - \dim_{\breve{\mathcal{I}}} \breve{\mathcal{K}}$.*

*(2) The cardinality of $Irr^{\max}(\breve{\mathcal{K}}\dot{w}\breve{\mathcal{K}} \cap [b])$ equals the leading coefficient of $F_{w,K,[b]}$.*

THEOREM 2.22. *Let $\breve{\mathcal{K}}$ be a $\sigma$-stable parahoric subgroup of $G(\breve{F})$ and $x \in {}^K\tilde{W}$. Then*

*(1) $\dim_{\breve{\mathcal{K}}}(\breve{\mathcal{K}} \cdot_\sigma \breve{\mathcal{I}}\dot{x}\breve{\mathcal{I}} \cap [b]) = \deg F_{x,[b]}$.*

*(2) The cardinality of $Irr^{\max}(\breve{\mathcal{K}} \cdot_\sigma \breve{\mathcal{I}}\dot{x}\breve{\mathcal{I}} \cap [b])$ equals the leading coefficient of $F_{x,[b]}$.*

Note that $F_{w,K,[b]}$ only depends on the Iwahori-Weyl group $\tilde{W}$ together with $\sigma : \tilde{W} \to \tilde{W}$, the element $w \in \tilde{W}$, the subset $K$ of $\tilde{\mathbb{S}}$ and straight $\sigma$-twisted conjugacy classes $\mathcal{O}_{[b]}$ of $\tilde{W}$. Thus one may relate the questions on the emptiness/non-emptiness pattern, dimension, the number of irreducible components of maximal dimension, etc., of $\breve{\mathcal{I}}\dot{w}\breve{\mathcal{I}} \cap [b]$ in a ramified group to the questions in an unramified group. This allows us to translate some results in the unramified groups (proved by other methods) to results in the ramified group (where the methods in the unramified groups might not apply).

2.8.3. *Relation with affine Deligne-Lusztig varieties.* Now we discuss some relations between the intersection $\breve{\mathcal{K}}\dot{w}\breve{\mathcal{K}} \cap [b]$ and the affine Deligne-Lusztig varieties.

Let $\breve{\mathcal{K}}$ be a $\sigma$-stable parahoric subgroup of $G(\breve{F})$ and $w \in {}^K\tilde{W}^K$. Let $b \in G(\breve{F})$. The affine Deligne-Lusztig variety is introduced by Rapoport in [**Ra05**]

$$X_{K,w}(b) = \{g\breve{\mathcal{K}} \in G(\breve{F})/\breve{\mathcal{K}}; g^{-1}b\sigma(g) \in \breve{\mathcal{K}}\dot{w}\breve{\mathcal{K}}\}.$$

If $\breve{\mathcal{K}}$ is the Iwahori subgroup $\breve{\mathcal{I}}$, we simply write $X_w(b)$ instead of $X_{\breve{\mathcal{I}},w}(b)$. If $\breve{\mathcal{K}}$ is a special maximal parahoric subgroup, then we may write

$$X_{K,\mu}(b) = \{g\breve{\mathcal{K}} \in G(\breve{F})/\breve{\mathcal{K}}; g^{-1}b\sigma(g) \in \breve{\mathcal{K}}\epsilon^\mu\breve{\mathcal{K}}\}$$

for the corresponding affine Deligne-Lusztig variety. We have

THEOREM 2.23. *Let $\breve{\mathcal{K}}$ be a $\sigma$-stable parahoric subgroup of $G(\breve{F})$ and $w \in {}^K\tilde{W}^K$. Let $b \in G(\breve{F})$. Then*

$$\dim_{\breve{\mathcal{K}}}(\breve{\mathcal{K}}\dot{w}\breve{\mathcal{K}} \cap [b]) - \dim X_{\breve{\mathcal{K}},w}(b) = \langle 2\rho, \nu_b\rangle.$$

PROOF. We first consider the case where $\breve{\mathcal{K}} = \breve{\mathcal{I}}$. In this case, the Deligne-Lusztig reduction method still works for $\dim X_w(b)$ (see [**He14**, Proof of Theorem 6.1]). The reduction steps are the same. The only difference lies in the basic step. When $w$ is a minimal length element in its $\sigma$-twisted conjugacy class, then

$$\dim X_w(b) = \begin{cases} \ell(w) - \ell(\mathcal{O}_b), & \text{if } \dot{w} \in [b]; \\ -\infty, & \text{if } \dot{w} \notin [b]. \end{cases}$$

Thus $\dim X_w(b)$ equals the degree of $\sum_{\mathcal{O}' \in \tilde{W}/\tilde{W}_\sigma, f(\mathcal{O}')=f(\mathcal{O}_b)} q^{\ell(\mathcal{O}')-\ell(\mathcal{O}_b)} F_{w,\mathcal{O}'}$ and

$$\dim_{\breve{\mathcal{I}}}(\breve{\mathcal{I}}\dot{w}\breve{\mathcal{I}} \cap [b]) - \dim X_w(b) = \ell(\mathcal{O}_b) = \langle 2\rho, \nu_b\rangle.$$

In the general case, $X_{\breve{\mathcal{K}},w}(b)$ equals the image of $\sqcup_{w' \in W_K w W_K} X_{w'}(b)$ under the natural projection map $\pi : G(\breve{F})/\breve{I} \to G(\breve{F})/\breve{\mathcal{K}}$. Note that each fiber of

$$\sqcup_{w' \in W_K w W_K} X_{w'}(b) \to X_{\breve{\mathcal{K}},w}(b)$$

is isomorphic to $\breve{\mathcal{K}}/\breve{\mathcal{I}}$. Thus

$$\begin{aligned} \dim X_{\breve{\mathcal{K}},w}(b) &= \max_{w' \in W_K w W_K} \dim X_{w'}(b) - \dim \breve{\mathcal{K}}/\breve{\mathcal{I}} \\ &= \max_{w' \in W_K w W_K} \dim_{\breve{\mathcal{I}}}(\breve{\mathcal{I}}\dot{w}'\breve{\mathcal{I}} \cap [b]) - \dim \breve{\mathcal{K}}/\breve{\mathcal{I}} - \langle 2\rho, \nu_b\rangle \\ &= \deg F_{K,w,[b]} - \dim \breve{\mathcal{K}}/\breve{\mathcal{I}} - \langle 2\rho, \nu_b\rangle \\ &= \dim_{\breve{\mathcal{K}}}(\breve{\mathcal{K}}\dot{w}\breve{\mathcal{K}} \cap [b]) - \langle 2\rho, \nu_b\rangle. \end{aligned}$$

□

**2.9. Non-emptiness pattern.** In this subsection, we discuss several cases that explicit non-emptiness patterns are known.

2.9.1. *Special maximal parahoric.* We first consider the intersection $\breve{\mathcal{K}}\epsilon^\mu\breve{\mathcal{K}} \cap [b]$, where $\breve{\mathcal{K}}$ is a special maximal parahoric subgroup. If $G = GL_n$, then the intersection is nonempty if and only if the Mazur's inequality is satisfied (see §2.7). Kottwitz and Rapoport [**KR03**] conjecture that it is still the case for any reductive group $G$ (and special maximal parahoric subgroups $\breve{\mathcal{K}}$).

THEOREM 2.24. *Let $\breve{\mathcal{K}}$ be a special maximal parahoric subgroup of $G(\breve{F})$. Let $\mu$ be a dominant coweight and $[b] \in B(G)$. Then $[b] \cap \breve{\mathcal{K}}\epsilon^\mu\breve{\mathcal{K}} \neq \emptyset$ if and only if $\mu \geqslant \nu_b$ and $\kappa([b]) = \kappa(\epsilon^\mu)$.*

The "only if" side is proved by Rapoport and Richartz in [**RR96**]. The "if" side is proved for type A and C in [**KR03**]. It is then proved by Lucarelli [**Luc04**] for classical split groups and then by Gashi [**Ga10**] for unramified cases. The general case is obtained in [**He14**, Theorem 7.1], combining the relation between the class polynomials and the non-emptiness pattern of such intersection, the relation between Iwahori-Weyl groups for ramified and unramified groups, and Gashi's result in the unramified case.

2.9.2. *Reduction to quasi-split, adjoint groups.* Now we consider the intersection $\breve{\mathcal{I}}\dot{w}\breve{\mathcal{I}} \cap [b]$. We first discuss the reduction from arbitrary group to a quasi-split, adjoint group.

As explained in [**GHN15**, §2.2], it suffices to consider the case where $G$ is adjoint. Let $H$ be the quasi-split inner form of $G$. Then $G(\breve{F}) = H(\breve{F})$ and $\breve{I}$ is the Iwahori subgroup of $H(\breve{F})$ as well. We identify the Iwahori-Weyl group of $H$ with $\tilde{W}$. The difference is the group automorphisms on $\tilde{W}$ induced from the Frobenius morphisms. We have $\sigma_G = \mathrm{Ad}(\gamma) \circ \sigma_H \in \mathrm{Aut}(\tilde{W})$ for some $\gamma \in \Omega$. The map $g \mapsto g\gamma$ gives a bijection from the $\sigma_G$-conjugacy class $[g]_G$ to the $\sigma_H$-conjugacy class $[g\gamma]_H$. We have

$$\dim_{\breve{\mathcal{I}}}(\breve{\mathcal{I}}\dot{w}\breve{\mathcal{I}} \cap [g]_G) = \dim_{\breve{\mathcal{I}}}(\breve{\mathcal{I}}\dot{w}\dot{\gamma}\breve{\mathcal{I}} \cap [g\gamma]_H).$$

In the sequel, we denote by $\varsigma$ the group automorphism $\sigma_H$ on $\tilde{W}$.

2.9.3. *$P$-alcove elements.* For quasi-split groups, an important notion is the $P$-alcove element, first introduced by Görtz, Haines, Kottwitz, and Reuman [**GHKR10**] for split group and then generalized in [**GHN15**] to quasi-split groups. We recall the definition.

Assume that $G$ is quasi-split. Let $P = MN$ be a standard $\sigma$-stable parabolic subgroup of $G$ and $x \in W_0$. We say that $w \in \tilde{W}$ is a $(P, x)$-alcove element if

- $x^{-1} w \sigma(x) \in \tilde{W}_M$;
- $\dot{x} N(\breve{F}) \dot{x}^{-1} \cap \dot{w}\breve{\mathcal{I}}\dot{w}^{-1} \subset \breve{\mathcal{I}}$.

The reason to put the above two requirement on $w$ is that we would like to reduce the study of $\breve{\mathcal{I}}\dot{w}\breve{\mathcal{I}}$ to the study of a Iwahori double coset in a Levi subgroup of $G$. In particular, we need that $w$ is $\sigma$-conjugate to an element in the Iwahori-Weyl group of a standard Levi subgroup. This is where the first condition comes from. The second condition is more involved. Roughly speaking, we have the decomposition

$$\breve{\mathcal{I}} = (\breve{\mathcal{I}} \cap N(\breve{F}))(\breve{\mathcal{I}} \cap M(\breve{F}))(\breve{\mathcal{I}} \cap N^-(\breve{F})),$$

where $N^-$ is the unipotent radical of the opposite parabolic subgroup. And we need this decomposition to be "compatible" with $\breve{\mathcal{I}}\dot{w}\breve{\mathcal{I}}$ so that we are able to eventually get rid of $\breve{\mathcal{I}} \cap N(\breve{F})$ after $\sigma$-conjugation by a suitable element.

The upshot is the following result, first obtained for split groups by Görtz, Haines, Kottwitz, and Reuman [**GHKR10**] and then generalized to quasi-split groups in [**GHN15**].

THEOREM 2.25. *Let $G$ be a quasi-split group. Let $w \in \tilde{W}$ be a $(P,x)$-alcove element. Then any element in $\breve{\mathcal{I}}\dot{w}\breve{\mathcal{I}}$ is $\sigma$-conjugate by $\breve{\mathcal{I}}$ to an element in*

$$(\breve{\mathcal{I}} \cap \dot{x}M(\breve{F})\dot{x}^{-1})w\sigma(\breve{\mathcal{I}} \cap \dot{x}M(\breve{F})\dot{x}^{-1}) \subset \dot{x}M(\breve{F})\sigma(\dot{x}^{-1}).$$

2.9.4. *Basic $\sigma$-conjugacy classes.* We say a $\sigma$-conjugacy class $[b]$ of $G(\breve{F})$ is *basic* if the associated Newton point is central, i.e., $\langle 2\rho, \nu_b \rangle = 0$. Use the notion of $P$-alcove elements, we have a complete description of the non-emptiness pattern of $\breve{\mathcal{I}}\dot{w}\breve{\mathcal{I}} \cap [b]$ for basic $\sigma$-conjugacy class $[b]$.

In [**GHN15**], we introduced the "no Levi obstruction" (NLO) condition. We say that there is *no Levi obstruction* if for any pair $(P,x)$ such that $w$ is a $(P,x)$-alcove element, there exists an element $b_J \in [b] \cap M(\breve{F})$ such that $\kappa_M(x^{-1}w\sigma(x)) = \kappa_M(b_J)$. By Theorem 2.25, "no Levi obstruction" is a necessary condition for $\breve{\mathcal{I}}\dot{w}\breve{\mathcal{I}} \cap [b] \neq \emptyset$. In [**GHN15**], it is proved that the notion of $P$-alcove elements is compatible with the Deligne-Lusztig reduction, and as a consequence, "no Levi obstruction" is also a sufficient condition. The following result is [**GHN15**, Theorem A].

THEOREM 2.26. *Let $G$ be a quasi-split group. Let $w \in \tilde{W}$ and $[b] \in B(G)$ be a basic $\sigma$-conjugacy class. Then $\breve{\mathcal{I}}\dot{w}\breve{\mathcal{I}} \cap [b] \neq \emptyset$ if and only if there is no Levi obstruction.*

Let us make the "no Levi obstruction" condition more explicit.

A *critical strip* of the apartment $V$ by definition is the subset $\{v; -1 < \langle \alpha, v \rangle < 0\}$ for a given positive finite root in the reduced root datum $\mathscr{R}$. We remove all the critical strips from $V$ and call each connected component of the remaining subset of $V$ *a Shrunken Weyl chamber.*

Let $w \in \tilde{W}$. Let $x$ by the unique element in $W_0$ such that $w(\mathfrak{a}) \subset x\mathcal{C}^-$. Let $\eta_\sigma(w)$ be the unique element in $W_0$ such that $x^{-1}w\sigma(x)\sigma(p) \in \eta_\sigma(w)\mathcal{C}$ for any sufficiently regular element $p \in \mathcal{C}^-$. In the quasi-split case, $\sigma$ preserves the antidominant chamber $\mathcal{C}^-$. The element $w$ can be written as $xt^\lambda y$ with $x \in W_0$ and $t^\lambda y(\mathfrak{a}) \subset \mathcal{C}^-$, then $x^{-1}w\sigma(x)\sigma(\mathcal{C}^-) = t^\lambda y\sigma(x)\mathcal{C}^-$. Thus $\eta_\sigma(w) = y\sigma(x)$. In general, we have $\sigma = \mathrm{Ad}(\gamma) \circ \varsigma$ for some $\gamma \in \Omega$ and $x^{-1}w\sigma(x)\sigma(p) = x^{-1}w\gamma\varsigma(x)\gamma^{-1}\gamma(\varsigma(p)) = x^{-1}w\gamma\varsigma(x)\varsigma(p)$. Therefore

$$\eta_\sigma(w) = \eta_\varsigma(w\gamma). \tag{2.3}$$

We have the following explicit description of non-emptiness pattern, first conjectured by Görtz, Haines, Kottwitz and Reuman in [**GHKR10**, Conjecture 1.1.1] for split groups. The quasi-split case is proved in [**GHN15**, Theorem B] and the general case follows from §2.9.2, the equality (2.3) and the result for the quasi-split case.

THEOREM 2.27. *Assume that $G$ is simple. Let $w \in \tilde{W}$ such that $w\sigma(\mathfrak{a})$ is contained in a Shrunken Weyl chamber. Then for any basic $\sigma$-conjugacy* *class* $[b]$, $\breve{\mathcal{I}}\dot{w}\breve{\mathcal{I}} \cap [b] \neq \emptyset$ *if and only if* $\kappa(w) = \kappa(b)$ *and that* $\eta_\sigma(w)$ *is not contained in the relative Weyl group* $W_M$ *for any proper standard* $\varsigma$*-stable Levi subgroup* $M$ *of quasi-split inner form of* $G$.

2.9.5. *Translation elements.* We have similar results for some nonbasic $\sigma$-conjugacy classes as well.

Let $G$ be a residually split group and $\mu$ be a dominant coweight. For any $x \in W_0$, we denote by $\mathcal{C}_x^\sharp$ the Shrunken Weyl chamber inside $x(\mathcal{C}^-)$. We set

$$\mathcal{C}_{\mu,x}^{**} = t^{x(\mu)}(\mathcal{C}_x^\sharp) = \mathcal{C}_x^\sharp - x(\mu) \subset x(\mathcal{C}^-)$$

and call it a *very Shrunken Weyl chamber* with respect to $\mu$. Then

THEOREM 2.28. *Assume that* $G$ *is simple. Let* $G$ *be a residually split group and* $\mu$ *be a dominant coweight. Let* $w \in \tilde{W}$ *such that* $w(\mathfrak{a})$ *is contained in a very Shrunken Weyl chamber with respect to* $\mu$. *If* $\breve{\mathcal{I}}\dot{w}\breve{\mathcal{I}} \cap [\epsilon^\mu] \neq \emptyset$, *then* $k(w) = k(\epsilon^\mu)$ *and* $\eta_\sigma(w)$ *is not contained in the relative Weyl group* $W_M$ *for any proper standard Levi subgroup* $M$ *of* $G$.

Here the proof is similar to the proof of the "only if" part of Theorem 2.27. We skip the details. We will also see in the next subsection that the converse is true and we have an explicit dimension formula.

**2.10. Dimension formula.** In this subsection, we discuss several cases that an explicit dimension formula is known.

2.10.1. *Defect.* Let us first recall the notion of *defect* introduced by Kottwitz in [**Ko06**]. Let $b \in G(\breve{F})$. By definition, the defect $\mathrm{def}(b)$ of $b$ is the difference between the $F$-rank of $G$ and the $F$-rank of $J_b$, where $J_b = \{g \in G(\breve{F}); g^{-1} b \sigma(g) = b\}$ is the $\sigma$-centralizer of $b$. It can be reformulated in another way.

Recall that $V$ is the apartment associated to the maximal $\breve{F}$-split torus $S$ defined over $F$. Then $\dim V$ equals the $F$-rank of $G$. Let $w \in \mathcal{O}_b$ and $V_w = \{v \in V; w\sigma(v) = v + \nu_w\}$. By [**Ko06**, §1.9], $\dim V_w$ equals the $F$-rank of $J_b$. Thus

$$\mathrm{def}(b) = \dim V - \dim V_w.$$

Note that the right hand side only depend on the Iwahori-Weyl group $\tilde{W}$, the group automorphism $\sigma : \tilde{W} \to \tilde{W}$ and a straight $\sigma$-twisted conjugacy class $\mathcal{O}_b$. This formula will be used in our reduction from ramified groups to unramified groups.

2.10.2. *Special maximal parahoric.* We start with the special maximal parahoric subgroup case.

THEOREM 2.29. *Let* $\breve{\mathcal{K}}$ *be a special maximal parahoric subgroup of* $G(\breve{F})$. *Let* $\mu$ *be a dominant coweight and* $[b] \in B(G)$. *If* $\nu_b \leqslant \mu$, *then*

$$\dim_{\breve{\mathcal{K}}}(\breve{\mathcal{K}}\epsilon^\mu\breve{\mathcal{K}} \cap [b]) = \langle \rho, \mu + \nu_b \rangle - \frac{1}{2} def_G(b),$$

$$\dim X_{K,\mu}(b) = \langle \rho, \mu - \nu_b \rangle - \frac{1}{2} def_G(b).$$

The dimension formula of $X_{K,\mu}(b)$ is conjectured by Rapoport in [**Ra05**, Conjecture 5.10]. For split groups, the conjectural formula is obtained by Görtz, Haines, Kottwitz and Reuman [**GHKR06**] and Viehmann [**Vie06**]. The conjectural formula for quasi-split unramified groups is obtained independently by Zhu [**Zhx**] and Hamacher [**Ha15**]. The general case can be obtained by combining Theorem 2.21 (which reduce the study of dimension formula to the study of degree of class polynomials), the relation between Iwahori-Weyl groups for ramified and unramified groups, and the results of Zhu and Hamacher in the unramified case.

2.10.3. *Virtual dimension.* Now we consider the case where $\breve{\mathcal{K}} = \breve{\mathcal{I}}$ is the Iwahori subgroup and we give an upper bound of the dimension.

THEOREM 2.30. *For any $w \in \tilde{W}$ and $[b] \in B(G)$, define the virtual dimension*

$$d_w([b]) = \frac{1}{2}(\ell(w) + \ell(\eta_\sigma(w)) - def_G(b)) + \langle \rho, \nu_b \rangle.$$

*Then*

$$\dim_{\breve{\mathcal{I}}}(\breve{\mathcal{I}} \dot{w} \breve{\mathcal{I}} \cap [b]) \leqslant d_w([b]).$$

PROOF. As explained in §2.9.2, it suffices to consider the case where $G$ is a inner form of a quasi-split, adjoint group $H$. We have $\sigma_G = \mathrm{Ad}(\gamma) \circ \sigma_H \in \mathrm{Aut}(\tilde{W})$ for some $\gamma \in \Omega$. We have $\dim_{\breve{\mathcal{I}}}(\breve{\mathcal{I}} \dot{w} \breve{\mathcal{I}} \cap [b]_G) = \dim_{\breve{\mathcal{I}}}(\breve{\mathcal{I}} \dot{w} \gamma \breve{\mathcal{I}} \cap [b\gamma]_H)$. We also have $d_w([b]_G) = d_{w\gamma}([b\gamma]_H)$ since $\eta_{\sigma_G}(w) = \eta_{\sigma_H}(w\gamma)$.

Suppose that $w\gamma \in W_0 t^\mu W_0$, where $\mu$ is a dominant coweight. By Theorem 2.23, [**He14**, Theorem 10.3], and Theorem 2.29, we have

$$\begin{aligned}
&\dim_{\breve{\mathcal{I}}}(\breve{\mathcal{I}} \dot{w} \gamma \breve{\mathcal{I}} \cap [b\gamma]_H) = \dim X^H_{w\gamma}(b\gamma) + \langle 2\rho, \nu^H_{b\gamma} \rangle \\
&\leqslant \dim X^H_\mu(b\gamma) + \frac{1}{2}(\ell(w\gamma) + \ell(\eta_{\sigma_H}(w))) - \langle \rho, \mu \rangle + \langle 2\rho, \nu^H_{b\gamma} \rangle \\
&= \langle \rho, \mu - \nu^H_{b\gamma} \rangle - \frac{1}{2}\mathrm{def}_H(b\gamma) + \frac{1}{2}(\ell(w\gamma) + \ell(\eta_{\sigma_H}(w))) - \langle \rho, \mu \rangle + \langle 2\rho, \nu^H_{b\gamma} \rangle \\
&= d_{w\gamma}([b\gamma]_H).
\end{aligned}$$

□

In several cases, we are able to get an equality $\dim X_w(b) = d_w([b])$.

2.10.4. *Basic $\sigma$-conjugacy classes.* Let $[b]$ be a basic $\sigma$-conjugacy class. Recall that in Theorem 2.27, we give a criterion on $\breve{\mathcal{I}} \dot{w} \breve{\mathcal{I}} \cap [b] \neq \emptyset$ for $w \in \tilde{W}$ such that $w\sigma(\mathfrak{a})$ is contained in the Shrunken Weyl chamber. Combining Theorem 2.23, [**He14**, Theorem 12.1] and Theorem 2.30, we have

THEOREM 2.31. *Assume that $G$ is simple. Let $w \in \tilde{W}$ such that $w\sigma(\mathfrak{a})$ is contained in the Shrunken Weyl chamber and that $(\eta_\sigma(w))$ is not contained in the relative Weyl group $W_M$ for any proper standard $\varsigma$-stable Levi subgroup $M$ of $G$. Then for any basic $\sigma$-conjugacy class of $G(\breve{F})$ with $\kappa([b]) = \kappa(w)$, we have*

$$\dim_{\breve{\mathcal{I}}}(\breve{\mathcal{I}} \dot{w} \breve{\mathcal{I}} \cap [b]) = d_w([b]).$$

The dimension formula for the affine Deligne-Lusztig varieties $X_w(b)$ with $w$ and $b$ as above is conjectured by Görtz, Haines, Kottwitz, and Reuman in [**GHKR10**, Conjecture 1.1.3] and is previously verified in [**He14**, §12] for residually split group.

2.10.5. *Translation elements.* Now we consider the translation elements. We assume furthermore that $G$ is residually split. We first relate the intersection $\breve{\mathcal{I}}\dot{w}\breve{\mathcal{I}} \cap [\epsilon^\mu]$ with $\breve{\mathcal{I}}\dot{w}'\breve{\mathcal{I}} \cap [1]$ for some $w'$.

Let $H$ be a closed subgroup of $G$. A bounded subset of $V$ of $H(\breve{F})$ is called admissible if it is stable under the right action of $H(\breve{F}) \cap \breve{\mathcal{I}}_n$ for some $n \in \mathbb{N}$. In this case, we say that $V$ is locally closed in $H(\breve{F})$ if $V/(H(\breve{F}) \cap \breve{\mathcal{I}}_n)$ is locally closed in $H(\breve{F})/(H(\breve{F}) \cap \breve{\mathcal{I}}_n)$. We define

$$\dim_{H(\breve{F}) \cap \breve{\mathcal{I}}} V = \dim V/(H(\breve{F}) \cap \breve{\mathcal{I}}_n) - \dim(H(\breve{F}) \cap \breve{\mathcal{I}})/(H(\breve{F}) \cap \breve{\mathcal{I}}_n).$$

We also need the following result of Görtz, Haines, Kottwitz and Reuman [**GHKR06**, Theorem 6.3.1].

THEOREM 2.32. *Let $G$ be a split group. For any $w \in \tilde{W}$ and $\mu \in X_*(T)$, there is an equality*

$$\dim X_w(\epsilon^\mu) = \max_{x \in W_0} \dim_{\dot{x}U(\breve{F})\dot{x}^{-1} \cap \breve{\mathcal{I}}}(\breve{\mathcal{I}}\dot{w}\breve{\mathcal{I}}\epsilon^{-x(\mu)} \cap \dot{x}U(\breve{F})\dot{x}^{-1}) + \langle \rho, \mu - \bar{\mu} \rangle,$$

*where $U$ is the subgroup of $G$ generated by the (finite) positive root subgroups.*

Now we are able to compare the dimension of affine Deligne-Lusztig varieties for various $b$ in the torus $S$ for a residually split group.

THEOREM 2.33. *Let $G$ be a residually split group. Let $w \in \tilde{W}$ and $\mu \in X_*(T)_\Gamma$ such that $\ell(w) = \ell(wt^{-\mu}) + \ell(t^\mu)$. Then*

$$\dim X_w(t^\mu) \geqslant \dim X_{wt^{-\mu}}(1).$$

PROOF. By Theorem 2.19 and the remark after Theorem 2.22, it suffices to prove the case where $G$ is split.

By Theorem 2.32, there exists $x \in W_0$ such that

$$\dim X_{wt^{-\mu}}(1) = \dim_{\dot{x}U(\breve{F})\dot{x}^{-1} \cap \breve{\mathcal{I}}}(\breve{\mathcal{I}}\dot{w}\breve{\mathcal{I}} \cap \dot{x}U(\breve{F})\dot{x}^{-1}).$$

Again by Theorem 2.32,

$$\begin{aligned}\dim X_w(\epsilon^\mu) &= \dim X_w(\epsilon^{x^{-1}(\mu)}) \\ &\geqslant \dim_{\dot{x}U(\breve{F})\dot{x}^{-1} \cap \breve{\mathcal{I}}}(\breve{\mathcal{I}}\dot{w}\breve{\mathcal{I}}\epsilon^{-\mu} \cap \dot{x}U(\breve{F})\dot{x}^{-1}) + \langle \rho, x^{-1}(\mu) - \bar{\mu} \rangle.\end{aligned}$$

Here the first equality follows from the fact that $\epsilon^\mu$ and $\epsilon^{x^{-1}(\mu)}$ are in the same $\sigma$-conjugacy class.[3]

[3]This is where the assumption that $G$ is residually split is used. At present, I do not know how to modify the argument so that it works for quasi-split groups as well.

Since $\ell(w) = \ell(wt^{-\mu}) + \ell(t^{\mu})$, we have the following commutative diagram

$$\begin{array}{ccc}
\breve{\mathcal{I}}\dot{w}\epsilon^{-\mu}\breve{\mathcal{I}} \times_{\breve{\mathcal{I}}} \breve{\mathcal{I}}\epsilon^{\mu}\breve{\mathcal{I}}\epsilon^{-\mu} & \xrightarrow[\cong]{m} & \breve{\mathcal{I}}\dot{w}\breve{\mathcal{I}}\epsilon^{-\mu} \\
\Big\downarrow \cong & & \Big\| \\
\breve{\mathcal{I}}\dot{w}\epsilon^{-\mu}\breve{\mathcal{I}} \times_{\breve{\mathcal{I}} \cap \epsilon^{\mu}\breve{\mathcal{I}}\epsilon^{-\mu}} \epsilon^{\mu}\breve{\mathcal{I}}\epsilon^{-\mu} & \xrightarrow[\cong]{m} & \breve{\mathcal{I}}\dot{w}\breve{\mathcal{I}}\epsilon^{-\mu}.
\end{array}$$

Here $m$ is the multiplication map.

The restriction to $\dot{x}U(\breve{F})\dot{x}^{-1}$ gives an injective map

$$\begin{aligned}
&(\breve{\mathcal{I}}\dot{w}\epsilon^{-\mu}\breve{\mathcal{I}} \cap \dot{x}U(\breve{F})\dot{x}^{-1}) \times_{\breve{\mathcal{I}} \cap \epsilon^{\mu}\breve{\mathcal{I}}\epsilon^{-\mu} \cap \dot{x}U(\breve{F})\dot{x}^{-1}} (\epsilon^{\mu}\breve{\mathcal{I}}\epsilon^{-\mu} \cap \dot{x}U(\breve{F})\dot{x}^{-1}) \\
&\quad \to \breve{\mathcal{I}}\dot{w}\breve{\mathcal{I}}\epsilon^{-\mu} \cap \dot{x}U(\breve{F})\dot{x}^{-1}.
\end{aligned}$$

By [**GHKR06**, Lemma 2.13.1],

$$\begin{aligned}
&\dim_{\dot{x}U(\breve{F})\dot{x}^{-1} \cap \breve{\mathcal{I}}}(\breve{\mathcal{I}}\dot{w}\breve{\mathcal{I}}\epsilon^{-\mu} \cap \dot{x}U(\breve{F})\dot{x}^{-1}) \\
&\geqslant \dim_{\dot{x}U(\breve{F})\dot{x}^{-1} \cap \breve{\mathcal{I}}}(\breve{\mathcal{I}}\dot{w}\epsilon^{-\mu}\breve{\mathcal{I}} \cap \dot{x}U(\breve{F})\dot{x}^{-1}) \\
&\quad + \dim((\epsilon^{\mu}\breve{\mathcal{I}}\epsilon^{-\mu} \cap \dot{x}U(\breve{F})\dot{x}^{-1})/(\breve{\mathcal{I}} \cap \epsilon^{\mu}\breve{\mathcal{I}}\epsilon^{-\mu} \cap \dot{x}U(\breve{F})\dot{x}^{-1})) \\
&= \dim_{\dot{x}U(\breve{F})\dot{x}^{-1} \cap \breve{\mathcal{I}}}(\breve{\mathcal{I}}\dot{w}\epsilon^{-\mu}\breve{\mathcal{I}} \cap \dot{x}U(\breve{F})\dot{x}^{-1}) + \sum_{\alpha \in R_+} \min\{\langle x\alpha, \mu\rangle, 0\} \\
&= \dim_{\dot{x}U(\breve{F})\dot{x}^{-1} \cap \breve{\mathcal{I}}}(\breve{\mathcal{I}}\dot{w}\epsilon^{-\mu}\breve{\mathcal{I}} \cap \dot{x}U(\breve{F})\dot{x}^{-1}) + \langle \rho, x^{-1}(\mu) - \bar{\mu}\rangle \\
&= \dim X_{wt^{-\mu}}(1).
\end{aligned}$$

□

We have the following result on the non-emptiness pattern and dimension formula for the intersection $\breve{\mathcal{I}}\dot{w}\breve{\mathcal{I}} \cap [\epsilon^{\mu}]$. For split groups, it is conjectured by Görtz, Haines, Kottwitz, and Reuman in [**GHKR10**, Conjecture 9.5.1 (b)].

THEOREM 2.34. *Let $G$ be a residually split, simple group. Let $\mu$ be a dominant coweight. Let $w \in \tilde{W}$ such that $\kappa(w) = \kappa(\epsilon^{\mu})$ and that $w(\mathfrak{a})$ is contained in a very Shrunken Weyl chamber with respect to $\mu$. If $\eta_\sigma(w)$ is not contained in the relative Weyl group $W_M$ for any proper standard Levi subgroup $M$ of $G$. Then*

$$\dim_{\breve{\mathcal{I}}}(\breve{\mathcal{I}}\dot{w}\breve{\mathcal{I}} \cap [\epsilon^{\mu}]) = d_w([\epsilon^{\mu}]).$$

*In particular, $\breve{\mathcal{I}}\dot{w}\breve{\mathcal{I}} \cap [\epsilon^{\mu}] \neq \emptyset$.*

PROOF. By Theorem 2.30, $\dim_{\breve{\mathcal{I}}}(\breve{\mathcal{I}}w\breve{\mathcal{I}} \cap [\epsilon^{\mu}] \leqslant d_w([\epsilon^{\mu}]))$.

Let $x \in W_0$ such that $w(\mathfrak{a}) \subset x(\mathcal{C}^-)$. Let $\mu'$ be the coweight with $w^{-1}t^{x(\mu)}w = t^{\mu'}$. By definition, $\ell(wt^{-\mu'}) = \ell(t^{-x(\mu)}w) = \ell(w) - \ell(t^{\mu})$ and $wt^{-\mu'}(\mathfrak{a}) \subset x(\mathcal{C}^-)$ is in the Shrunken Weyl chamber. Thus $\eta_\sigma(w) = \eta_\sigma(wt^{-w^{-1}x(\mu)})$. Note that $[\epsilon^{\mu}] = [\epsilon^{\mu'}]$. By Theorem 2.23, Theorem 2.33 and Theorem 2.31,

$$\begin{aligned}
&\dim_{\breve{\mathcal{I}}}(\breve{\mathcal{I}}\dot{w}\breve{\mathcal{I}} \cap [\epsilon^{\mu'}]) \\
&= \dim X_w(\epsilon^{\mu'}) + \langle 2\rho, \mu\rangle \geqslant \dim X_{wt^{-\mu'}}(1) + \langle 2\rho, \mu\rangle \\
&= d_{wt^{-\mu'}}([1]) + \langle 2\rho, \mu\rangle = \frac{1}{2}(\ell(wt^{-\mu'}) + \ell(\eta_\sigma(wt^{-\mu'}))) + \langle 2\rho, \mu\rangle \\
&= \frac{1}{2}(\ell(w) + \ell(\eta_\sigma(w)) + \langle \rho, \bar{\mu}\rangle = d_w([\epsilon^\mu]). \qquad \square
\end{aligned}$$

**2.11. The Kottwitz-Rapoport conjecture.**

2.11.1. *Admissible set* $\mathrm{Adm}(\mu)$. In fact, we are not only interested in the intersection of a $\sigma$-conjugacy class with a single $\breve{\mathcal{I}}$-double coset, but also interested in the intersection of a $\sigma$-conjugacy class with a certain union of $\breve{\mathcal{I}}$-double cosets. Such union of $\breve{\mathcal{I}}$-double cosets arises in the study of Shimura varieties. Let us first look at the following example [**Ha05**, §4.4].

EXAMPLE 2.35. Let $R$ be a $\mathbb{Z}_p$-algebra. We consider the pairs $(\mathcal{F}_0, \mathcal{F}_1)$ of locally free rank 1 submodules of $R^2$ such that the diagram commutes

$$\begin{array}{ccccc}
R \oplus R & \xrightarrow{\begin{bmatrix} p & 0 \\ 0 & 1 \end{bmatrix}} & R \oplus R & \xrightarrow{\begin{bmatrix} 1 & 0 \\ 0 & p \end{bmatrix}} & R \oplus R \\
\uparrow & & \uparrow & & \uparrow \\
\mathcal{F}_0 & \longrightarrow & \mathcal{F}_1 & \longrightarrow & \mathcal{F}_0
\end{array}$$

This functor represents a closed subscheme of $\mathbb{P}^1_{\mathbb{Z}_p} \times \mathbb{P}^1_{\mathbb{Z}_p}$, which we denote by $M^{loc}$. This is the "local model" of Shimura variety associated to the fake unitary group $U(1,1)$.

In the generic fiber, $M^{loc}_{\mathbb{Q}_p} \cong \mathbb{P}^1_{\mathbb{Q}_p}$.

In the special fiber, $M^{loc}_{\mathbb{F}_p}$ is the union of two $\mathbb{P}^1_{\mathbb{F}_p}$ meeting in a point and we may regard $M^{loc}_{\mathbb{F}_p}$ as the union of the three Schubert cells in the affine flag variety of $GL_2(\mathbb{F}_p((\epsilon)))$. These two open Schubert cells correspond to the elements $t^{[1,0]}$ and $t^{[0,1]}$ in the Iwahori-Weyl group of $GL_2$ and the closed Schubert cell correspond to the element $t^{[1,0]}(12)$, which is the unique length-zero element in $t^{[1,0]}W_a = t^{[0,1]}W_a$.

In general, for any coweight $\mu$, we define the admissible set

$$\mathrm{Adm}(\mu) = \{w \in \tilde{W}; w \leqslant t^{x(\mu)} \text{ for some } x \in W_0\}.$$

Similarly, for any parahoric subgroup $K$ of $G$, we set $\mathrm{Adm}(\mu)^K = W_K \mathrm{Adm}(\mu) W_K$ and denote by $\mathrm{Adm}(\mu)_K$ the image of $\mathrm{Adm}(\mu)^K$ in $W_K \backslash \tilde{W} / W_K$. The admissible set is an interesting combinatorial object. We refer to [**HHx**] and [**Hax**] for some nice properties on $\mathrm{Adm}(\mu)$.

The set $\mathrm{Adm}(\mu)_K$ is expected to parametrize a natural stratification of the special fiber of the local model of a Shimura variety with level $K$ structure. This has been established for Shimura varieties of PEL type at places of tame, parahoric reduction by Pappas and Zhu [**PZ13**].

2.11.2. *The union $M_K$ of $\breve{\mathcal{K}}$-double cosets.* We consider the following set

$$M_K = \sqcup_{w \in \mathrm{Adm}(\mu)_K} \breve{\mathcal{K}} \dot{w} \breve{\mathcal{K}}.$$

We discuss the relation of $M_K$ with the admissible subsets of $G(\breve{F})$ in §2.3 and §2.4.

As a $\breve{\mathcal{K}} \times \breve{\mathcal{K}}$-double cosets, we have

THEOREM 2.36. *Let $\breve{\mathcal{K}}$ be a parahoric subgroup of $G(\breve{F})$. Then*

$$\breve{\mathcal{K}} \backslash M_K / \breve{\mathcal{K}} = \mathrm{Adm}(\mu)_K.$$

This is just a reformulation of definition.

Next, as a subset of $G(\breve{F})$ stable under the $\sigma$-twisted conjugation action of $\breve{\mathcal{K}}$,

THEOREM 2.37. *Let $\breve{\mathcal{K}}$ be a parahoric subgroup of $G(\breve{F})$. Then*

$$M_K = \sqcup_{w \in \mathrm{Adm}(\mu) \cap {}^K \tilde{W}} \breve{\mathcal{K}} \cdot_\sigma \breve{\mathcal{I}} \dot{w} \breve{\mathcal{I}}.$$

This follows from Theorem 2.5 and the fact that $\mathrm{Adm}(\mu)^K \cap {}^K \tilde{W} = \mathrm{Adm}(\mu) \cap {}^K \tilde{W}$ proved in [**Hex**, Theorem 6.1] (see [**HHx**] for another proof).

As a consequence, we may compare the set $M_K$ for different parahoric subgroups.

COROLLARY 2.38. *Let $\breve{\mathcal{K}}' \subset \breve{\mathcal{K}}$ be parahoric subgroups. Then*

$$M_K = \breve{\mathcal{K}} \cdot_\sigma M_{K'}.$$

2.11.3. $\mathrm{Adm}(\mu)$ *and* $B(G, \mu)$. Now we discuss the relation of $M_K$ with the $\sigma$-conjugacy classes of $G(\breve{F})$.

We denote by $B(G, \mu)$ the subset of $B(G)$ consisting of $\sigma$-conjugacy classes $[b]$ of $G(\breve{F})$ such that $\kappa(b) = \kappa(\mu)$ and the Newton point $\nu_b$ is less than or equal to the $\varsigma$-average of $\mu$ in the dominance order. The relation between $M_K$ and $B(G)$ is as follows.

THEOREM 2.39. *Let $\breve{\mathcal{K}}$ be a parahoric subgroup of $G(\breve{F})$. Let $[b] \in B(G)$. Then $[b] \cap M_K \neq \emptyset$ if and only if $[b] \in B(G, \mu)$.*

This result is conjectured by Kottwitz and Rapoport in [**KR03**] and [**Ra05**]. The "only if" part is a group-theoretic version of Mazur's inequality. The case where $G$ is an unramified group and $\breve{\mathcal{K}}$ is a hyperspecial maximal subgroup, is proved by Rapoport and Richartz in [**RR96**, Theorem 4.2]. Another proof is given by Kottwitz in [**Ko03**]. The case where $G$ is an unramified group and $\breve{\mathcal{K}}$ is an Iwahori subgroup, is proved in [**Ra05**, Notes added June 2003, (7)]. The "if" part is the "converse to Mazur's inequality" and is proved by Wintenberger in [**Wi05**] in case $G$ is quasi-split. The general case of both directions is proved in [**Hex**, Theorem A].

2.11.4. *The closure relation on $B(G, \mu)$.* In §2.6, we have discussed the closure relation between the $\sigma$-conjugacy classes of $G(\breve{F})$. For $[b], [b'] \in$

$B(G,\mu)$, $[b'] \subset \overline{[b]}$ if and only if $\nu_{b'} \leqslant \nu_b$ in the dominance order (here $\kappa(b') = \kappa(b)$ is automatically satisfied).

Now we show that $M_K = \sqcup_{[b] \in B(G,\mu)} M_K \cap [b]$ is a weak stratification. This is the group-theoretic analogy of the Grothendieck conjecture on the Newton strata of Shimura varieties.

THEOREM 2.40. *For* $[b], [b'] \in B(G,\mu)$, $\overline{M_K \cap [b]} \cap [b'] \neq \emptyset$ *if and only if* $\nu_{b'} \leqslant \nu_b$.

REMARK 2.41. The proof is similar to the proof of [**HRx**, Theorem 5.6] on Shimura varieties. The statement in loc.cit relies on several axioms on the Shimura varieties. The statement here, on the other hand, is valid unconditionally.

PROOF. If $\overline{M_K \cap [b]} \cap [b'] \neq \emptyset$, then $\overline{[b]} \cap [b'] \neq \emptyset$ and thus $\nu_{b'} \leqslant \nu_b$.

Now suppose that $\nu_{b'} \leqslant \nu_b$. Since $\kappa(b') = \kappa(b)$, by Theorem 2.12, $\mathcal{O}_{[b']} \preceq \mathcal{O}_{[b]}$. By [**Hex**, Theorem 3.3], there exists a $\sigma$-straight element $w \in \mathrm{Adm}(\mu) \cap \mathcal{O}_{[b]}$. By Theorem 2.17, $\breve{\mathcal{I}} \dot{w} \breve{\mathcal{I}} \subset M_K \cap [b]$. Thus $\overline{\breve{\mathcal{I}} \dot{w} \breve{\mathcal{I}}} \subset \overline{M_K \cap [b]}$. By the definition of $\preceq$, there exists an element $w' \in \mathcal{O}_{[b']}$ with $w' \leqslant w$. Thus $\dot{w}' \in [b']$ and $\dot{w}' \in \overline{\breve{\mathcal{I}} \dot{w} \breve{\mathcal{I}}} \subset \overline{M_K \cap [b]}$. In other words, $\overline{M_K \cap [b]} \cap [b'] \neq \emptyset$. □

**2.12. Application.** The study of the admissible subsets of $G(\breve{F})$ and their intersections we discussed above has found important applications in the study of Shimura varieties. They serve as the group-theoretic model of the some characteristic subsets (Newton strata, Kottwitz-Rapoport strata, Ekedahl-Oort strata, etc) in the reduction modulo $p$ of a Shimura variety with parahoric level structure. These subsets have been studied intensively in the last two decades. We refer to the survey articles by Rapoport [**Ra05**], by Haines [**Ha05**], and to the new preprint [**HRx**] on a group-theoretic approach to study these characteristic subsets.

In Table 5, we give a very rough comparison between the admissible subsets of $G(\breve{F})$ we discussed above and the characteristic subsets of Shimura varieties.

Let $(\mathbf{G}, \{h\})$ be a Shimura datum. Let $\mathbf{K} = K^p K_p$ be an open compact subgroup of $\mathbf{G}(\mathbb{A}_f)$, where $K = K_p$ is a parahoric subgroup of $\mathbf{G}(\mathbf{Q}_p)$. Let $G = \mathbf{G} \otimes_{\mathbb{Q}} \mathbb{Q}_p$. Let $\mu$ be the dominant coweight corresponding to $\{h\}$. We regard $M_K$ as the group-theoretic model for the special fiber $Sh_K$ of the Shimura variety $Sh_{\mathbf{K}}$.

The stratifications

$$M_K = \sqcup_{w \in \mathrm{Adm}(\mu)_K} \breve{\mathcal{K}} \dot{w} \breve{\mathcal{K}}, \qquad M_K = \sqcup_{[b] \in B(G,\mu)} (M_K \cap [b])$$

are the group-theoretic analogy of the Kottwitz-Rapoport stratification and the Newton stratification of the corresponding Shimura varieties. The stratification

$$M_K = \sqcup_{w \in \mathrm{Adm}(\mu) \cap {}^K \tilde{W}} \breve{\mathcal{K}} \cdot_\sigma \breve{\mathcal{I}} \dot{w} \breve{\mathcal{I}}$$

TABLE 5. Group-theoretic analogy of Shimura varieties

| Group-theoretic side | Shimura variety side |
|---|---|
| $M_K$ | $Sh_K$ |
| $\breve{\mathcal{K}}\dot{w}\breve{\mathcal{K}}$ for $w \in W_K\backslash\tilde{W}/W_K$ | Kottwitz-Rapoport stratum $KR_{K,w}$ |
| $M_K \cap [b]$ for $[b] \in B(G)$ | Newton stratum $S_{K,[b]}$ |
| $\breve{\mathcal{K}} \cdot_\sigma \breve{\mathcal{I}}\dot{w}\breve{\mathcal{I}}$ for $w \in {}^K\tilde{W}$ | EKOR stratum $EKOR_{K,w}$ |
| $M_K = \sqcup_{w \in \mathrm{Adm}(\mu)_K} \breve{\mathcal{K}}\dot{w}\breve{\mathcal{K}}$ | Non-emptiness of KR strata (in the natural range) |
| $M_K = \sqcup_{[b] \in B(G,\{\mu\})}(M_K \cap [b])$ | Non-emptiness of Newton strata |
| $M_K = \sqcup_{w \in \mathrm{Adm}(\mu) \cap {}^K\tilde{W}} \breve{\mathcal{K}} \cdot_\sigma \breve{\mathcal{I}}\dot{w}\breve{\mathcal{I}}$ | Non-emptiness of EKOR strata |
| $\breve{\mathcal{K}}\dot{w}\breve{\mathcal{K}} = \sqcup_{x \in {}^K\tilde{W} \cap W_K w W_K} \breve{\mathcal{K}} \cdot_\sigma \breve{\mathcal{I}}\dot{w}\breve{\mathcal{I}}$ | Each KR stratum is a union of EKOR strata |
| $\overline{\breve{\mathcal{K}}\dot{w}\breve{\mathcal{K}}} = \sqcup_{w' \leqslant w} \breve{\mathcal{K}}\dot{w}'\breve{\mathcal{K}}$ | Closure relation between KR strata |
| $\overline{M_K \cap [b]} \cap [b'] \neq \emptyset \iff \nu_{b'} \leqslant \nu_b$ | Closure relation between Newton strata |
| $\overline{\breve{\mathcal{K}} \cdot_\sigma \breve{\mathcal{I}}\dot{w}\breve{\mathcal{I}}} = \sqcup_{w' \preceq_{K,\sigma}} \breve{\mathcal{K}} \cdot_\sigma \breve{\mathcal{I}}\dot{w}'\breve{\mathcal{I}}$ | Closure relation between EKOR strata |
| $\dim_{\breve{\mathcal{K}}} \breve{\mathcal{K}}\dot{w}\breve{\mathcal{K}} = \max\{\ell(x); x \in W_K w W_K \cap {}^K\tilde{W}\}$ | Conjectural dimension for KR strata |
| $\dim_{\breve{\mathcal{K}}} \breve{\mathcal{K}} \cdot_\sigma \breve{\mathcal{I}}\dot{w}\breve{\mathcal{I}} = \ell(w)$ | Conjectural dimension for EKOR strata |
| For $\sigma$-straight $w$, $\breve{\mathcal{K}} \cdot_\sigma \breve{\mathcal{I}}\dot{w}\breve{\mathcal{I}} \subset [\dot{w}]$ | Special EKOR stratum in a single Newton stratum |
| For $\breve{\mathcal{K}}' \subset \breve{\mathcal{K}}$, $M_K = \breve{\mathcal{K}} \cdot_\sigma M_{K'}$ | Change of parahoric |

is the group-theoretic analogy of the Ekedahl-Kottwitz-Oort-Rapoport stratification, a stratification for Shimura varieties with parahoric level structure which interpolates between the Kottwitz-Rapoport stratification of Shimura varieties with Iwahori level structure and the Ekedahl-Oort stratification [**Oo01**], [**Vie14**] of Shimura varieties with hyperspecial level structure.

We do not have a simple dimension formula for Newton strata. However, we have the following diagram

$$\begin{array}{ccc} & \{g \in G(\breve{F}); g^{-1}b\sigma(g) \in M_K\} & \\ \swarrow^{p_1} & & \searrow^{p_2} \\ X(\mu, b)_K & & M_K \cap [b] \end{array}$$

Here the map $p_1$ is the projection map and the map $p_2$ is a $J_b$-torsor. Note that $J_b$ is a discrete group. This suggests that there might be a way to "locally" identify $M_K \cap [b]$ with $\{g \in G(\breve{F}); g^{-1}b\sigma(g) \in M_K\}$. Notice that

$$\dim_{\breve{\mathcal{K}}}(M_K \cap [b]) - \dim X(\mu, b)_K = \langle 2\rho, \nu_b \rangle,$$

where $X(\mu, b)_K = \{g\breve{\mathcal{K}} \in G(\breve{F})/\breve{\mathcal{K}}; g^{-1}b\sigma(g) \in M_K\}$ is a union of affine Deligne-Lusztig varieties. The number $\langle 2\rho, \nu_b \rangle$ is exactly the conjectural dimension of central leaves.

## 3. Part III. Cocenters of affine Hecke algebras

**3.1. Motivation: group algebras.** To explain some basic idea of the cocenter-representation duality, we start with a naive example.

Let $G$ be a finite group and $V$ be a finite dimensional complex representation of $G$. We define the character of $V$ by $\chi_V(g) = Tr(g, V)$ for $g \in G$. Then $\chi_V$ is a class function, i.e., $\chi_V(g) = \chi_V(g')$ if $g$ and $g'$ are conjugate in $G$.

Let $V_1, \cdots, V_k$ be the irreducible representations of $G$ (up to isomorphism) and $\mathcal{O}_1, \cdots, \mathcal{O}_l$ be the conjugacy classes of $G$. It is well known that $k = l$ and the matrix $(\chi_{V_i}(g_j))_{1 \leqslant i,j \leqslant k}$ is invertible, where $g_j$ is a representative of $\mathcal{O}_j$. This matrix is called the *character table* of $G$.

We reformulate it in a different way.

Let $H_1 = \mathbb{C}[G]$ be the group algebra of $G$. Let $[H_1, H_1]$ be the commutator of $H_1$, the subspace of $H_1$ spanned by $[h, h'] := hh' - h'h$ for all $h, h' \in H_1$. We call the quotient space $\bar{H}_1 = H_1/[H_1, H_1]$ the *cocenter* of $H_1$.

It is easy to see that

(1) For any $g, g'$ in a given conjugacy class $\mathcal{O}$ of $G$, the image of $g$ and $g'$ in $\bar{H}_1$ are the same. We denote it by $[\mathcal{O}]$.

(2) $\{[\mathcal{O}_1], \cdots, [\mathcal{O}_k]\}$ is a basis of $\bar{H}_1$.

Let $R(H_1) = R(G)$ be the Grothendieck group of finite dimensional complex representations of $H_1$. Then $\{V_1, \cdots, V_k\}$ is a basis of $R(H_1)$. The trace map

$$Tr : H_1 \to R(H_1)^*, \qquad g \mapsto (V \mapsto Tr(g, V))$$

factors through $Tr : \bar{H}_1 \to R(H_1)^*$. Moreover,

$$Tr : \bar{H}_1 \to R(H_1)^* \text{ is an isomorphism of vector spaces.}$$

We call it the *cocenter-representation duality* of the group $G$.

**3.2. Hecke algebras.** Let $(W, \mathbb{S})$ be a Coxeter system. Fix a set of indeterminates $\mathbf{c} = \{c(s); s \in \mathbb{S}\}$ such that $c(s) = c(t)$ if $s$ and $t$ are conjugate in $W$, and let $\Lambda = \mathbb{Z}[c(s); s \in \mathbb{S}]$.

The generic Hecke algebra $\mathcal{H} = \mathcal{H}(W, \mathbf{c})$ is the $\Lambda$-algebra generated by $\{T_w; w \in \tilde{W}\}$ subject to the relations:

(1) $T_w \cdot T_{w'} = T_{ww'}$, if $\ell(ww') = \ell(w) + \ell(w')$;
(2) $(T_s + 1)(T_s - c(s)) = 0$, $s \in \mathbb{S}$.

Note that the definition of Hecke algebra works for extended affine Weyl groups as well.

Let $\mathbf{k}$ be a field. If we assign an element $q_s \in \mathbf{k}$ to $c(s)$ for $s \in \mathbb{S}$, then we can regard $\mathbf{k}$ as a $\Lambda$-module and $H_{\mathbf{q}} = \mathcal{H} \otimes_\Lambda \mathbf{k}$ is the specialization of $\mathcal{H}$. In particular, if $q_s = 1$ for all $s \in \mathbb{S}$, then we obtain the group algebra $H_1 = \mathbf{k}[W]$.

We are interested in the cocenter of $\mathcal{H}$, the representations of $\mathcal{H}$ and their relation via the trace map.

One difficulty is that given two elements $w$ and $w'$ in the same conjugacy class of $W$, the image of $T_w$ and $T_{w'}$ in $\bar{\mathcal{H}}$ may not be the same. In fact, we should look at the "strongly conjugate" elements or the elements in the same $\approx$-equivalence class.

PROPOSITION 3.1. *Let $w, w' \in W$.*

*(1) If $w \approx w'$, then the image of $T_w$ and $T_{w'}$ in $\bar{\mathcal{H}}$ are the same.*

*(2) If $w \sim w'$ and $q_s \neq 0$ for all $s \in \mathbb{S}$, then the image of $T_w$ and $T_{w'}$ in $\bar{H}_{\mathbf{q}}$ are the same.*

PROOF. (1) It suffices to prove the case where $w \xrightarrow{s} w'$ and $\ell(w) = \ell(w')$. Without loss of generality, we may assume furthermore that $sw < w$. Then $T_w = T_s T_{sw}$ and $T_{w'} = T_{sw} T_s$. Hence the image of $T_w$ and $T_{w'}$ are the same.

(2) It suffices to prove the case where $w' = xwx^{-1}$ and $\ell(xw) = \ell(w) + \ell(x) = \ell(w') + \ell(x)$. Then we have $T_x T_w = T_{xw} = T_{w'x} = T_{w'} T_x$. Since $q_s \neq 0$ for all $s \in \mathbb{S}$, $T_x$ is invertible in $H_{\mathbf{q}}$. Thus $T_w = T_x^{-1} T_{w'} T_x \equiv T_{w'} \mod [H_{\mathbf{q}}, H_{\mathbf{q}}]$. □

PROPOSITION 3.2. *We denote by $W_{\min}$ the set of elements of $W$ minimal length in their conjugacy class. Suppose that the* **Red-Min** *Property holds for every conjugacy class of $W$. Then for any $w \in W$, the image of $T_w$ in $\bar{\mathcal{H}}$ is a linear combination of the image of $T_x$ for $x \in W_{\min}$.*

PROOF. We argue by induction on $w$.

If $w \in W_{\min}$, then the statement automatically holds for $w$.

If $w \notin W_{\min}$, then by the **Red-Min** property, there exists $w' \in W$ and $s \in \mathbb{S}$ such that $w' \approx w$ and $sw's < w'$. By Proposition 3.1 (1),

$$T_w \equiv T_{w'} \mod [\mathcal{H}, \mathcal{H}].$$

Since $sw's < w$,

$$\begin{aligned} T_{w'} &= T_s T_{sw's} T_s \equiv T_{w'} T_s^2 = (c(s) - 1) T_{sw's} T_s + c(s) T_{sw's} \\ &= (c(s) - 1) T_{sw'} + c(s) T_{sw's} \mod [\mathcal{H}, \mathcal{H}]. \end{aligned}$$

Now the statement for $w$ follows from inductive hypothesis on $sw'$ and on $sw's$. □

THEOREM 3.3. *Let $W$ be a finite Coxeter group or an extended affine Weyl group. Suppose that $q_s \neq 0$ for all $s \in \mathbb{S}$. Then*

*(1) For any conjugacy class $\mathcal{O}$ of $W$, the image of $T_w$ in $\bar{H}_{\mathbf{q}}$ is independent of the choice of $w \in \mathcal{O}_{\min}$. We denote the image by $T_{\mathcal{O}}$.*

*(2) The set* $\{T_{\mathcal{O}}; \mathcal{O} \in W/W_{\Delta}\}$ *spans* $\bar{H}_{\mathbf{q}}$.

**3.3. Finite Hecke algebras.** In this section, we assume that $W$ is a finite Coxeter group and $H_{\mathbf{q}}$ is a finite Hecke algebra with nonzero parameters.

By Tits' deformation theorem [**GP00**, Theorem 7.4.6 & 7.4.7], for generic parameter $\mathbf{c} = \{c(s); s \in \mathbb{S}\}$, the Hecke algebra $H_{\mathbf{c}}$ is isomorphic to the group algebra $H_1$, and hence the number of irreducible representations of $H_{\mathbf{c}}$ equals the number of irreducible representations of $H_1$, and hence equals the number of conjugacy classes of $W$.

By comparing the number of irreducible representations of $H$ and the number of conjugacy classes of $W$, one deduces that $\{T_{\mathcal{O}}\}$ is in fact a basis of $\bar{H}$. Therefore,

THEOREM 3.4. *Let $W$ be a finite Coxeter group and $H_{\mathbf{c}}$ be the Hecke algebra of $W$ with generic parameter* $\mathbf{c}$. *Then $Tr : \bar{H}_{\mathbf{c}} \to R(H_{\mathbf{c}})^*$ is an isomorphism.*

This leads to the definition and study of "character table" of finite Hecke algebras in [**GP93**].

EXAMPLE 3.5. Let $W = S_3$ be the Weyl group of type $A_2$. There are three conjugacy classes of $W$: $\{1\}, \{s_1, s_2\}, \{s_1 s_2, s_2 s_1\}$. For generic parameter $q \in \mathbf{k}^{\times}$, there are three irreducible representations of $H_{\mathbf{q}}$: trivial representation, Steinberg representation and a 2-dimensional irreducible representation $\pi$. The character table is given as follows:

| $A_2$ | $triv$ | $St$ | $\pi$ |
|---|---|---|---|
| $T_{s_1 s_2}$ | $q^2$ | $1$ | $-q$ |
| $T_{s_1}$ | $q$ | $-1$ | $q-1$ |
| $1$ | $1$ | $1$ | $2$ |

EXAMPLE 3.6. Let $W$ be the Weyl group of type $C_2$. There are three conjugacy classes of $W$: $\{1\}, \{s_1, s_2 s_1 s_2\}, \{s_2, s_1 s_2 s_1\}, \{s_1 s_2, s_2 s_1\}, \{s_1 s_2 s_1 s_2\}$. For generic parameter $\mathbf{q} = (q_1, q_2) \in \mathbf{k}^{\times} \times \mathbf{k}^{\times}$, there are five irreducible representations of $H_{\mathbf{q}}$. We label these five modules by the bipartitions which parameterized the corresponding representations of the finite Weyl group: $2 \times 0$, $11 \times 0$, $0 \times 2$, $0 \times 11$, and $1 \times 1$. The character table is given as follows:

| $C_2$ | $2 \times 0$ | $11 \times 0$ | $0 \times 2$ | $0 \times 11$ | $1 \times 1$ |
|---|---|---|---|---|---|
| $T_{s_1 s_2}$ | $q_1 q_2$ | $-q_2$ | $-q_1$ | $1$ | $0$ |
| $T_{s_1 s_2 s_1 s_2}$ | $(q_1 q_2)^2$ | $q_2^2$ | $q_1^2$ | $1$ | $-2q_1 q_2$ |
| $T_{s_1}$ | $q_1$ | $-1$ | $q_1$ | $-1$ | $q_1 - 1$ |
| $T_{s_2}$ | $q_2$ | $q_2$ | $-1$ | $-1$ | $q_2 - 1$ |
| $1$ | $1$ | $1$ | $1$ | $1$ | $2$ |

**3.4. Representations of affine Hecke algebras and of $p$-adic groups.** In this subsection, we recall the relations between the representations of affine Hecke algebras and of $p$-adic groups. The main reference of this subsection is Vignéras' talk at ICM2002 [**Vig02**]. This serves as one of the main motivation to our study of cocenter-representation duality of affine Hecke algebras.

Let $G$ be a connected reductive group over a non-archimedean local field $F$ with residual field $\mathbb{F}_q$ and let $\mathcal{I}$ be an Iwahori subgroup of $G(F)$. Let $\mathbf{k}$ be an algebraically closed field of characteristic $l \neq p$. Let $H_{\mathbf{k}}(G, \mathcal{I}) = \mathrm{End}_{\mathbf{k}[G]} \mathbf{k}[\mathcal{I} \backslash G(F)]$ be the Iwahori Hecke $\mathbf{k}$-algebra of $G(F)$. It is isomorphic to the extended affine Hecke algebra associated to $G$ with parameter $q \in \mathbf{k}$.

For any $\mathbf{k}$-representation $V$ of $G(F)$, the space $V^{\mathcal{I}}$ of $\mathcal{I}$-fixed points is a right $H_{\mathbf{k}}(G, \mathcal{I})$-module. Moreover,

THEOREM 3.7. *The map $V \mapsto V^{\mathcal{I}}$ gives a bijection between the irreducible $\mathbf{k}$-representations of $G$ with $V^{\mathcal{I}} \neq 0$ and the simple right $H_{\mathbf{k}}(G, \mathcal{I})$-modules.*

For complex representations, the equivalence of category is well-known. For modular representations, it is due to Vignéras [**Vig98**].

The case $l = p$ is very different. However, there is still a conjectural relation between the representations of $G$, and of $H_{\mathbf{k}}(G, \mathcal{I})$ (with zero parameter) and of the pro-$p$-Iwahori Hecke algebra $H_{\mathbf{k}}(G, \mathcal{I}_p)$. For $GL_2(\mathbb{Q}_p)$, it is proved by Barthel and Livné in [**BL94**] and [**BL95**].

**3.5. Affine Hecke algebras.** We recall the definition of affine Hecke algebras. Let $\mathscr{R} = (X^*, X_*, R, R^\vee, \Pi)$ be a based root datum and $\tilde{W} = X_* \rtimes W_0$ be the extended affine Weyl group associated to $\mathscr{R}$ (see §1.7). Let $\mathbf{c} = \{c(s); s \in \tilde{\mathbb{S}}\}$ such that $c(s) = c(t)$ if $s$ and $t$ are conjugate in $\tilde{W}$, and let $\Lambda = \mathbb{Z}[c(s); s \in \tilde{\mathbb{S}}]$. We define the extended affine Hecke algebra $\tilde{\mathcal{H}}$ and its specialization $\tilde{H}_{\mathbf{q}}$ in a similar way as in §3.2. We fix a set of nonzero parameters $\mathbf{q}$ in an algebraically closed field $\mathbf{k}$ and we may simply write $\tilde{H}$ instead of $\tilde{H}_{\mathbf{q}}$. The basis $\{T_w\}_{w \in \tilde{W}}$ gives the Iwahori-Matsumoto presentation of $\tilde{H}$. It is related to the quasi-Coxeter structure of $\tilde{W}$.

Another presentation we need for $\tilde{H}$ is the Bernstein-Lusztig presentation:

$$\tilde{H} = \oplus_{x \in X_*, w \in W_0} \mathbf{k}\, \theta_x T_w.$$

We refer to [**Lu89**] for the definition of $\theta_x$. This presentation is related to the semi-product $\tilde{W} = X_* \rtimes W_0$. It plays an important role in the study of representations of affine Hecke algebras, especially the (parabolic) induction and restriction functors.

For any $J \subset \Pi$, let $\tilde{H}_J$ be the parabolic subalgebra of $\tilde{H}$ generated by $\{\theta_x T_w; x \in X, w \in W_J\}$. This is the extended affine Hecke algebra for the parabolic subgroup $\tilde{W}_J = X_* \rtimes W_J$, for some parameters.

We may then define the (parabolic) induction and restriction functors between the Grothendieck groups of the finite dimension $\mathbf{k}$-representations of $\tilde{H}$ and $\tilde{H}_J$.

$$i_J : R(\tilde{H}_J) \to R(\tilde{H}), \qquad r_J : R(\tilde{H}) \to R(\tilde{H}_J).$$

We have the central character $\chi_t$ of $\tilde{H}_J$, here $t$ runs over a complex torus $T^J$ associated to $J$. If $\sigma \in R(\tilde{H}_J)$, then $\sigma\chi_t \in R(\tilde{H}_J)$. Following Bernstein, Deligne and Kazhdan, we say that a form $f \in R(\tilde{H})^*$ is *good* if for any $J \subset \Pi$ and $\sigma \in R(\tilde{H}_J)$, the function $t \mapsto f(i_J(\sigma\chi_t))$ is a regular function on $T^J$.

The following result is obtained by Bernstein, Deligne and Kazhdan [**BDK86**] and [**Kaz86**].

THEOREM 3.8. *Let $\mathbb{H} = \mathbb{C}_c^\infty(G(F))$ be the (complex) Hecke algebra of a $p$-adic group $G(F)$. Then*
*(1) The image of $Tr : \bar{\mathbb{H}} \to R(\mathbb{H})^*$ is the set of good forms.*
*(2) The map $Tr : \bar{\mathbb{H}} \to R(\mathbb{H})^*$ is injective.*

Part (1) is referred to as trace Paley-Wiener Theorem and Part (2) is referred to as the density Theorem. The proofs rely on p-adic analysis.

Our goal is to understand for which parameters, the trace Paley-Wiener Theorem and the density Theorem holds for affine Hecke algebras and to have an explicit basis of the cocenter.

We have seen in Theorem 3.3 that the set $\{T_{\mathcal{O}}; \mathcal{O} \in \tilde{W}/\tilde{W}_\Delta\}$ forms a basis of $\overline{\tilde{H}}$. We also have the Bernstein-Lusztig presentation of these elements. It is obtained in [**HN15**, Theorem B].

THEOREM 3.9. *Let $i_J : \tilde{H}_J \to \tilde{H}$ be the inclusion and $\bar{i}_J : \overline{\tilde{H}_J} \to \overline{\tilde{H}}$ the induced map. Let $(J, x, K, C) \in \tilde{\Gamma}$ and $\mathcal{O}$ be the associated $\delta$-conjugacy class of $\tilde{W}$, then*

$$T_{\mathcal{O}} = \bar{i}_J(T^J_{Cx}),$$

*where $T^J_{Cx}$ is the image of $T^J_{ux}$ in $\overline{\tilde{H}_J}$ for any minimal length element $u \in C$.*

Note that the relation between Iwahori-Matsumoto basis and Bernstein-Lusztig basis of $\tilde{H}$ is complicated. The relation between the minimal length elements in $\mathcal{O}$ (with respect to the length function of $\tilde{W}$) and the minimal length elements in $C$ (with respect to the length function of $\tilde{W}_J$) is also complicated. It is amazing that these two complexities cancel each other out. This leads to the above matching between Iwahori-Matsumoto basis and Bernstein-Lusztig basis of the cocenter of $\tilde{H}$.

Compared to the finite Hecke algebras, it is more difficult to prove that $\{T_{\mathcal{O}}\}$ is linearly independent in the cocenter of extended affine Hecke algebras. The reason is that both $\overline{\tilde{H}}$ and $R(\tilde{H})$ are of infinite rank and thus one can't compare their rank directly to make the conclusion. For equal parameter case, linearly independence is proved in [**HN14**] using Lusztig's $J$-ring. The general case is proved in [**CHx**] using the cocenter-representation duality that we are going to discuss.

**3.6. Cocenter-representation duality for affine Hecke algebras.** In this section, we fix a nonzero parameter $\mathbf{q}$ in $\mathbb{C}$ and we simply write $\tilde{H}$ instead of $\tilde{H}_{\mathbf{q}}$.

3.6.1. *Elliptic quotient.* Elliptic representation theory, introduced by Arthur [**Ar93**], studies the Grothendieck group of certain representations of a Lie-theoretic group modulo those induced from proper parabolic subgroups. The elliptic theory of representations of semisimple $p$-adic groups and Iwahori-Hecke algebras is further studied intensively, e.g., Schneider-Stuhler [**SS97**], Bezrukavnikov [**Be98**], Reeder [**Re01**], Opdam-Solleveld [**OS09**].

For an affine Hecke algebra $\tilde{H}$ (of a given nonzero parameters in $\mathbb{C}$), the *elliptic quotient* is defined to be

$$R(\tilde{H})_{ell} = R(\tilde{H})_{\mathbb{C}} / \sum_{J \subsetneq \Pi} i_J(R(\tilde{H}_J)_{\mathbb{C}}).$$

Opdam and Solleveld in [**OS09**] studied the affine Hecke algebras for positive parameters and showed that

THEOREM 3.10. *Let* $\mathbf{q}$ *be a positive parameter function on* $\tilde{H}$. *Then*

*(1) The dimension of* $R(\tilde{H})_{ell}$ *is at most the number of elliptic conjugacy classes of* $\tilde{W}$.

*(2) The inequivalent discrete series forms an orthogonal set of* $R(\tilde{H})_{ell}$.

In particular, one has an upper bound of the number of irreducible discrete series for affine Hecke algebra of positive parameters. A lower bound can be obtained by counting the central characters of the discrete series. This leads to the classification of irreducible discrete series for affine Hecke algebras of positive parameters in [**OS10**].

The method of Opdam-Solleveld is analytic, passing from $\tilde{H}$ to its Schwartz algebra, a certain topological completion of $\tilde{H}$. For Hecke algebras of $p$-adic groups, the elliptic quotient and its relation with the trace map was also studied in [**BDK86**] and analytic methods were used to obtain an upper bound of the dimension of the elliptic quotient.

3.6.2. *Rigid cocenter and rigid quotient.* Here is our motivation to develop a different method to study elliptic representation theory for affine Hecke algebras.

First, we would like to understand the affine Hecke algebras for arbitrary parameters (not just the positive parameters) and for representations over a field of positive characteristic. It is desirable to have a more algebraic method.

Second, we would like to put the elliptic quotient in the framework of "cocenter-representation duality". Namely, the elliptic quotient $R(\tilde{H})_{ell}$ corresponds to a subspace of $\overline{\tilde{H}}$ via the trace map $Tr : \overline{\tilde{H}} \times R(\tilde{H})_{\mathbb{C}} \to \mathbb{C}$. What is this subspace? Results in [**BDK86**] indicates that this subspace is very complicated and may not have a nice explicit description.

The idea in [**CHx**] is to replace the elliptic quotient by the so-called rigid quotient. For simplicity, we only consider the extended affine Hecke algebras associated to semisimple root data. The reductive root data can be reduced to semisimple ones via [**CHx**, §8.1].

We define the *rigid cocenter*

$$\overline{\tilde{H}}^{\mathrm{rig}} = \mathrm{span}\{T_{\mathcal{O}}; \nu_{\mathcal{O}} = 0\}$$

and the *rigid quotient*

$$R(\tilde{H})_{\mathrm{rig}} = R(\tilde{H})_{\mathbb{C}} / \langle i_J(\sigma) - i_J(\sigma\chi_t); J \subset \Pi, \sigma \in R(\tilde{H}_J), t \in T^J \rangle,$$

the quotient of $R(\tilde{H})$ by the difference of induced modules. Both the rigid cocenter and the rigid quotient are finite dimensional since the root datum is semisimple.

It is proved in [**CHx**, Theorem 1.1] that

THEOREM 3.11. *(1) For generic parameters, the trace map* $Tr : \bar{H} \times R(\tilde{H})_{\mathbb{C}} \to \mathbb{C}$ *induces a perfect pairing*

$$Tr : \overline{\tilde{H}}^{\mathrm{rig}} \times R(\tilde{H})_{\mathrm{rig}} \to \mathbb{C}.$$

*In particular,* $\dim R(\tilde{H})_{\mathrm{rig}} = \dim \overline{\tilde{H}}^{\mathrm{rig}}$ *equals the number of conjugacy classes in* $\tilde{W}$ *whose Newton points equal to zero.*

*(2) For arbitrary nonzero parameters, the induced map* $Tr : \overline{\tilde{H}}^{\mathrm{rig}} \to R(\tilde{H})^*_{\mathrm{rig}}$ *is surjective.*

As some consequences,

- Trace Paley-Wiener Theorem: For arbitrary nonzero parameters, the image of the map $Tr : \overline{\tilde{H}} \to R(\tilde{H})^*$ is $R(\tilde{H})^*_{good}$.
- Density Theorem: For generic parameters, the map $Tr : \overline{\tilde{H}} \to R(\tilde{H})^*$ is injective.
- Basis Theorem: The set $\{T_{\mathcal{O}}; \mathcal{O} \in \tilde{W}/\tilde{W}_{\Delta}\}$ forms a basis of $\overline{\tilde{H}}$.

We also have the following deformation theorem.

THEOREM 3.12. *For generic parameters* $\mathbf{q}$*, there exists a basis* $\{V_{\pi,\mathbf{q}}\}$ *of* $R(\tilde{H})_{\mathbb{C}}$ *such that for any* $w \in \tilde{W}$ *and any* $\pi$*, the action of* $T_w$ *on* $V_{\pi,\mathbf{q}}$ *depends analytically on* $\mathbf{q}$.

For affine Hecke algebra with positive parameters, a similar result was obtained by Opdam and Solleveld in [**OS10**] using Schwartz algebras. The idea of our proof (without the restriction on positive parameters) is to use Lusztig's graded affine Hecke algebras [**Lu89**] and the deformation theorem for graded affine Hecke algebras obtained in [**CH16**]. The passage from affine Hecke algebras to graded affine Hecke algebras is analytic. This is the reason that we have the analytic deformation theorem here. However, we expect that there exists a family of representations depending algebraically on $\mathbf{q}$.

3.6.3. *Class polynomials.* Now suppose that all the parameters $q_s$ are equal. We write $q = q_s$ for any $s \in \tilde{\mathbb{S}}$. For any $w \in \tilde{W}$, there exists $F_{w,\mathcal{O}} \in \mathbb{Z}[q]$ for each conjugacy class $\mathcal{O}$ of $\tilde{W}$ such that

$$T_w \equiv \sum_{\mathcal{O}} F_{w,\mathcal{O}} T_{\mathcal{O}} \mod [\tilde{H}, \tilde{H}]. \tag{3.1}$$

The polynomials $F_{w,\mathcal{O}}$ are called the *class polynomials* and can be computed inductively on $w$ as we discussed in §2.8.2. Note that the set $\{T_{\mathcal{O}}\}$ is a basis of $\overline{\tilde{H}}$. Thus the polynomials $F_{w,\mathcal{O}}$ are uniquely determined by the equality 3.1 and thus is independent of the choice of the reduction procedure in §2.8.2.

We also have that

$$Tr(T_w, V) = \sum_{\mathcal{O}} F_{w,\mathcal{O}} Tr(T_{\mathcal{O}}, V). \tag{3.2}$$

for any finite dimensional representation $V$ of $\tilde{H}$. Therefore, the character value of any element in $\tilde{H}$ is known if one knows the character value of all the minimal length elements, together with the class polynomials. Moreover, by the density Theorem for generic parameter $q$, the polynomials $F_{w,\mathcal{O}}$ are also determined by the equality 3.2. This gives a representation-theoretic definition of the class polynomials.

As we have seen in §2.8.2, the class polynomials (especially the leading term of the class polynomial), have found important application in the arithmetic geometry. It is desirable to have an explicit formula for the class polynomials. This is a challenging problem. Some computations for type $\tilde{A}_2$ is done by Yang [**Yax**].

### 3.7. The "Modular case".

3.7.1. *Naive kernel conjecture.* Now we move to the "modular case". Here instead of complex representations, we consider representations over an algebraically closed field $\mathbf{k}$ of positive characteristic. We also consider the case where the parameter is a root of unity. In this situation, one can't expect to have a perfect pairing as in Theorem 3.11 (1). However, we expect that the cocenter-representation duality fails in a "controllable" way and thus provide useful information on the representations of affine Hecke algebras in the "modular case".

We still have a surjective map $Tr : \overline{\tilde{H}}^{\text{rig}} \to R(\tilde{H})^*_{rig}$, but it fails to be injective. It is interesting to see whether the whole kernel comes from the non-semisimplicity of the parahoric subalgebras.

Let us make it precise. Let $K \subset \tilde{\mathbb{S}}$. If $W_K$ is finite, then we say that the subalgebra $H_K$, generated by $T_w$ for $w \in W_K$, is a *parahoric subalgebra* of $\tilde{H}$. In this case, $H_K$ is a finite Hecke algebra. We set $\Omega(K) = \{\tau \in \Omega; \tau K \tau^{-1} = K\}$ and define the *extended parahoric subalgebra*

$$H_K^\sharp = H_K \rtimes \mathbf{k}[\Omega(K)].$$

We have a natural map $\bar{H}_K^\sharp \to \overline{\tilde{H}}^{\mathrm{rig}}$. The composition $\bar{H}_K^\sharp \to \overline{\tilde{H}}^{\mathrm{rig}} \to R(\tilde{H})^*_{\mathrm{rig}}$ is not injective in general. It contains $\ker(\bar{H}_K^\sharp \to R(H_K^\sharp)^*)$ in its kernel.

Now we introduce the *naive kernel*. Let $\ker^{naive}(Tr : \overline{\tilde{H}}^{\mathrm{rig}} \to R(\tilde{H})^*_{\mathrm{rig}})$ be the sum of the image in $\overline{\tilde{H}}^{\mathrm{rig}}$ of $\ker(\bar{H}_K^\sharp \to R(H_K^\sharp)^*)$, where $H_K$ runs over all the parahoric subalgebras of $\tilde{H}$.

CONJECTURE 3.13. *The naive kernel* $\ker^{naive}(Tr : \overline{\tilde{H}}^{\mathrm{rig}} \to R(\tilde{H})^*_{\mathrm{rig}})$ *equals the kernel of the trace map* $Tr : \overline{\tilde{H}}^{\mathrm{rig}} \to R(\tilde{H})^*_{\mathrm{rig}}$.

If this conjecture holds, then one may reduce the study of parametrization of irreducible modular representations of extended affine Hecke algebras to the study of parametrization of irreducible modular representations of extended finite Hecke algebras. The latter case has been studied extensively by Geck and Jacon [**GJ11**]. By combining all these together, one obtain the rank of the rigid quotient and the elliptic quotient of $R(\tilde{H})$.

In particular, if all the extended parahoric subalgebras of $\tilde{H}$ are semisimple, then the parameterization of irreducible modules of $\tilde{H}$ are independent of the change of parameters $\mathbf{q}$. If $q_s = q \in \mathbb{C}^\times$ for all $s \in \tilde{\mathbb{S}}$, then the condition that all the extended parahoric subalgebras are semisimple is equivalent to the condition that $q$ is not a root of the Poincaré polynomial of $\tilde{H}$. The classification of irreducible representations in the case where $q$ is not a root of unity is obtained by Kazhdan and Lusztig [**KL87**]. It is proved later by Xi [**Xi07**] that the Kazhdan-Lusztig classification remains valid if $q$ is not a root of Poincaré polynomial.

Now we provide some examples.

EXAMPLE 3.14. Let $G = SL_3$. Then $\Omega = \{1\}$. By Example 1.22, the rigid cocenter $\overline{\tilde{H}}^{\mathrm{rig}}$ has a basis $\{T_1 = 1, T_{s_1}, T_{s_1 s_2}, T_{s_0 s_1}, T_{s_0 s_2}\}$. There are three maximal (extended) parahoric subalgebras: $H_K$ for $K = \{1,2\}, \{0,1\}, \{0,2\}$. They are finite Hecke algebras associated to the group $S_3$.

The Poincare polynomial $P_{S_3}(q) = \Phi_3(q)\Phi_2(q)$, where $\Phi_i$ is the $i$-th cyclotomic polynomial. If $q$ is not a root of $P_{S_3}(q)$, then the three parahoric Hecke algebras are semisimple. In particular, the map $Tr : \bar{H}_K \to R(H_K)^*$ is injective and the naive kernel is trivial. The conjectural rank of $R(\tilde{H})_{\mathrm{rig}}$ equals the dimension of $\bar{\mathcal{H}}^{\mathrm{rig}}$, which is 5.

If $\Phi_2(q) = 0$, then the kernel of the map $Tr : \bar{H}_{\{1,2\}} \to R(H_{\{1,2\}})^*$ is spanned by $T_{s_1} + 1$. Note that the image of $T_{s_1} + 1, T_{s_2} + 1, T_{s_0} + 1$ in $\overline{\tilde{H}}^{\mathrm{rig}}$ are the same. Thus the naive kernel is spanned by the image of $T_{s_1} + 1$ in $\overline{\tilde{H}}^{\mathrm{rig}}$. The conjectural rank of $R(\tilde{H})_{\mathrm{rig}}$ equals the dimension of $\overline{\tilde{H}}^{\mathrm{rig}}$ minus 1, which is 4.

If $\Phi_3(q) = 0$, then the kernel of the map $Tr : \bar{H}_{\{1,2\}} \to R(H_{\{1,2\}})^*$ is spanned by $(q+1)T_{s_1 s_2} + (q+2)T_{s_1} + 1$. The naive kernel is spanned by the image of $(q+1)T_{s_1s_2}+(q+2)T_{s_1}+1$, $(q+1)T_{s_0s_2}+(q+2)T_{s_0}+1$, $(q+1)T_{s_0s_1}+(q+2)T_{s_0}+1$ in $\overline{\tilde{H}}^{\text{rig}}$. The conjectural rank of $R(\tilde{H})_{\text{rig}}$ equals the dimension of $\overline{\tilde{H}}^{\text{rig}}$ minus 3, which is 2.

Here the characteristic of $\mathbf{k}$ does not affect the conjectural dimension of $R(\tilde{H})_{\text{rig}}$. The following table lists the conjectural rank of $R(\tilde{H})_{\text{rig}}$ for various choice of $q$ and $char(\mathbf{k})$.

| $SL_3$ | $char(\mathbf{k}) \neq 3$ | $char(\mathbf{k}) = 3$ |
|---|---|---|
| $\Phi_3(q)\Phi_2(q) \neq 0$ | 5 | 5 |
| $\Phi_2(q) = 0$ | 4 | 4 |
| $\Phi_3(q) = 0$ | 2 | 2 |

EXAMPLE 3.15. Let $G = PGL_3$. Then $\Omega = \mu_3$. By Example 1.21, the rigid cocenter $\overline{\tilde{H}}^{\text{rig}}$ has a basis $\{T_1 = 1, T_{s_1}, T_{s_1s_2}, \tau, \tau^2\}$, where $\tau$ is a non-trivial element in $\Omega$. There are four maximal extended parahoric subgroups: $H_K$ for $K = \{1,2\}, \{0,1\}, \{0,2\}$ and $H_{\emptyset}^{\sharp}$, the group algebra of $\Omega$. The first three are conjugate by $\Omega$ and we only need to consider the contribution of $\ker(\bar{H}_K^{\sharp} \to R(H_K^{\sharp})^*)$ for $K = \{1,2\}$ and $K = \emptyset$ to the naive kernel.

If $\Phi_2(q) = 0$ or $\Phi_3(q) = 0$, then the kernel of $Tr : \bar{H}_{\{1,2\}} \to R(H_{\{1,2\}})^*$ is 1-dimensional and this makes an one-dimensional contribution of the naive kernel.

If $char(\mathbf{k}) = 3$, then the only irreducible representation of $\Omega$ is trivial and the kernel of $Tr : \bar{H}_{\emptyset}^{\sharp} \to R(H_{\emptyset}^{\sharp})^*$ is 2-dimensional. This makes a two-dimensional contribution of the naive kernel.

The following table lists the conjectural rank of $R(\tilde{H})_{\text{rig}}$ for various choice of $q$ and $char(\mathbf{k})$.

| $PGL_3$ | $char(\mathbf{k}) \neq 3$ | $char(\mathbf{k}) = 3$ |
|---|---|---|
| $\Phi_3(q)\Phi_2(q) \neq 0$ | 5 | 3 |
| $\Phi_2(q) = 0$ | 4 | 2 |
| $\Phi_3(q) = 0$ | 4 | 2 |

Here the characteristic of $\mathbf{k}$ does affect the conjectural dimension of $R(\tilde{H})_{\text{rig}}$. It is also interesting to compare this example with the previous example to see how the different isogeny classes affect the conjectural rank of $R(\tilde{H})_{\text{rig}}$.

3.7.2. *Rigid determinant.* Recall that in §3.3, we discussed the character table for finite Hecke algebras. In particular, we may compute the determinant of the character table. It is a polynomial of the generic parameters $\mathbf{q}$ and it is determined by the finite Hecke algebra (up to sign depending on the ordering of the irreducible representations).

We may define the rigid character table for extended affine Hecke algebras (with generic complex parameters) as well. Here we use a family of representations, depending analytically on the parameters, whose images in $R(\tilde{H})_{\text{rig}}$ form a basis. The existence of such family is proved in Theorem

3.12. We define the rigid determinant as the determinant of the rigid character table. It is again a polynomial of the parameters $\mathbf{q}$ and it is determined up to scalar.

We denote by $\det^{\mathrm{rig}}_{\spadesuit}$ the rigid determinant of an extended affine Hecke algebra of type $\spadesuit$ and by $\det_{\heartsuit}$ for the determinant of the character table of an (extended) finite Hecke algebra of type $\heartsuit$.

CONJECTURE 3.16. *Let $\tilde{H}$ be an extended affine Hecke algebra with generic parameter $\mathbf{q}$. Then the rigid determinant $\det^{\mathrm{rig}} \in \mathbf{k}[\mathbf{q}]$ is a factor of $\Pi_K \det_{H_K^\sharp}$, where $H_K$ runs over all the parahoric subalgebras of $\tilde{H}$.*

REMARK 3.17. In fact, in the conjecture we only need to take the product of the maximal extended parahoric subalgebras. And we also expect that there is an explicit formula for the rigid determinant, as an alternating product of the determinants of the character tables for various of (partially) extended parahoric subalgebras, as we will see in the examples below. However, due to its complexity, we do not formulate the (conjectural) explicit formula here.

Also this conjecture predicts that the pairing between the rigid cocenter and the rigid quotient is a perfect pairing if all the extended parahoric subalgebras are semisimple (i.e., the determinant of the character tables are invertible). This coincides with the prediction from the naive kernel conjecture 3.13.

Now we give two examples to support the rigid determinant conjecture. We also provide explicit formulas for the rigid determinants in these examples.

EXAMPLE 3.18. Let $G = Sp(4)$ and let $\tilde{H}$ be the affine Hecke algebra attached to the affine diagram of type $C_2$ with three parameters

$$q_0 \Longrightarrow q_1 \Longleftarrow q_2$$

The rigid cocenter $\overline{\tilde{H}}^{\mathrm{rig}}$ has a basis $\{T_1 = 1, T_{s_0}, T_{s_1}, T_{s_2}, T_{s_0s_1}, T_{s_0s_2}, T_{s_1s_2}, T_{s_0s_1s_0s_1}, T_{s_1s_2s_1s_2}\}$. There are three maximal parahoric subalgebras: $H_K$ for $K = \{1,2\}, \{0,1\}, \{0,2\}$. Here $T_{s_i}$ for $i = 0,1,2$ each appears in the cocenters of two out of three maximal parahoric subalgebras. We have

$$\begin{aligned}
\det{}_{C_2(q_0,q_1)} &= (1+q_0)^2(1+q_1)^2(1+q_0q_1)(q_0+q_1),\\
\det{}_{A_1(q_0)\times A_1(q_2)} &= (1+q_0)^2(1+q_2)^2,\\
\det{}_{C_2(q_1,q_2)} &= (1+q_1)^2(1+q_2)^2(1+q_1q_2)(q_1+q_2),
\end{aligned}$$

The rigid determinant is computed in [**CHx**, Example 7.9]. We have

$$\det{}^{\mathrm{rig}}_{\tilde{C}_2(q_0,q_1,q_2)} = \frac{\det_{C_2(q_0,q_1)} \det_{A_1(q_0)\times A_1(q_2)} \det_{C_2(q_1,q_2)}}{\det_{A_1(q_0)} \det_{A_1(q_1)} \det_{A_1(q_2)}}.$$

EXAMPLE 3.19. Let $G = PSp(4)$ and let $\tilde{H}$ be the affine Hecke algebra attached to the affine diagram of type $C_2$ with three parameters

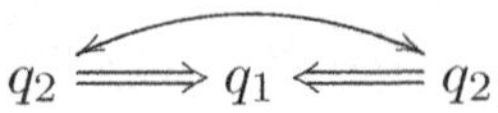


The rigid cocenter $\bar{\tilde{H}}^{\text{rig}}$ has a basis $\{T_1 = 1, T_{s_1}, T_{s_2}, T_{s_0 s_2}, T_{s_1 s_2}, T_{s_1 s_2 s_1 s_2}, \tau, T_{s_1}\tau, T_{s_0}\tau\}$. There are four maximal extended parahoric subalgebras: $H_K^\sharp$ for $K = \{1,2\}, \{0,1\}, \{0,2\}, \{1\}$. The first two are conjugate by $\Omega$. The elements $T_{s_1}, T_{s_2}, \tau$ each appears twice in the cocenters of the three isomorphism classes of maximal parahoric subalgebras. We have

$$\begin{aligned} \det{}_{C_2(q_1,q_2)} &= (1+q_1)^2(1+q_2)^2(1+q_1q_2)(q_1+q_2), \\ \det{}_{(A_1(q_2)\times A_1(q_2))\rtimes\mu_2} &= 4(1+q_2)^4, \\ \det{}_{A_1(q_1)\times\mu_2} &= 4(1+q_1). \end{aligned}$$

The rigid determinant is computed in [**CHx**, Example 7.10]. We have

$$\det{}^{\text{rig}}_{\tilde{C}_2\rtimes\mu_2(q_1,q_2)} = \frac{\det{}_{C_2(q_1,q_2)} \det{}_{(A_1(q_2)\times A_1(q_2))\rtimes\mu_2} det_{A_1(q_1)\times\mu_2}}{8\det{}_{A_1(q_1)} \det{}_{A_1(q_2)} \det{}_{\mu_2}}.$$

**3.8. 0-Hecke algebras.** If all the parameters equal to zero, then we say that the corresponding Hecke algebra a 0-*Hecke algebra*. In this subsection, we discuss the cocenter and representations of affine 0-Hecke algebras. The purpose is two-folded:

- Affine 0-Hecke algebras serve as models for the pro-$p$ Iwahori-Hecke algebras, which play an important role in the study of mod-$p$ representations of $p$-adic groups.
- For affine 0-Hecke algebras, the cocenter-representation duality fails. The good news is that it fails in a "controllable" way. This serves as an evidence of the naive kernel conjecture 3.13.

3.8.1. *The cocenter.* Let $W$ be a finite or (extended) affine Weyl group. As we have seen in §3.3 and §3.6, the cocenter of $H_{\mathbf{q}}$ (for nonzero parameters $\mathbf{q}$) has a standard basis indexed by $W_{\min}/\sim$. For 0-Hecke algebras, the situation is different.

By Proposition 3.2 the induction argument still works and the cocenter of $H_0$ is spanned by the image of $T_w$, where $w$ runs over the elements in $W_{\min}$. However, for $w \sim w'$, $T_w$ and $T_{w'}$ may have different image in $\bar{H}_0$.

By Proposition 3.1 (1), if $w \approx w'$, then the image of $T_w$ and $T_{w'}$ in $\bar{H}_0$ are the same. For any $\approx$-equivalence class $\Sigma$ of $W_{\min}$, we denote by $T_\Sigma$ the image of $T_w$ in $\bar{H}_0$ for any $w \in \Sigma$. Then

THEOREM 3.20. *The set* $\{T_\Sigma; \Sigma \in W_{\min}/\approx\}$ *forms a* $\mathbb{Z}$*-basis of* $\bar{H}_0$.

For finite Hecke algebras, it is proved in [**He15**, Theorem 5.5], where the basis theorem of $\bar{H}_{\mathbf{q}}$ (for nonzero parameters $\mathbf{q}$) is used. A different and simpler approach is found in [**HNx**, §2.4], which works for both finite and affine Hecke algebras.

3.8.2. *Finite* 0*-Hecke algebras.* In this subsection, we assume that $W$ is a finite Coxeter group. Recall that

$$\Gamma = \{(J, C); J \subset \mathbb{S}, C \text{ is an elliptic conjugacy class of } W_J\}.$$

It is proved in Proposition 1.10 that the set of equivalence classes of $\Gamma$ is in natural bijection with the set of conjugacy classes of $W$. The following result is proved in [**He15**, Corollary 3.2].

PROPOSITION 3.21. *The map*

$$\Gamma \to W_{\min}/\approx, \qquad (J, C) \mapsto C_{\min}$$

*gives a bijection from* $\Gamma$ *to the* $\approx$*-equivalence classes of* $W_{\min}$.

The irreducible representations of $H_0$ are easy to construct. For any $J \subset \mathbb{S}$, let $\lambda_J$ be the one-dimensional representation of $H_0$ defined by

$$\lambda_J(T_s) = \begin{cases} -1, & \text{if } s \in J; \\ 0, & \text{if } s \notin J. \end{cases}$$

By [**No79**], the set $\{\lambda_J\}_{J \subset \mathbb{S}}$ is the set of all the irreducible representations of $H_0$.

The following result [**He15**, Proposition 5.6] describes how the cocenter-representation duality fails for $H_0$.

PROPOSITION 3.22. *The trace map* $\bar{H}_0 \to R(H_0)$ *is surjective and the kernel is spanned by* $T_\Sigma - T_{\Sigma'}$, *where* $\Sigma, \Sigma'$ *are* $\approx$*-equivalence classes of* $W_{\min}$ *of the same support.*

Now let us compare the trace function for $H_{\mathbf{q}}$ for generic parameters $\mathbf{q}$ and $H_0$. Let $\Sigma \in W_{\min}/\approx$ and $(J, C) \in \Gamma$ be the associated pair. Then the trace function of $H_0$ remembers the support $J$ of the element $T_\Sigma$, but ignores the difference between various elliptic conjugacy classes of $W_J$. On the other hand, the trace function of $H_{\mathbf{q}}$ remembers the elliptic conjugacy classes, but ignore the difference between the various subsets of $\mathbb{S}$ that are conjugate to each other.

3.8.3. *Affine* 0*-Hecke algebras.* Recall that $\tilde{\Gamma}$ is the set of standard quadruples defined in §1.8.3 for ordinary conjugation action. It is proved in Theorem 1.19 that the set of equivalence classes of $\tilde{\Gamma}$ is in natural bijection with the set of conjugacy classes of $\tilde{W}$.

For any $\Sigma \in \tilde{W}_{\min}/\approx$ corresponding to a standard quadruple $(J, x, K, C)$, we set $J_\Sigma = J$. We set

$$\overline{\tilde{H}_0}^{\text{rig}} = \oplus_{\Sigma \in \tilde{W}_{\min}/\approx, J_\Sigma = \Pi} \mathbb{Z} T_\Sigma, \qquad \overline{\tilde{H}_0}^{\text{nrig}} = \oplus_{\Sigma \in \tilde{W}_{\min}/\approx, J_\Sigma \subsetneqq \Pi} \mathbb{Z} T_\Sigma$$

Let $K \subset \tilde{\mathbb{S}}$ with $W_K$ finite. Recall that $\Omega(K) = \{\tau \in \Omega; \tau(K) = K\}$. Let $\chi \in \operatorname{Hom}_{\mathbb{Z}}(\Omega(K), \mathbf{k}^\times)$. We extend $\chi$ to a 1-dimensional $H_{a,0} \rtimes \Omega(K)$-module, where $T_s^J$ acts by $-1$ if $s \in K$ and by 0 if $s \in \tilde{\mathbb{S}} \smallsetminus K$. Then we set

$$\pi_{K,\chi} = \tilde{\mathcal{H}}_0 \otimes_{H_{a,0} \rtimes \Omega(K)} \chi.$$

We say that $(K,\chi) \sim (K',\chi')$ if they are conjugate by an element in $\Omega$. It is easy to see that if $(K,\chi) \sim (K',\chi')$, then $\pi_{K,\chi}$ is isomorphic to $\pi_{K',\chi'}$.

We set

$$R(\tilde{\mathcal{H}}_0)_{\text{rig}} = \oplus_{(K,\chi)/\sim; K\subset\tilde{\mathbb{S}} \text{ with } W_K \text{ finite}} \mathbb{Z}\pi_{K,\chi}.$$

The following result is proved in [**HNx**, Proposition 5.1].

THEOREM 3.23. *For* $M \in R(\tilde{\mathcal{H}}_0)$, $Tr(\overline{\tilde{H}_0}^{\text{nrig}}, M) = 0$ *if and only if* $M \in R(\tilde{\mathcal{H}}_0)_{\text{rig}}$.

In other words, the trace map $Tr : \overline{\tilde{H}_0} \to R(\tilde{\mathcal{H}}_0)$ induces a map

$$Tr : \overline{\tilde{H}_0}^{\text{rig}} \to R(\tilde{\mathcal{H}}_0)^*_{\text{rig}}.$$

This map is surjective, but not injective. Similar to the finite 0-Hecke algebra case, the trace map on the rigid cocenter remembers the support $K$ of the element $T_\Sigma$, but ignore the difference between various elliptic conjugacy classes of $W_K$. Thus the above Theorem provides an evidence to the naive kernel conjecture 3.13.

In [**Vig05**], Vignéras introduces the Bernstein-Lusztig basis $\{E_w; w \in \tilde{W}\}$ of $\tilde{\mathcal{H}}_0$. She also gives the definition of the supersingular modules of $\tilde{H}_0$. By definition, a module $M$ of $\tilde{H}_0$ is *supersingular* if $E_w M = 0$ for $w \in \tilde{W}$ with $\ell(w) \gg 0$.

We have the following characterization of supersingular modules.

PROPOSITION 3.24. *Let* $M \in R(\tilde{\mathcal{H}}_0)$. *The following conditions are equivalent:*

*(1)* $M$ *is supersingular.*

*(2)* $M \in \oplus_{(K,\chi)/\sim; K\subset\tilde{\mathbb{S}} \text{ with } W_K, W_{\tilde{\mathbb{S}}-K} \text{ finite}} \mathbb{Z}\pi_{K,\chi}$.

*(3)* $Tr(\overline{\tilde{H}_0}^{nrig} + \iota(\overline{\tilde{H}_0}^{nrig}), M) = 0$, *where* $\iota$ *is the involution on* $\tilde{\mathcal{H}}_0$ *defined in* [**Vig05**, Corollary 2].

Here the equivalence between (1) and (2) is first obtained by Vignéras in [**Vig14**]. The criterion (3) for the supersingular modules is given in [**HNx**, Proposition 5.4] and a new proof of the equivalence between (1) and (2) is also given in loc.cit.

## References


[Ar93] J. Arthur, *On elliptic tempered characters*, Acta Math. **171** (1993), 73–138.

[BDK86] J. Bernstein, P. Deligne, D. Kazhdan, *Trace Paley-Wiener theorem for reductive p-adic groups*, J. d'Analyse Math. **47** (1986), 180–192.

[Be98] R. Bezrukavnikov, *Homological properties of representations of p-adic groups related to geometry of the group at infinity*, Ph.D. thesis, Tel Aviv University, 1998.

[BL94] L. Barthel and R. Livné, *Irreducible modular representations of GL(2) of a local field*, Duke Math. J. **75** (1994), 261–292.

[BL95] L. Barthel, R. Livné, *Modular representations of GL(2) of a local field: the ordinary, unramified case*, J. Number Theory **55** (1995), 1–27.

[BR08] C. Bonnafé and R. Rouquier, *Affineness of Deligne-Lusztig varieties for minimal length elements*, J. Algebra **320** (2008), no. 3, 1200–1206.

[BSx] B. Bhatt and P. Scholze, *Projectivity of the Witt vector affine Grassmannian*, arXiv:1507.06490.

[BT72] F. Bruhat and J. Tits, *Groupes rèductifs sur un corps local, I*, Publ. Math. IHES **41** (1972), 5–276.

[CH16] D. Ciubotaru and X. He, *The cocenter of graded affine Hecke algebras and the density theorem*, J. Pure Appl. Algebra **220** (2016), 382–410.

[CHx] D. Ciubotaru and X. He, *Cocenters and representations of affine Hecke algebras*, arXiv:1409.0902, to appear in JEMS.

[DL76] P. Deligne and G. Lusztig, *Representations of reductive groups over finite fields*, Ann. of Math. (2) **103** (1976), no. 1, 103–161.

[Ga10] Q. Gashi, *On a conjecture of Kottwitz and Rapoport*, Ann. Sci. École Norm. Sup. (4) **43** (2010), no. 6, 1017–1038.

[GH10] U. Görtz and X. He, *Dimensions of affine Deligne-Lusztig varieties in affine flag varieties*, Documenta Math. **15** (2010), 1009–1028.

[GH15] U. Görtz and X. He, *Basic loci in Shimura varieties of Coxeter type*, Cambridge Journal of Mathematics **3** (2015), 323–353.

[GHKR06] U. Görtz, T. Haines, R. Kottwitz, D. Reuman, *Dimensions of some affine Deligne-Lusztig varieties*, Ann. Sci. École Norm. Sup. (4) **39** (2006), 467–511.

[GHKR10] U. Görtz, T. Haines, R. Kottwitz, D. Reuman, *Affine Deligne-Lusztig varieties in affine flag varieties*, Compositio Math. **146** (2010), 1339–1382.

[GHN15] U. Görtz, X. He, S. Nie, *P-alcoves and nonemptiness of affine Deligne-Lusztig varieties*, Ann. Sci. Ècole Norm. Sup. **48** (2015), 647–665.

[GJ11] M. Geck and N. Jacon, *Representations of Hecke algebras at roots of unity*, Springer-Verlag London, Ltd., London, 2011.

[GKP00] M. Geck, S. Kim, and G. Pfeiffer, *Minimal length elements in twisted conjugacy classes of finite Coxeter groups*, J. Algebra **229** (2000), no. 2, 570–600.

[GM97] M. Geck and J. Michel, *"Good" elements of finite Coxeter groups and representations of Iwahori-Hecke algebras* Proc. London Math. Soc. (3) **74** (1997), 275–305.

[GP93] M. Geck and G. Pfeiffer, *On the irreducible characters of Hecke algebras*, Adv. Math. **102** (1993), no. 1, 79–94.

[GP00] M. Geck and G. Pfeiffer, *Characters of finite Coxeter groups and Iwahori-Hecke algebras*, London Mathematical Society Monographs. New Series, vol. 21, The Clarendon Press Oxford University Press, New York, 2000.

[Ha05] T. Haines, *Introduction to Shimura varieties with bad reduction of parahoric type*, Clay Mathematics Proceedings **4** (2005), 583–642.

[Hax] T. Haines, *Dualities for root systems with automorphisms and applications to non-split groups*, arXiv:1604.01468.

[HHx] T. Haines and X. He, *Vertexwise criteria for admissibility of alcoves*, arXiv:1411.5450, to appear in Amer. J. Math.

[Ha15] P. Hamacher, *The dimension of affine Deligne-Lusztig varieties in the affine Grassmannian of unramified groups*, Int. Math. Res. Notices **23** (2015), 12804–12839.

[HL12] X. He and G. Lusztig, *A generalization of Steinberg's cross-section*, J. Amer. Math. Soc. **25** (2012), 739–757.

[HRx] X. He and M. Rapoport, *Stratifications in the reduction of Shimura varieties*, arXiv:1509.07752, to appear in manuscripta mathematica.

[He07a] X. He, *The G-stable pieces of the wonderful compactification*, Trans. Amer. Math. Soc. **359** (2007), 3005–3024.

[He07b] X. He, *Minimal length elements in some double cosets of Coxeter groups*, Adv. Math. **215** (2007), 469–503.

[He11] X. He, *Closure of Steinberg fibers and affine Deligne-Lusztig varieties*, Int. Math. Res. Notices **14** (2011), 3237–3260.

[He14] X. He, *Geometric and homological properties of affine Deligne-Lusztig varieties*, Ann. Math. **179** (2014), 367–404.

[He15] X. He, *Centers and cocenters of* 0*-Hecke algebras*, Representations of reductive groups, 227–240, Prog. Math. Phys., 312, Birkhäuser/Springer, Cham, 2015.

[Hex] X. He, *Kottwitz-Rapoport conjecture on unions of affine Deligne-Lusztig varieties*, arXiv:1408.5838, to appear in Ann. Sci. Ècole Norm. Sup.

[Hexx] X. He, *Minimal length elements in conjugacy classes of extended affine Weyl group*, arXiv:1004.4040, unpublished note.

[HN12] X. He and S. Nie, *Minimal length elements of finite Coxeter groups*, Duke Math. J. **161** (2012), 2945–2967.

[HN14] X. He and S. Nie, *Minimal length elements of extended affine Weyl group*, Compos. Math. **150** (2014), 1903–1927.

[HN15] X. He and S. Nie, *$P$-alcoves, parabolic subalgebras and cocenters of affine Hecke algebras*, Selecta. Math. **21** (2015), 995–1019.

[HNx] X. He and S. Nie, *Cocenters and representations of affine* 0*-Hecke algebras*, arXiv:1502.02184.

[IM65] N. Iwahori and H. Matsumoto, *On some Bruhat decomposition and the structure of the Hecke rings of p-adic Chevalley groups*, Inst. Hautes Études Sci. Publ. Math. (1965), no. 25, 5–48.

[Kaz86] D. Kazhdan, *Representations groups over close local fields*, Journal d'Analyse Mathematique **47** (1986), 175–179.

[KL87] D. Kazhdan and G. Lusztig, *Proof of the Deligne Langlands conjecture for Hecke algebras*, Invent. Math. **87** (1987), 153–215.

[Ko85] R. Kottwitz, *Isocrystals with additional structure*, Compositio Math. **56** (1985), no. 2, 201–220.

[Ko97] R. Kottwitz, *Isocrystals with additional structure. II*, Compositio Math. **109** (1997), no. 3, 255–339.

[Ko03] R. Kottwitz, *On the Hodge-Newton decomposition for split groups*, Int. Math. Res. Not. **26** (2003), 1433–1447.

[Ko06] R. Kottwitz, *Dimensions of Newton strata in the adjoint quotient of reductive groups*, Pure Appl. Math. Q. **2** (2006), 817–836.

[KR03] R. Kottwitz and M. Rapoport, *On the existence of F-crystals*, Comment. Math. Helv. **78** (2003), 153–184.

[Luc04] C. Lucarelli, *A converse to Mazur's inequality for split classical groups*, J. Inst. Math. Jussieu **3** (2004), no. 2, 165–183.

[Lu89] G. Lusztig, *Affine Hecke algebras and their graded version*, J. Amer. Math. Soc. **2** (1989), 599–635.

[Lu03] G. Lusztig, *Hecke algebras with unequal parameters*, CRM Monograph Series, 18. American Mathematical Society, Providence, RI, 2003. vi+136 pp.

[Lu04a] G. Lusztig, *Parabolic character sheaves. I*, Mosc. Math. J. **4** (2004), 153–179, 311.

[Lu04b] G. Lusztig, *Parabolic character sheaves. II*, Mosc. Math. J. **4** (2004), 869–896, 981

[Lu10] G. Lusztig, *Parabolic character sheaves. III*, Mosc. Math. J. **10** (2010), 603–609.

[Lu11] G. Lusztig, *From conjugacy classes in the Weyl group to unipotent classes*, Represent. Theory **15** (2011), 494–530.

[Ma73] B. Mazur, *Frobenius and the Hodge filtration (estimates)*, Ann. of Math. (2) **98** (1973), 58–95.

[Ni13] S. Nie, *Minimal length elements of Coxeter groups*, J. Algebra **384** (2013), 135–142.

[No79] P. N. Norton, 0-*Hecke algebras*, J. Austral. Math. Soc. Ser. A **27** (1979), no. 3, 337–357.

[Oo01] F. Oort, *A stratification of a moduli space of abelian varieties*, in: Moduli of abelian varieties (Texel Island, 1999), 345–416, Progr. Math., 195, Birkhäuser, Basel, 2001.

[OR08] S. Orlik and M. Rapoport, *Deligne-Lusztig varieties and period domains over finite fields*, J. Algebra **320** (2008), no. 3, 1220–1234.

[OS09] E. Opdam and M. Solleveld, *Homological algebra of affine Hecke algebras*, Adv. Math. **220** (2009), 1549–1601.

[OS10] E. Opdam and M. Solleveld, *Discrete series characters for affine Hecke algebras and their formal degrees*, Acta Math. **205** (2010), 105–187.

[PZ13] G. Pappas and X. Zhu, *Local models of Shimura varieties and a conjecture of Kottwitz*, Invent. Math. **194** (2013), 147–254.

[Ra05] M. Rapoport, *A guide to the reduction modulo p of Shimura varieties*, Astérisque (2005), no. 298, 271–318.

[Re01] M. Reeder, *Euler-Poincaré pairings and elliptic representations of Weyl groups and p-adic groups*, Compositio Math. **129** (2001), 149–181.

[RR96] M. Rapoport and M. Richartz, *On the classification and specialization of $F$-isocrystals with additional structure*, Compositio Math. **103** (1996), no. 2, 153–181.

[Rox] S. Rostami, *Conjugacy classes of non-translations in affine Weyl groups and applications to Hecke algebras*, Trans. Amer. Math. Soc. **368** (2016), 621–646.

[SS97] P. Schneider and U. Stuhler, *Representation theory and sheaves on the Bruhat-Tits building*, Publ. I.H.E.S. **85** (1997), 97–191.

[Ti57] J. Tits, *Sur les analogues algébriques des groupes semi-simples complexes*, in Colloque d'algébre supérieure, 261–289, 1957.

[Vie06] E. Viehmann. *The dimension of affine Deligne-Lusztig varieties*, Ann. Sci. École Norm. Sup. (4) **39** (2006), 513–526.

[Vie14] E. Viehmann, *Truncations of level 1 of elements in the loop group of a reductive group*, Ann. of Math. (2) **179** (2014), 1009–1040.

[Vig98] M.-F. Vignéras, *Induced representations of reductive p-adic groups in characteristic $\lambda \neq p$*, Selecta Math. **4** (1998), 549–623.

[Vig02] M.-F. Vignéras, *Modular representations of p-adic groups and of affine Hecke algebras*, Proceedings of the International Congress of Mathematicians, Vol. II (Beijing, 2002), Higher Ed. Press, Beijing, 2002, 667–677.

[Vig05] M.-F. Vignéras, *Pro-p-Iwahori Hecke ring and supersingular $\bar{\mathbb{F}}_p$-representations*, Math. Ann. **331** (2005), no. 3, 523–556. Erratum in 333 (2), 699–701.

[Vig14] M.-F. Vignéras, *The pro-p-Iwahori-Hecke algebra of a reductive p-adic group III*, J. Inst. Math. Jussieu (2015), 1–38.

[Wi05] J.-P. Wintenberger, *Existence de $F$-cristaux avec structures supplémentaires*, Adv. Math. **190** (2005), 196–224.

[Xi07] N. Xi, *Representations of affine Hecke algebras and based rings of affine Weyl groups*, J. Amer. Math. Soc. **20** (2007), 211–217.

[Yax] Z. Yang, *Class polynomials for some affine Hecke algebras*, J. Algebra **452** (2016), 502–548.

[Zhx] X. Zhu, *Affine Grassmannians and the geometric Satake in mixed characteristic*, arXiv:1407.8519.



Department of Mathematics, University of Maryland, College Park, MD 20742

*E-mail address*: xuhuahe@math.umd.edu

Current Developments in Mathematics, 2015

# Green function, mean field equation and Painlevé VI equation

Chang-Shou Lin

Abstract. Recently jointly with C.L. Chai, Z.J. Chen, T.J. Kuo and C.L. Wang, we have developed a theory to connect different subjects such as (multiple) Green functions, mean field equations, Lamé equation, hyperelliptic curves, modular forms and Painlevé VI equation. Among others in those joint works, the premodular forms $Z_{r,s}^{(n)}(\tau)$ occupies the central role. We will discuss its role from the aspects of differential geometry (Green function), conformal geometry (mean field equations) and the classical integral Lamé equation. The key issue is to study the function properties of $Z_{r,s}^{(n)}(\tau)$ such as the simple zero properties and the asymptotics as $\tau \to +\infty$. Unexpectedly, Painlevé VI is the essential tool for the proof of these properties. The Okamoto transformation for Painlevé VI is used to set up the induction step. This is the key for many difficult computation. We also show how to apply the Painlevé properties to study the degenerate curves for branch points of the hyperelliptic curve $Y^{(n)}(\tau)$, $n = 2$.

Contents



## 1. Introduction

Throughout the paper, we fix the following notations $\omega_0 = 0$, $\omega_1 = 1$, $\omega_2 = \tau$, $\omega_3 = 1 + \tau$, $\Lambda_\tau = \mathbb{Z} + \tau\mathbb{Z}$, and $E_\tau \doteqdot \mathbb{C}/\Lambda_\tau$ where $\tau \in \mathbb{H} = \{\tau | \text{Im}\tau > 0\}$. For a point $a \in \mathbb{C}$, we often denote it by $[a]$ to emphasize that this point

$[a]$ is considered to be the point $a + \Lambda_\tau$ in the torus $E_\tau$. We also define $E_\tau[2] \doteqdot \left\{\frac{\omega_i}{2} \middle| i = 0,1,2,3\right\}$ to be the set of the lattice point and 2-torsion points in $E_\tau$.

We recall $\wp(z|\tau)$, $\zeta(z|\tau)$ and $\sigma(z)$ to be defined by:

$$\wp(z|\tau) = \frac{1}{z^2} + \sum_{\omega \in \Lambda_\tau \setminus \{0\}} \left( \frac{1}{(z-\omega)^2} - \frac{1}{\omega^2} \right),$$

$$\zeta'(z|\tau) = -\wp, \text{ and } \zeta(z) = \frac{\sigma'(z)}{\sigma(z)}.$$

Here $\zeta(z|\tau)$ is odd and quasi-periodic:

$$\zeta(z + \omega_i|\tau) = \zeta(z) + \eta_i(\tau), \quad i = 1,2,3.$$

We refer [**1**] for the introduction to Weierstrass elliptic functions. Let $G(z|\tau)$ be the Green function on $E_\tau$:

$$\begin{cases} \Delta G(z) = \delta_0 - \frac{1}{|E_\tau|}, \\ \int G(z) \mathrm{d}x\mathrm{d}y = 0, \end{cases} \tag{1.1}$$

where $\delta_0$ is the Dirac measure at 0 and $|E_\tau|$ is the area of $E_\tau$.

In this article, we want to study critical points of $G^{(n)}(z|\tau)$ on $E_\tau^n = E_\tau \times \cdots \times E_\tau$, where

$$G^{(n)}(z|\tau) = n \sum_{i=1}^{n} G(z_i) - \sum_{i<j} G(z_i - z_j), \tag{1.2}$$

under the deformation of $\tau$.

There are two types of critical points. A critical point $(a_1, \cdots, a_n)$ is called *trivial* if $\{a_1, \cdots, a_n\} = \{-a_1, \cdots, -a_n\}$, otherwise, it is called *non-trivial*. For each $n \in \mathbb{N}$, $G^{(n)}$ has $(2n+1)$ trivial critical points. Locally each trivial critical point depends on $\tau$ holomorphically. A trivial critical point $a$ is called *degenerate* if the Hessian of $G^{(n)}(z)$ at $z = a$ is zero. We call the set of $\tau$ where trivial critical points $a(\tau)$ are degenerate as the degenerate curve in $\mathbb{H}$. The fundamental issue is whether the degenerate curve is smooth or not. This problem seems not trivial at all. It is unexpected that we could answer this questions by applying the Painlevé property of some Painlevé VI equation.

Indeed the study of critical points are related to mean field equations, Lamé equations, hyper-elliptic curves, modular forms and Painlevé VI equation. This is the purpose of the article to survey the links with these subjects.

In section 2, we discuss the connection with the mean field equations. A critical point $a = (a_1, \cdots, a_n)$ of $G^{(n)}$ is related to the blowup set of a sequence of blowing up solutions of mean field equations. Indeed, our motivation for this project has been originated from studying the mean field equation at critical parameters. The mean field equation considered in section 2 is also an integrable system by the Liouville theorem. From this point of view, the mean field equation is related to the Lamé equation. See section 5.

The set of critical points for $n = 1$ are completely understand. See [**27**] and [**10**]. However even for $n = 2$, the number of non-trivial critical points is still unsettled. In section 2, we will discuss the cases $n = 1$ and $n = 2$ for this issue.

In section 3, we consider the classical Lamé equation and the hyper-elliptic curve $Y_n(\tau)$ which is embedded in $E_\tau^n$, $n$ copies of $E_\tau$. The structure of the closure of $Y_n(\tau)$ has been thoroughly studied in [**6**] and [**27**]. Via the embedding into $E_\tau^n$, we naturally introduce the addition map $\sigma_n$ from $\overline{Y_n}(\tau)$ to $E_\tau$. In [**27**], the formula of degree $\sigma_n = \frac{n(n+1)}{2}$ has been obtained. This implies the function field $K(\overline{Y_n}(\tau))$ is a finite extension of $K(E_\tau)$ and

$$[K(\overline{Y_n}(\tau)) : K(E_\tau)] = \frac{n(n+1)}{2}.$$

The primitive element of this extension is denoted by $z_n(a)$ and we prove

$$z_n(a) = \zeta(\sum_{i=1}^{n} a_i) - \sum_{i=1}^{n} \zeta(a_i).$$

Therefore, there is a minimal polynomial $W_n(X)$ such that $W_n(z_n(a)) \equiv 0$ in $\overline{Y_n}(\tau)$. Set

$$Z_{r,s}^{(n)}(\tau) = W_n(Z_{r,s}(\tau)),$$

where $Z_{r,s}(\tau) = \zeta(r + s\tau \mid \tau) - r\eta_1(\tau) - s\eta_2(\tau)$. We note that the pair $(r, s)$ is the monodromy data of the Lamé equation.

In section 4, we see that $Z_{r,s}^{(n)}(\tau)$ is modular form of weight $\frac{n(n+1)}{2}$ w.r.t $\Gamma(N)$ provided that $(r, s)$ is a $N-$torsion point. The question:

Is every zero of $Z_{r,s}^{(n)}(\tau)$ simple?

had arisen in [**27**]. Interestingly, this question is answered affirmatively in [**9**]. The proof is to apply the Painlevé property of some Painlevé equation.

In section 6, we briefly overview the basic theory of Painlevé VI equation and its elliptic form. We refer the readers to [**23**] for the excellent introduction of Painlevé VI equations. Two important features of Painlevé VI are: Painlevé property and the isomonodromic deformation. Clearly, Painlevé property could be applied only if the solutions has explicit expressions. In literature, the Hitchin theorem is the first result in this direction (except for the Picard solutions). In § 6, we give two proofs of this remarkable result: one uses the isomodromic deformation, the other one is to apply the Okamoto transformation. By the Hitchin theorem, we immediately see how the algebraic solutions related to the Hecke form $Z_{r,s}(\tau)$. The Hecke form $Z_{r,s}(\tau)$ is a modular form if $(r, s)$ is $N-$torsion point. Hecke proved that $Z_{r,s}(\tau)$ is the Eisenstein series of weight 1 up to the constant $N$.

The Green function $G(z|\tau)$ has either three or five critical points. This was proved in [**28**]. Indeed $G(z|\tau_0)$ has five critical points iff there exists a real pair $(r, s) \in \mathbb{R}^2 \backslash \frac{1}{2}Z^2$ such that $Z_{r,s}(\tau_0) = 0$ holds. Denote this set of $\tau$

mod $\mathrm{SL}_2(z)$ by $\Omega^{(5)}$, and the interior of its complimentary set mod $\mathrm{SL}_2(Z)$ by $\Omega^{(3)}$. In [**10**], we prove both $\Omega^{(3)}$ and $\Omega^{(5)}$ are simply connected. In § 6, we give a survey of this result.

This simple zero property of $Z_{r,s}^{(n)}(\tau)$ can be proved by applying the Painlevé VI equation. To do it, we need a generalization of Hitchin theorem. This is the main subject discussed in § 7. In § 7, we also discuss the explicit expression of non-completely reducible solutions for $n = 2$. Form it, we prove the smoothness of degenerate curve for the trivial critical points of $G^{(2)}(z|\tau)$. So, we have seen that both completely reducible solutions or non-completely reducible solution (the so-called Riccati type solutions) are useful for studying critical points of $G^{(n)}(z|\tau)$.

In the last section, we consider

$$M_N^{(n)}(\tau) = \prod_{(r,s)\in Q_N} Z_{r,s}^{(n)}(\tau).$$

We calculate the vanishing order $\nu_\infty(M_N^{(n)})$ of $M_N^{(n)}(\tau)$ at $\tau = \infty$ by using the Okamoto transformation. The function $M_N^{(n)}(\tau)$ is a modular form of weight $\frac{n(n+1)|Q_N|}{2}$ of level 1. The Okamoto transformation provides an induction process on $n$. Using the formula of $\nu_\infty(M_N^{(n)})$, we solve the conjecture proposed by Dahmen and Beukers for the number of Lamé equations, under scalar equivalence, whose monodromy group is the dihedral group $D_N$ of order $2N$.

## 2. Green function and mean field equation

**2.1. Critical points of multiple Green function $E_\tau^N$.** My original motivation is to understand the following problem: Consider $a = (a_1, \cdots, a_N) \in E_\tau \times \cdots \times E_\tau$ satisfies

$$\sum_{\ell=0}^{3} n_\ell \nabla G(a_i - \frac{\omega_\ell}{2}) = \sum_{j\neq i}^{N} \nabla G(a_i - a_j), \quad 1 \le i \le N, \tag{2.1}$$

where $n_\ell \in \mathbb{N} \cup \{0\}$ and $N = \sum_{\ell=0}^{3} n_\ell$. Obviously, $a$ can be considered as a critical point of the Green function $G^{(\vec{n})}(z)$ on $E_\tau^N$:

$$G^{(\vec{n})}(z) = \sum_{i=1}^{N}\sum_{\ell=0}^{3} n_\ell G(z_i - \frac{\omega_\ell}{2}) - \sum_{i<j}^{N} G(z_i - z_j),$$

where $\vec{n} = (n_0, n_1, n_2, n_3)$ and $z = (z_1, \ldots, z_n) \in E_\tau^N$.

Equation (2.1) is often related to the bubbling phenomena of a nonlinear elliptic PDE. For example, we consider

$$\Delta u + e^u = 8\pi \sum_{i=0}^{3} n_i \delta\left(\frac{\omega_i}{2}\right) \text{ in } E_\tau, \tag{2.2}$$

where $n_i \in \mathbb{N} \cup \{0\}$, $\delta(p)$ is the Dirac measure at $p$. We call $u$ is a solution of (3.1) if $u \in C^\infty(E_\tau \backslash E_\tau[2])$ satisfies

$$\Delta u + e^u = 0 \text{ in } E_\tau \backslash E_\tau[2]$$

and

$$u(z) = -2n_i \log |z - q| + O(1) \text{ as } z \to q \in E_\tau[2].$$

In conformal geometry, a solution $u$ of (1.2) is equivalent to saying that the Gaussian curvature of the new metric $\frac{1}{2}e^u|dz|^2$ is equal to 1, but it has conic singularity at $\frac{\omega_i}{2}$. In this paper, we always assume $n_i$ are positive integers.

Suppose $u_k$ is a sequence of solutions of (3.1). A point $a \in E_\tau$ is called a blowup point if there are $x_k$ such that $x_k \to a$ and $u_k(x_k) \to +\infty$. Now assume $u_k$ blows up at $a_1, \cdots, a_m$. Assume $a_i \notin E_\tau[2]$. Then we have the following result from PDE:

THEOREM 2.1. [**11, 12, 13**] *Suppose $\{a_1, \ldots, a_m\}$ is the blowup set of a sequence of solutions of (2.2) and $a_i \notin E_\tau[2]$. Then*

*(i) $m = N = \sum_{i=0}^3 n_i$, and*
*(ii) $a_1, \cdots, a_N$ satisfies the balance condition (2.1).*

The fact (i) is due to the quantization property of the entire solution of (2.2), a non-trivial result from PDE theory and the balance condition (2.1) is derived from the well-known Pohozaev identity.

**2.2. Non-trivial critical points of Green function on $E_\tau$.** Let $n = (1, 0, 0, 0)$. Then $N = 1$ and (2.1) is reduced to

$$\nabla G(a) = 0, \tag{2.3}$$

i.e., $a$ is a critical point of the Green function $G(z)$ on $E_\tau$. The Green function $G(z)$ is **even**, i.e., $G(-z) = G(z)$, and has singularity only at 0, the lattice point. Thus, $G$ has at least three trivial critical points at half periods $\frac{\omega_1}{2}$, $\frac{\omega_2}{2}$ and $\frac{\omega_3}{2}$.

A critical point $a$ of $G$ is said *non-trivial* if $a \notin E_\tau[2]$. The corresponding mean field equation is

$$\Delta u + e^u = 8\pi\delta_0 \quad \text{in } E_\tau. \tag{2.4}$$

Then we have the following connection of (2.3) and (2.4):

THEOREM 2.2. [**28**] *The Green function $G(z)$ of $E_\tau$ has a non-trivial critical point iff (2.4) has a solution.*

Naturally, we ask: How many non-trivial critical points at $E_\tau$? This question was answered in [**28**]:

THEOREM 2.3. [**28**] *For each $\tau$, $G$ has at most one pair of non-trivial critical points $\pm a$.*

In [**28**], Theorem 2.3 was proved by using Alexander-Bol's inequality and the isoparametrical inequality. Later, Bergweiler and Eremenko [**3**] gave a proof by the dynamic arguments. If $\tau = ib, b > 0$, then $G$ has no non-trivial critical points and then (2.4) has no solutions. But for $\tau = e^{i\pi/3}$, $G$ has only one pair of non-trivial critical points $\pm\frac{\omega_3}{3}$ and (2.4) has an unique solution.

We remark that the proof of Theorem 2.3 is a consequence of the following PDE result:

(2.5) *For any* $\tau \in \mathbb{H}$, (2.4) *has at most one even solution.*

This uniqueness for (2.4) is a highly non-trivial result. See [**28**]

In order to understand the set of critical points, one might deform $\tau$ starting from the imaginary axis. During this deformation process, we want to study the following questions:

(a) Where the bifurcation of critical point might occur?
(b) What is the geometry of those $\tau$ such that the Green function has non-trivial critical points.

The question (a) can be answered by Theorem 2.4.

THEOREM 2.4. *The bifurcation of critical points of Green function could occurs only at trivial critical points (half periods).*

REMARK 2.5. Suppose $G(z|\tau)$ has non-trivial critical points. Then $G(z|\tau'), \tau' = \gamma\cdot\tau, \gamma \in \mathrm{SL}(2,\mathbb{Z})$, also has non-trivial critical points. This is due to that fact $E_\tau$ and $E_{\tau'}$ are holomorphically equivalent.

Let $\Omega_5 = \{\tau \in \mathbb{H} | G(z|\tau) \text{ has non-trivial critical points}\}$. By the remark, $\Omega_5$ is invariant by the action of $\mathrm{SL}(2,\mathbb{Z})$. For some reason from the analytic point of view, we consider $\tau \in F_2$:

$$F_2 = \{\tau \in \mathbb{H} | 0 \le \operatorname{Re}\tau \le 1, |\tau - \frac{1}{2}| \ge \frac{1}{2}\}.$$

It is not difficult to see that $F_2$ is the closure of a fundamental domain of $\Gamma_0(2)$, where

$$\Gamma_0(2) = \left\{\gamma \in \mathrm{SL}(2,\mathbb{Z}) | \gamma = \begin{pmatrix} a & b \\ c & d \end{pmatrix}, c \equiv 0 \mod 2\right\}.$$

For the simplicity, the set $\Omega_5 \cap F_2$ is still denoted by $\Omega_5$ if there are no confusions. Similarly, we denote $\Omega_3$ by

$$\Omega_3 = \{\tau \in F_2 | G(z|\tau) \text{ has no non-trivial critical points}\}.$$

Then we have the following theorem to describe the geometry of $\Omega_5$ and $\Omega_3$.

THEOREM 2.6. [**10**]

*(i)* $\Omega_5$ *is open and simply connected.*
*(ii)* $\partial\Omega_5 = \partial\Omega_3$ *consists of three smooth curves* $C_i$, $i = 1, 2,$ *and* $3$, *where* $C_i = \{\tau \in F_2 | \frac{\omega_i}{2}$ *is a degenerate critical point of* $G(z|\tau)\}$.
*(iii)* $C_i \cap C_j = \emptyset$

We remark that (iii) implies that for any torus the Green function has at most one degenerate half period. This seems new in literature.

The proof of Theorem 2.6 is interesting in itself, a combination of some PDE results together with a simple fact of theory of modular form with respect to $\mathrm{SL}(2,\mathbb{Z})$:

*Any modular form of weight 8 has only one zero at* $\tau = e^{\frac{i\pi}{3}}$.

The complete proof of Theorem 2.6 was given in [**10**]. In section 4, we will give a sketch of the proof after the discussion of connection of our problems with modular forms.

Naturally, for the general case we ask the similar questions to understand the solution structure of (2.1) when $\tau$ deforms in $\mathbb{H}$. Understandably, both questions (a) and (b) would be not easy to answer, because equation (2.1) is not an algebraic equation and the uniqueness of solutions of

$$\Delta u + e^u = 8\pi n \delta_0 \text{ in } E_\tau$$

no longer holds true for $n \geqslant 2$. Even through we still conjecture that Theorem 2.4 should be true for $n \geqslant 2$. But this is a very difficult question.

For question (b), we could say more than (a). Indeed, we will apply the Painlevé VI equations to prove the degenerate curves (in $\tau$) of trivial critical points are smooth. In section 7, we will use $n = 2$ as an example to explain how to apply the Painlevé VI equation for our purpose.

Equation (2.1) can be decomposed into two parts: an algebraic part and an non-algebraic part (one single equation). The equation for the algebraic part will give a hyper-elliptic structure, and it turns out this hyper-elliptic curve is identical with the spectral curve for the Treibich-Verdier potential in KdV theory. Let us to denote the hyper-elliptic curve by $Y^{(\vec{n})}$ and $X^{(\vec{n})} = Y^{(\vec{n})} \setminus \{\text{branch points}\}$. So, (2.1) is reduced to a single equation defined on $Y^{(\vec{n})}$. Like the case $n = (1, 0, 0, 0)$, all the branch points of $Y^{(\vec{n})}$ are solutions to this single equation. See the Remark 3.3. We call $a = (a_1, \ldots, a_N) \in Y^{(\vec{n})}$ is a nontrivial solution of (2.1) if $a$ is not a branch point.

Geometrically, this single non-algebraic equation yields the global monodromy data for some linear ODE. For the multiple Green function, this ODE is the Lamé equation, which is the main subject discussed in section 5.

## 3. Hyper-elliptic curves

**3.1. The Green function and $Y^{(n)}(\tau)$.** In this section, we want to introduce the hyper-elliptic curve $Y^{(n)}(\tau)$ associated with the multiple Green function through (2.1) and its embeddings into $E_\tau^n$. In this case, (2.1) becomes

$$n \nabla G(a_i) = \sum_{j \neq i}^{n} \nabla G(a_i - a_j), \; 1 \leq i \leq N = n. \tag{3.1}$$

Clearly, any solution $a = (a_1, \ldots, a_n)$ of (3.1) is a critical point of $G^{(n)}(z|\tau)$, where

$$G^{(n)}(z|\tau) = n\sum_{i=1}^{n} G(z_i|\tau) - \sum_{i<j}^{n} G(z_i - z_j|\tau)$$

for $z = (z_1, \ldots, z_n) \in E_\tau^n$.

First we recall $\zeta(z|\tau) = -\int \wp(z|\tau)dz$ and $\eta_i(\tau)$, $i = 1, 2$, are the quasi-periods. The Green function can be written in terms of Weierstrass elliptic functions, see (3.7) and (3.8) below. In particular,

$$\frac{\partial G(z|\tau)}{\partial z} = \zeta(z|\tau) - \eta(z|\tau), \tag{3.2}$$

where $\eta(z)$ is defined: if $z = r + s\tau$ for $r, s \in \mathbb{R}$, then

$$\eta(r + s\tau) = r\eta_1(\tau) + s\eta_2(\tau).$$

By using (3.2), (3.1) can be rewritten as:

PROPOSITION 3.1 ([**6**]). *Let $a \in E_\tau^n$. Then $a = (a_1, \cdots, a_n)$ satisfies (3.1) if and only if*

$$\sum_{j\neq i}^{n} (\zeta(a_i - a_j) + \zeta(a_j) - \zeta(a_i)) = 0, \ 1 \leq i \leq n, \ \textit{and} \tag{3.3}$$

$$\sum_{i=1}^{n} \nabla G(a_i) = 0. \tag{3.4}$$

The equivalence of (3.1) and (3.3)–(3.4) are straightforward computations. See [**6**]. The first equation, equation (3.3), is an algebraic equation by the addition theorem. Let us define $Y^{(n)}(\tau)$ is the quotient of the solution set of (3.3) by the permutation $S_n$:

$$\{a \in E_\tau^n | a_i \neq 0 \ \forall i \text{ and } a_i \neq a_j \ \forall i \neq j \text{ such that (3.3) holds}\}/S_n.$$

Then in [**6**] we could prove

THEOREM 3.2. [**6**]

(i) *$Y^{(n)}(\tau)$ is a hyper-elliptic curve, i.e., there exists $\ell_n(B) \in Q[g_2, g_3, B]$ of degree $2n + 1$ such that $C^2 = \ell_n(B)$ holds, where*

$$C = \wp'(a_i)\Pi_{j\neq i}(\wp(a_i) - \wp(a_j)) \text{ and } B = (2n-1)\sum_{i=1}^{n} \wp(a_i).$$

(ii) *Except for a finite set of $\tau$, $Y^{(n)}(\tau)$ is smooth.*

(iii) *The branch points are the set of $(a_1, \cdots, a_n)$ such that $\{a_1, \cdots, a_n\} = \{-a_1, \cdots, -a_n\}$.*

(iv) *Let $a = (a_1, \ldots, a_n) \in Y^{(n)}(\tau)$. Then $a$ is not branch point iff $\{a_1, \cdots, a_n\} \cap \{-a_1, \cdots, -a_n\} = \emptyset$.*

(v) *Let $\overline{Y^{(n)}}(\tau)$ be the closure in $E_\tau^n$. Then $\overline{Y^{(n)}}(\tau) = Y^{(n)}(\tau) \cup \{(0, \cdots, 0)\}$.*

(vi) $\overline{Y^{(n)}}(\tau)$ *is the spectral curve of the potential* $I_0(z) = n(n+1)\wp(z)$ *in the KdV theory.*

REMARK 3.3. By (iii), any component $a_i$ of a branch point $a = (a_1, \cdots, a_n)$ satisfies either $a_i \in E_\tau[2]$ (then $\nabla G(a_i) = 0$) or the pair $\pm a_i$ appears in $a$ (then $\nabla G(a_i) + \nabla G(-a_i) = 0$). Therefore, any branch point $a$ satisfies (3.4), i.e., any branch point is a solution to (3.1). By (i), there are $2n+1$ distinct branch (finite) points if $Y^n(\tau)$ is smooth. Thus, $a = (a_1, \ldots, a_n)$ is called a *non-trivial* solution of (3.1) if $a$ is not a branch point of $Y^{(n)}(\tau)$. By (iv), we have

$$\{a_1, \ldots, a_n\} \cap \{-a_1, -a_2, \ldots, -a_n\} = \emptyset$$

if $a$ is a non-trivial solution.

**3.2. The branch points of $Y^{(2)}(\tau)$.** Now we give some examples to compute these branch points. If $n = 1$, $Y^{(n)}(\tau) = E_\tau$, and the branch points are the three half periods. For $n = 2$, there are only five branch points which can be computed as follows:

PROPOSITION 3.4. *Suppose $n = 2$. Then for each $\tau \in \mathbb{H}$, there are only five branch points of $Y^{(2)}(\tau)$ :*

$$\left(\frac{\omega_i}{2}, \frac{\omega_j}{2}\right) \; 1 \le i \ne j \le 3, \; \textit{and} \; (q_\pm, -q_\pm), \tag{3.5}$$

*where $q_\pm$ satisfies $\wp(q_\pm) = \pm\sqrt{\frac{g_2(\tau)}{12}}$.*

PROOF. Since the three half periods $\frac{1}{2}\omega_i$, $i = 1, 2, 3$, are critical points of $G(z)$ and for any $i \ne j$, $\frac{1}{2}\omega_i - \frac{1}{2}\omega_j$ is the other half period, we see that $(a, b) = (\frac{1}{2}\omega_i, \frac{1}{2}\omega_j)$, $i \ne j$, $i, j \in \{1, 2, 3\}$ satisfies (3.1). We can also find a solution $(a, b)$ for (3.1) satisfying $a = -b$ and $a$ is not a half period (i.e., $\wp'(a) \ne 0$). In fact, since $\nabla G(z)$ is odd, (3.1) is equivalent to

$$\nabla G(a) + \nabla G(b) = 0, \quad \nabla G(a) - \nabla G(b) - \nabla G(a-b) = 0. \tag{3.6}$$

On the other hand, the Green function $G$ could be expressed in terms of well-known functions:

$$G(z) = -\frac{1}{2\pi}\left(\ln|\vartheta_1(z)| - \frac{\pi}{\operatorname{Im}\tau}(\operatorname{Im} z)^2\right) + C(\tau), \tag{3.7}$$

where

$$\vartheta_1(z;\tau) \doteqdot -i \sum_{n=-\infty}^{\infty} (-1)^n e^{(n+\frac{1}{2})^2\pi i\tau} e^{(2n+1)\pi i z}$$

is the odd theta function and $C(\tau)$ is a constant; see [**6**]. From (3.7), we have

$$\begin{aligned} -4\pi G_z(z) &= \zeta(z|\tau) - \eta_1(\tau)z + \frac{2\pi i \operatorname{Im} z}{\operatorname{Im}\tau} \\ &= \zeta(z|\tau) - r\eta_1(\tau) - s\eta_2(\tau), \end{aligned} \tag{3.8}$$

where $z = r + s\tau$, $r, s \in \mathbb{R}$. In view of (3.8), the second equation in (3.6) becomes

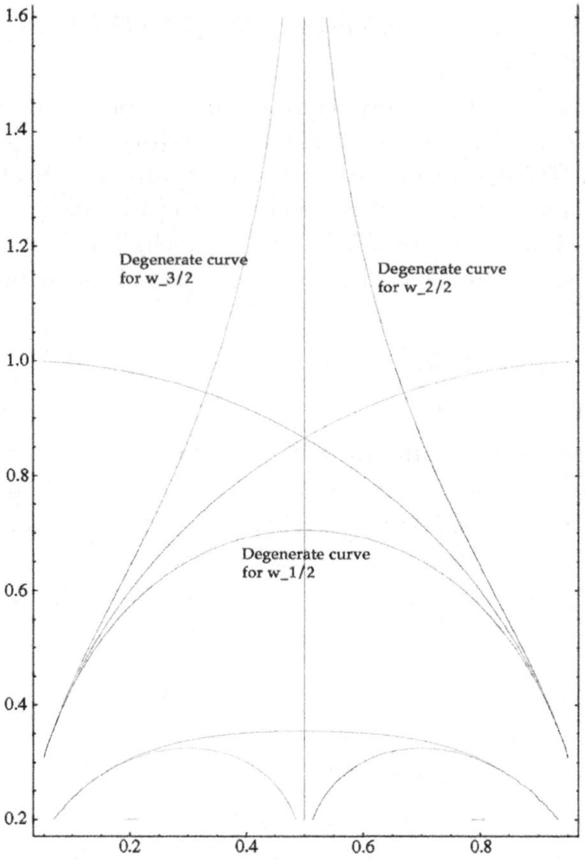


FIGURE 1.

$$\zeta(a) - \zeta(b) - \zeta(a-b) = 0. \tag{3.9}$$

If $a = -b$, then the first equation in (3.6) holds automatically, while (3.9) becomes $2\zeta(a) - \zeta(2a) = 0$. This together with $\zeta(2a) = 2\zeta(a) + \frac{1}{2}\frac{\wp''(a)}{\wp'(a)}$, implies that $a$ satisfies $\wp''(a) = 0$. Since $\wp''(z) = 6\wp^2(z) - \frac{g_2}{2}$, $a$ satisfies

$$\wp(a) = \wp(-a) = \pm\sqrt{\frac{g_2}{12}}. \qquad \square$$

REMARK 3.5. This proves $\wp(q_\pm) = \pm\sqrt{\frac{g_2}{12}}$ iff $\wp''(q_\pm) = 0$.

Let $F_2$ be the fundamental domain of $\Gamma_0(2)$:

$$F_2 = \left\{\tau \middle| 0 \le \mathrm{Re}\tau \le 1, |\tau - \frac{1}{2}| \ge \frac{1}{2}\right\},$$

For $n = 1$, let $C_i$ be the curve for those $\tau$ such that $\frac{\omega_i}{2}$ is a degenerate critical point of $G(z|\tau)$. Figure 1 is the numerical simulation of those degenerate curves, which is confirmed by Theorem 2.6.

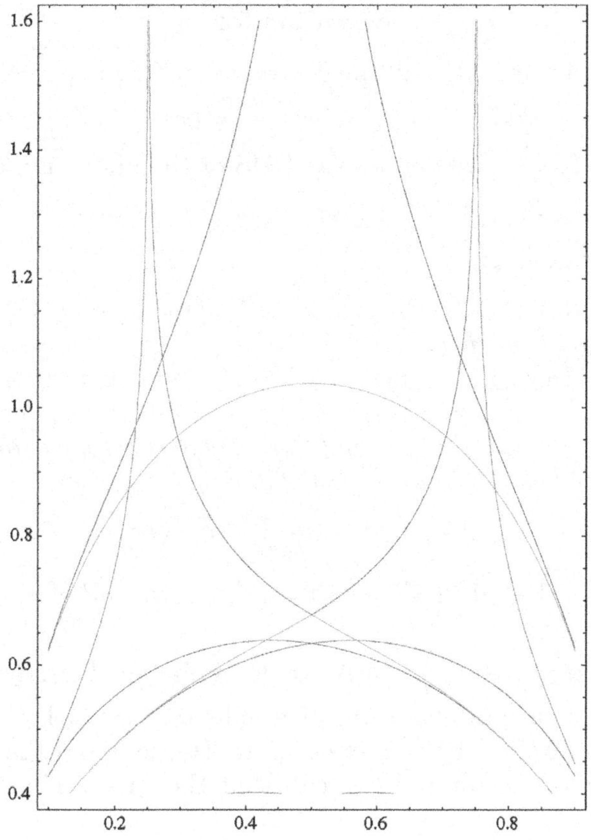


FIGURE 2.

For $n = 2$, we let $C_{ij}$ be the degenerate curve for $\left(\frac{\omega_i}{2}, \frac{\omega_j}{2}\right)$, i.e., $\left(\frac{\omega_i}{2}, \frac{\omega_j}{2}\right)$ is a degenerate critical point of the Green $G^{(2)}$ function on $E_\tau \times E_\tau$. Similarly, $C_\pm$ is the degenerate curve for $(q_\pm(\tau), -q_\pm(\tau))$.

THEOREM 3.6. *For $\tau \in F$, all the degenerate curves $C_{ij}$ and $C_\pm$ are smooth.*

By computer simulation, those curves are shown in Figure 2. The fundamental domain $F_2$ is divided into several regions by $C_{ij}$ and $C_\pm$, and in each region, the number of non-trivial solution to (3.1) should be the same. However it is still an open problem because this problem is related to the question (a) in section 2. So far, we could prove it only if the following conjecture holds true:

> *Except of a finite set of $\tau$, the bifurcation of critical points could occur only at branch points of $Y^{(n)}(\tau)$*

Theorem 3.6 is an application of the Painlevé VI equation (Riccati equation). A sketch of the proof of Theorem 3.6 is given in section 7. See Theorem 7.6.

## 4. Modular forms

**4.1. Hecke forms.** Recall the derivative of $G$: write $z = r+s\tau$, $r, s \in \mathbb{R}$

$$-4\pi G_z(z) = \zeta(r + s\tau) - r\eta_1(\tau) - s\eta_2(\tau).$$

Now fix $(r, s) \in \mathbb{C} \times \mathbb{C}$, we denote the RHS of the above by $Z_{r,s}(\tau)$:

$$Z_{r,s}(\tau) = \zeta(r + s\tau) - r\eta_1(\tau) - s\eta_2(\tau). \tag{4.1}$$

Some properties of $Z_{r,s}$:

PROPOSITION 4.1. *(i) If $(r, s) \in \mathbb{R}^2\backslash\{(0,0)\}$, then $Z_{r,s}(\tau)$ is a holomorphic function in $\tau \in \mathbb{H}$.*

*(ii) If $(r, s) \in \{(\pm\frac{1}{2}, \pm\frac{1}{2})\} \mod 1$, then $Z_{r,s}(\tau) \equiv 0$ in $\tau$.*

*(iii) If $(r, s) \in \mathbb{C} \times \mathbb{C}\backslash\mathbb{R} \times \mathbb{R}$, then $Z_{r,s}(\tau)$ has a simple pole at $\tau_0$ such that $r + s\tau_0 \in \Lambda_{\tau_0}$, where $\Lambda_{\tau_0}$ is the lattice generated by 1 and $\tau_0$.*

*(iv) $Z_{r,s}(\tau)$ satisfies the following modularity:*

$$Z_{r',s'}(\tau') = (c\tau + d)Z_{r,s}(\tau) \tag{4.2}$$

*if* $\gamma = \begin{pmatrix} a & b \\ c & d \end{pmatrix} \in \mathrm{SL}(2, \mathbb{Z})$, $(r, s) = (r', s') \cdot \gamma$, *and* $\tau' = \gamma \cdot \tau \div \frac{a\tau+b}{c\tau+d}$.

Hence if $(r, s) = \left(\frac{k_1}{N}, \frac{k_2}{N}\right)$ is a $N$-torsion point ($\gcd(k_1, k_2, N) = 1$), then by (4.2) $Z_{r,s}(\tau)$ is a modular form of weight one w.r.t. $\Gamma_N$. This was first proved by Hecke [**21**] on 1926. Furthermore, Hecke noted that $Z_{r,s}(\tau)$ is the Eisenstein series of weight 1. We recall that the Eisenstein series of weight 1 is defined by

$$\mathfrak{E}_1^N(\tau, s; k_1, k_2) := (\mathrm{Im}\tau)^s \sum_{(m,n)} (m\tau + n)^{-n}|(m\tau + n)|^{-2s},$$

where $(m, n)$ runs over $\mathbb{Z}^2$ under the condition $0 \neq (m, n) \equiv (k_1, k_2) \mod N$. It is known that $\mathfrak{E}_1^N(\tau, s; k_1, k_2)$ is a meromorphic function in the $s$-plane and holomorphic at $s = 0$. Set $\mathfrak{E}_1^N(\tau; k_1, k_2) := \mathfrak{E}_1^N(\tau, 0; k_1, k_2)$. It can be proved (see [**17**]) that if $(r, s)$ is a $N$-torsion point, then

$$Z_{r,s}(\tau) = N\mathfrak{E}_1^N(\tau; k_1, k_2) \text{ if } (r, s) \equiv \left(\frac{k_1}{N}, \frac{k_2}{N}\right) \mod \mathbb{Z}.$$

Recall $F_2$ is the fundamental domain of $\Gamma_0(2)$. The following result was proved in [**10**].

THEOREM 4.2. [**10**]

*(i) For any pair $(r, s) \notin \mathbb{C}^2\backslash\frac{1}{2}Z^2$, $Z_{r,s}(\tau) = \zeta(r + \zeta\tau) - (r\eta_1(\tau) + s\eta_2(\tau))$ has simple zeros only.*

*(ii) Suppose $(r, s) \in \mathbb{R}^2\backslash\{(\pm\frac{1}{2}, \pm\frac{1}{2})\}$. Then $Z_{r,s}(\tau)$ has a zero at $\tau \in F_2$ iff $\pm(r, s) \in \Delta_2 \mod 1$, where*

$$\Delta_2 = \left\{(r, s)|\, 0 \le r \le \frac{1}{2}, 0 \le s \le \frac{1}{2}, 0 \le r + s \le 1\right\}.$$

*Moreover, the zero is unique in $F_2$.*

REMARK 4.3. Since $Z_{m\pm r,n\pm s}(\tau) = \pm Z_{r,s}(\tau)$ for integers $m, n$, for any real pair $(r, s)$ we might assume $(r, s) \in \square = \{(r, s) | 0 \le r \le 1, 0 \le s \le \frac{1}{2}\}$. Then we decompose the rectangle $\square$ as the union of four triangles as follows. Thus Theorem 4.2 (ii) says that $Z_{r,s}(\tau)$ has exactly one zero if $(r, s) \in \Delta_2$,

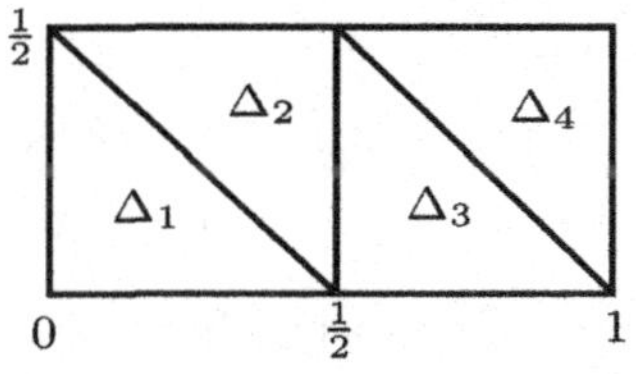


and has no zeros if $(r, s) \in \Delta_1 \cup \Delta_3 \cup \Delta_4 \bigcup_{i=1}^{4} \partial\Delta_i$. Let $\tau = \tau(r, s)$ be the unique zero of $Z_{r,s}(\tau)$ in $F_2$. Then $\tau(r, s)$ is a 1-1 map from $\Delta_2$ onto $\Omega_5$, a consequence of Theorem 2.3. Since $\Delta_2$ is simply connected, we have $\Omega_5$ is simply connected.

PROOF. We will prove (ii) and (i) for the case $(r, s) \in \mathbb{R}^2$. The general case will be proved in section 6.

To prove (ii), we should use the PDE result and the theory of modular form. The PDE result is about the non-existence of the following nonlinear equation

$$\Delta u + e^u = 8\pi\delta(0) \text{ in } E_\tau \tag{4.3}$$

if $E_\tau$ is a rectangle i.e. $\tau = ib$, $b > 0$, where $\delta(0)$ is the Dirac measure at $0 \in E_\tau$. See [28]. A function $u \in C^\infty(E_\tau \setminus \{0\})$ is said a solution to (4.3) if $\Delta u + e^u = 0$ in $E_\tau \setminus \{0\}$ and $u$ satisfies

$$u(z) = 4 \log |z| + O(1) \text{ as } z \to 0.$$

This non-existence of (4.3) for $\text{Re}\tau = 0$ implies that for any $(r, s) \in \mathbb{R}^2 \setminus \frac{1}{2}Z^2$,

$$Z_{r,s}(\tau) \neq 0 \quad \forall \tau = ib,\ b > 0. \tag{4.4}$$

Since $E_{\tau'} \approx E_\tau$ if $\tau' = \gamma \cdot \tau$, (4.4) implies

$$Z_{r,s}(\tau) \neq 0, \text{ for any } \tau \in \partial F \cap \mathbb{H}. \tag{4.5}$$

The second ingredient is to apply the theory of the modular forms with respect to $\text{SL}(2, \mathbb{Z})$. It is easy to see that the product $\prod_{(r,s):3\text{-torsion}} Z_{r,s}(\tau)$ is a modular form of weight 8 with respect to $\text{SL}(2, \mathbb{Z})$. There are only one modular form of weight 8 up to constants. This implies that $Z_{\frac{1}{3},\frac{1}{3}}(\tau)$ has only zero at $\tau = \rho = e^{\frac{\pi i}{3}}$, and the zero is a simple zero.

The following lemma shows that if $(r, s)$ is a $N$-torsion point with $N = 4$ or 6, then $Z_{r,s}(\tau) \neq 0\ \forall \tau \in \mathbb{H}$.

LEMMA 4.4. *For any* $\tau \in \mathbb{H}$,

*(i)* $\zeta(\frac{3}{4}\omega_1 + \frac{1}{4}\omega_2) \neq \frac{3}{4}\eta_1 + \frac{1}{4}\eta_2$.

*(ii)* $\zeta(\frac{1}{6}\omega_1 + \frac{1}{6}\omega_2) \neq \frac{1}{6}\eta_1 + \frac{1}{6}\eta_2$.

*(iii)* $\zeta(\frac{2}{6}\omega_1 + \frac{3}{6}\omega_2) \neq \frac{2}{6}\eta_1 + \frac{3}{6}\eta_2$.

The proof of Lemma 4.4 is a good exercise of the addition theorem. Now we can prove (ii) of Theorem 4.2.

**Step 1:** we compute the asymptotic of $Z_{r,s}(\tau)$ as $\tau$ tends to three cusp points $\infty, 0, 1$:

$$\lim_{\tau\to\infty} Z_{r,s}(\tau) = \begin{cases} 2\pi i(s-\frac{1}{2}) & \text{if } s \neq 0, \\ \pi \cot \pi r & \text{if } s = 0, \end{cases}$$

$$\lim_{\tau\to 0} Z_{r,s}(\tau) = \begin{cases} \frac{2\pi i}{\tau}(\frac{1}{2}-r) & \text{if } r \in (0,1), \\ \frac{\pi \cot \pi s}{\tau} & \text{if } r = 0, \end{cases}$$

$$\lim_{\tau\to 1} Z_{r,s}(\tau) = \begin{cases} \frac{2\pi i}{\tau-1}\left(\frac{1}{2}-(r+s)\right) & \text{if } r+s \in (0,1), \\ \frac{\pi \cot \pi s}{\tau-1} & \text{if } r+s = 0. \end{cases}$$

For the details of computation, we refer [**10**]. In particular if $(r,s) \notin \{(r,\frac{1}{2})|0 \leq r < 1\} \cup \{(\frac{1}{2},s)|0 \leq s < \frac{1}{2}\} \cup \{(r,s)|r+s = \frac{1}{2}\}$, then the limit of $Z_{r,s}(\tau)$ does not vanish as $\tau$ tends to any cusp $\in \{0,1,\infty\}$.

**Step 2:** By Lemma 4.4, $Z_{r,s}(\tau)$ has no zero at $(\frac{3}{4},\frac{1}{4})$, $(\frac{1}{6},\frac{1}{6})$ and $(\frac{2}{6},\frac{3}{6})$ which are contained in $\Delta_1$ and $\Delta_3 \cup \Delta_4 \cup \{(r,s)|r+s=1\}$. Since $\lim_{\tau\to\partial F} Z_{r,s}(\tau) \neq 0$ $\forall (r,s) \in \Delta_1 \cup \Delta_3 \cup \Delta_4 \cup \{(r,s)|r+s=\frac{1}{2}\}$ by Step 1 and the PDE result, we conclude that $Z_{r,s}(\tau)$ has no zeros in $F$ for $(r,s) \in \Delta_1 \cup \Delta_3 \cup \Delta_4 \cup \{(r,s)|r+s = \frac{1}{2}\}$. Consequently, for $(r,s) \in \bigcup_{i=1}^{4} \partial\Delta_i$, $Z_{r,s}(\tau)$ has no solutions.

For $(r,s) \in \Delta_2$, we have the number of zeros of $Z_{r,s}(\tau)$ = the number of zeros of $Z_{\frac{1}{3},\frac{1}{3}}(\tau)$. Then (ii) and (i) for real pairs $(r,s)$ follows easily. □

REMARK 4.5. The advantage of our method that $Z_{r,s}(\tau)$ for $(r,s) \notin Q^2$ is used to connect one modular form (with a $N_1$-torsion point $(r,s)$) with another modular form (with a $N_2$-torsion point). During the deformation, the zeros are contained within the interior of $F_2$.

**4.2. To construct premodular forms.** Next we will construct the modular form for $n \geq 2$. For the family of hyper-elliptic curves $Y^{(n)}(\tau)$, $\tau \in \mathbb{H}$, we could associate a "premodular form" $Z^{(n)}_{r,s}(\tau)$. Here $Z^{(n)}_{r,s}(\tau)$ is called a premodular form if $Z^{(n)}_{r,s}(\tau)$ is a modular form w.r.t $\Gamma(N)$ for any $N$-torsion point $(r,s)$.

THEOREM 4.6. [**27**] *For any pair* $(r,s) \in \mathbb{R}^2 \backslash \frac{1}{2}\mathbb{Z}^2$, *there exists a premodular form* $Z^{(n)}_{r,s}(\tau)$ *of weight* $\frac{n(n+1)}{2}$ *satisfying the following property:*

$$Z^{(n)}_{r_0,s_0}(\tau_0) = 0 \textit{ for some } (r_0,s_0) \notin \frac{1}{2}\mathbb{Z}^2 \textit{ iff there is an}$$

$$a = (a_1, \cdots, a_n) \in Y^{(n)}(\tau_0) \textit{ satisfying (3.4) and } \sum_{i=1}^{n} a_i \notin E_{\tau_0}[2],$$

*or equivalently, there exists a non-trivial solution a to (3.1) such that* $\sum_{i=1}^{n} a_i \notin E_{\tau_0}[2]$.

We will give a proof of Theorem 4.6 later after some necessary preparations. To construct $Z_{r,s}^{(n)}(\tau)$, we consider two maps $\sigma_n$ and $z_n$ on $Y^{(n)}(\tau)$:

(i) $\sigma_n(a) = \sum_{i=1}^{n} a_i$ is the addition map from $Y^{(n)}(\tau)$ to $E_\tau$ and

(ii) $z_n(a) = \zeta(\sum_{i=1}^{n} a_i) - \sum_{i=1}^{n} \zeta(a_i)$ is a meromorphic function from $\overline{Y^{(n)}}(\tau)$ to $\mathbb{P}^1 = \mathbb{C} \cup \{\infty\}$.

Recall $\overline{Y^{(n)}} = Y^{(n)} \cup \{(0,\dots,0)\}$. It implies that $z_n(a)$ has a (simple) pole at those $a$ such that either $a = (0,0,\dots,0)$ or $\sum a_i = 0 \mod \Lambda_\tau$. How many $a = (a_1,\dots,a_n)$ with $\sum a_i = 0 \mod \Lambda_\tau$? The answer is given by the following result.

THEOREM 4.7. **[27]** *Both* $\sigma_n$ *and* $z_n$ *have the degree* $\frac{n(n+1)}{2}$.

Theorem 4.7 is the main result in **[27]**. The part on $\sigma_n$ is proved by applying the cubics theorem in algebraic geometry. The degree formula implies that if $\sigma_0 \in E_\tau$ is not a critical value, then $\sigma_n^{-1}(\sigma_0)$ has $\frac{n(n+1)}{2}$ element. Through $\sigma_n$, the function field $K(\overline{Y^{(n)}}(\tau))$ of $\overline{Y^{(n)}}(\tau)$ is an extension of $K(E_\tau)$ with

$$[K(\overline{Y^{(n)}}(\tau)) : K(E_\tau)] = \frac{n(n+1)}{2}.$$

So we want to know what is the primitive element of this extension. It turns out $z_n$ is the primitive element. In fact we could prove there exists a monic polynomial $W_n(X;\sigma) \in Q[g_2, g_3, \wp(\sigma), \wp'(\sigma), X]$ where $g_2$ and $g_3$ are coefficients of $E_\tau : y^2 = 4x^3 - g_2 x - g_3$ such that

(iii) $\deg W = \frac{n(n+1)}{2}$,

(iv) $W_n(z_n(a); \sigma_n(a)) \equiv 0$ in $Y^{(n)}(\tau)$,

(v) $W_n$ is homogeneous in $g_2$, $g_3$, $\wp(\sigma)$, $\wp'(\sigma)$ and $X$ if we assign them the weights by 4,6,2,3,1 respectively.

The fact that $z_n$ is a primitive element is equivalent to the part of Theorem 4.7 concerning the degree of $z_n$, as a map from $\overline{Y^{(n)}}$ to $\mathbb{P}^1$. It is equivalent to prove that the values of $z_n(a)$ are distinct for $a \in \sigma_n^{-1}(\sigma_0)$ for generic $\sigma_0$, i.e.,

(4.6) if $\sigma_n(a) = \sigma_n(a')$ and $z_n(a) = z_n(a')$ for $a, a' \in Y^{(n)}(\tau)$ then $a = a'$.

The precise statement about (4.6) is the following lemma. See **[27]**.

LEMMA 4.8. **[27]** *Suppose* $a$ *is not a branch point of* $Y^{(n)}(\tau)$ *and* $\sigma_n(a) = \sigma_n(a')$ *for two distinct points* $a$, $a' \in Y^{(n)}(\tau)$. *If* $z_n(a) = z_n(a')$ *then* $a = a'$. *In particular, the preimage* $\sigma_n^{-1}(r + s\tau)$ *consists of* $\frac{n(n+1)}{2}$ *distinct elements in* $Y^{(n)}(\tau)$ *provided that* $(r,s) \in \mathbb{R}^2 \backslash \frac{1}{2}\mathbb{Z}^2$.

There are at least two ways to prove Lemma 4.8, one is our original proof in [**27**], which extend the classical method of the so-called second symmetric (tensor) product of the Lamé equation. Note that the classical second symmetric product is for solutions of the *same* equation, but our proof can be extended to solutions of *different* equations, but with the same *type*. Another method is to use the Painlevé VI, and Lemma 4.8 is a special case of a general theorem see [**9**].

Let us come back to our subject $W_n$. Once $W_n$ is known, we can construct our premodular form $Z^{(n)}_{r,s}(\tau)$ by

$$Z^{(n)}_{r,s}(\tau) = W_n\left(Z_{r,s}(\tau); \sigma_n\right), \quad \sigma_n = r + s\tau, \tag{4.7}$$

where $Z_{r,s}(\tau)$ is the Hecke form:

$$Z_{r,s}(\tau) = \zeta(r + s\tau|\tau) - r\eta_1(\tau) - s\eta_2(\tau).$$

We remark that $Z^{(n)}_{r,s}(\tau)$ could be defined for all $(r, s) \in \mathbb{C}^2 \backslash \frac{1}{2}Z^2$.

Recall that $Z_{r,s}(\tau)$ is a modular form of weight 1 w.r.t $\Gamma(N)$ if $(r, s)$ is a $N$-torsion point. This together with (v) implies that $Z^{(n)}_{r,s}(\tau)$ is a modular form of weight $\frac{n(n+1)}{2}$ w.r.t $\Gamma(N)$.

Proof of Theorem 4.6. Note that if $Z^{(n)}_{r_0,s_0}(\tau_0) = 0$ and $(r_0, s_0) \notin \frac{1}{2}\mathbb{Z}^2$ then $Z_{r_0,s_0}(\tau_0)$ is a root of $W(X; \sigma_n)$ and $\sigma_n = r_0 + s_0\tau_0$. By Lemma 4.8, $W(X; \sigma_n)$ has $\frac{n(n+1)}{2}$ distinct roots, which are $\{z_n(a)|a \in \sigma_n^{-1}(r_0 + s_0\tau_0)\}$. Therefore there exist $a \in Y^{(n)}(\tau)$ such that

$$r + s\tau = \sigma_n(a) = \sum_{i=1}^{n} a_i, \text{ and } Z_{r,s}(\tau_0) = z_n(a). \tag{4.8}$$

Recall the definition:

$$z_n(a) = \zeta\left(\sum a_i\right) - \sum_{i=1}^{n} \zeta(a_i).$$

So, the second equation of (4.8) implies

$$\sum_{i=1}^{n} \zeta(a_i) = r\eta_1 + s\eta_2 \tag{4.9}$$

To link (4.9) with (3.4), we recall

$$G_z(a_i) = \zeta(a_i) - (r_i\eta_1 + s_i\eta_2), \quad a_i = r_i + s_i\tau$$

Then (4.9) is identical with (3.4):

$$\begin{aligned}\sum_{i=1}^{n} G_z(a_i) &= \sum_{i=1}^{n} \zeta(a_i) - ((\sum r_i)\eta_1 + (\sum s_i)\eta_2) \\ &= \sum_{i=1}^{n} \zeta(a_i) - r\eta_1 - s\eta_2 = 0.\end{aligned}$$

Therefore $a$ satisfies (3.4). This proves the sufficient part.

Conversely if there is $a = (a_1, \ldots, a_n) \in Y^{(n)}(\tau_0)$ satisfying (3.4), then

$$0 = \sum_{i=1}^{n} G_z(a_i) = \sum_{i=1}^{n} \zeta(a_i) - (\sum r_i)\eta_1 - (\sum s_i)\eta_2.$$

Since $r + s\tau_0 = \sum_{i=1}^{n} a_i = (\sum r_i) + (\sum s_i)\tau_0$, we have $r = \sum r_i$ and $s = \sum s_i$, which implies $\sum_{i=1}^{n} \zeta(a_i) = r\eta_1 + s\eta_2$, and $z_n(a) = \zeta(\sum a_i) - \sum_{i=1}^{n} \zeta(a_i) = \zeta(r + s\tau_0) - (r\eta_1 + s\eta_2) = Z_{r,s}(\tau_0)$. Therefore, $Z_{r,s}^{(n)}(\tau_0) = W(Z_{r,s}(\tau_0)) = W(z_n(a), \sigma_n(a)) = 0$. □

REMARK 4.9. If a real pair $(r_0, s_0) \notin \frac{1}{2}Z^2$ and $Z_{r_0,s_0}^{(n)}(\tau_0) = 0$ for some $\tau_0$, then by Theorem 4.6, $r_0 + s_0\tau_0$ is a non-trivial critical point of $G^n(z|\tau_0)$. Since $Z_{r,s}(\tau)$ is holomorphic in $\tau$ by the Rouche theorem, the zero $\tau$ of $Z_{r,s}(\tau)$ is well-defined as a (real) analytic function of $(r, s)$. Thus $Z_{r,s}^{(n)}(\tau)$ has a zero near $\tau_0$ for any $(r, s)$ near $(r_0, s_0)$. The local behavior of $\tau(r, s)$ is related the question (a) in section 2. For example if we could prove $\tau(r, s)$ is an open mapping, then question (a) might be answered.

In §5, we see that the zeros of $Z_{r,s}^{(n)}(\tau)$ also consists of those tori such that (3.1) has non-trivial solutions, or equivalently, the mean field equation,

$$\Delta u + e^u = 8\pi n\delta_0 \quad \text{in } E_\tau, \tag{4.10}$$

has at least one solution. See Theorem 5.2.

The most important property of $Z_{r,s}^{(n)}(\tau)$, as a holomorphic function in $\mathbb{H}$, is the following *simple zero property*.

THEOREM 4.10. [**9**] *For $(r, s) \in \mathbb{C}^2 \backslash \frac{1}{2}\mathbb{Z}^2$, $Z_{r,s}^{(n)}(\tau)$ has simple zeros only.*

Theorem 4.10 was conjectured in [**27**]. Recently, Chen-Kuo-Lin proves this highly non-trivial result by applying the Painlevé VI equations. We will briefly discuss the proof in §7.

We remark that the explicit expression of $W_n(X; \sigma)$ is very difficult to compute even for small $n$. For $n = 2$ and 3, it was obtained by Dahmen [**15**], who used the classical formula (the so-called Frobenius-Stickelberger formula) to obtain it and for $n = 4$, it was obtained by Wang and the author [**27**] by applying the addition map $\sigma_n$ and the setting above. Here we list some known examples from [**27**] and [**15**].

- For $n = 2$,

$$W^{(2)}(X; \sigma) = X^3 - 3\wp(\sigma)X - \wp'(\sigma), \text{ and}$$
$$Z_{r,s}^{(2)}(\tau) = Z_{r,s}^{(3)}(\tau) - 3\wp(\sigma)Z_{r,s}(\tau) - \wp'(\sigma).$$

Obviously, $Z_{r,s}^{(2)}(\tau)$ is a modular form of weight 3 w.r.t. $\Gamma(N)$, if $(r, s)$ is a $N$-torsion point.

- For $n = 3$, $W^{(3)}(X;\sigma)$ is a polynomial of degree 6:

$$W^{(3)}(X;\sigma) = X^6 - 15\wp(\sigma)X^4 - 20\wp'(\sigma)X^3 + (\frac{27}{4}g_2 - 45\wp^2(\sigma))X^2 - 12\wp(\sigma)\wp'(\sigma)X - \frac{5}{4}\wp'(\sigma)^2$$

- For $n = 4$, $W^{(4)}(X;\sigma)$ is a polynomial of degree 10:

$$\begin{aligned} W^{(4)}(X;\sigma) = X^{10} &- 45\wp(\sigma)X^8 - 120\wp'(\sigma)X^7 + (\frac{399}{4}g_2 - 630\wp^2(\sigma))X^6 \\ &- 504\wp(\sigma)\wp'(\sigma)X^5 - \frac{15}{4}(280\wp^3(\sigma) - 49g_2\wp(\sigma) - 115g_3)X^4 \\ &+ 15(11g_2 - 24\wp^2(\sigma))\wp'(\sigma)X^3 \\ &- \frac{9}{4}(140\wp^4(\sigma) - 245g_2\wp^2(\sigma) + 190g_3\wp(\sigma) + 21g_2^2)X^2 \\ &- (40\wp^3(\sigma) - 163g_2\wp(\sigma) + 125g_3)\wp'(\sigma)X \\ &+ \frac{3}{4}(25g_3 - 3\wp^2(\sigma))\wp'(\sigma)^2. \end{aligned}$$

## 5. Generalized Lamé equations

**5.1. Mean field equation and the Lamé equation.** The classic Lamé equation is a Heun ordinary differential equation with four regular singular points (one of them is $\infty$):

$$p(x)\frac{d^2y}{dx^2} + \frac{1}{2}p'(x)\frac{dy}{dx} - (n(n+1)x + B)y = 0, \quad x \in \mathbb{C} \tag{5.1}$$

where $p(x) = 4x^3 - g_2(\tau)x - g_3(\tau)$. By using the two-one projection $x = \wp(z|\tau)$ from $E_\tau$ to $\mathbb{C} \cup \{\infty\}$, (5.1) is transferred to

$$y''(z) = (n(n+1)\wp(z) + B)y(z), \quad z \in E_\tau. \tag{5.2}$$

Classically, (5.1) is called the algebraic form of the Lamé equation, and (5.2) is called the elliptic form of (5.1). In this paper, the Lamé equation is often referred as the elliptic form. For (5.2), $n$ is called the index and $B$ is called the accessory parameter of the Lamé equation.

The Lamé equation has been thoroughly studied since 19th century, in particular the case $n \in \frac{1}{2}\mathbb{Z}$. When $n \in \frac{1}{2}\mathbb{Z}$, the monodromy representation is generally reduced to a homomorphism from $\pi_1(E_\tau)$ to $GL(2,\mathbb{C})$. We refer the readers to [**9**] for the full discussion of the Lamé equation with $n \in \frac{1}{2}\mathbb{Z}$. For the rest of this subsection, we consider $n \in \mathbb{N}$, and (5.1) is called an integral Lamé equation.

From the geometric aspects, there is an important link between the Lamé equation and the mean field equation:

$$\Delta u + e^u = 8\pi n\delta_0 \quad \text{in } E_\tau. \tag{5.3}$$

where $n \in \mathbb{N}$. Equation (5.3) is considered to be an integrable system in the following sense: for any solution $u(z)$ of (5.3), there exists an meromorphic function $f(z)$ on $\mathbb{C}$ such that

$$u(z) = \log \frac{8|f'(z)|^2}{(1+|f(z)|^2)^2}, \quad z \in E_\tau. \tag{5.4}$$

This meromorphic function $f$ is called a developing map of $u$. For $n \in \mathbb{R}^+$, $f(z)$ is multi-valued. However if $n \in \frac{1}{2}\mathbb{Z}$ then $f(z)$ becomes a single-valued function. To recover $f(z)$, we note that $u_{zz} - \frac{1}{2}u_z^2$ is meromorphic because

$$(u_{zz} - \frac{1}{2}u_z^2)_{\bar{z}} = 4(u_z e^u - (e^u)_z) = 0.$$

Since $u_{zz} - \frac{1}{2}u_z^2$ has singularities only at 0, we have

$$u_{zz} - \frac{1}{2}u_z^2 = -2(n(n+1)\wp(z) + B) \tag{5.5}$$

for some complex number B.

On the other hand, by the formula (5.4), we have

$$u_{zz} - \frac{1}{2}u_z^2 = \frac{f'''}{f'} - \frac{3}{2}(\frac{f''}{f'})^2.$$

Thus if $f_1$ and $f_2$ are two developing maps of $u$, then the classical Schwarz theorem says that

$$f_2(z) = \frac{af_1(z)+b}{cf_2(z)+d} \quad \text{for some } \gamma = \begin{pmatrix} a & b \\ c & d \end{pmatrix} \in \mathrm{SL}(2,\mathbb{C}).$$

Due to the restriction (5.4), we have $\gamma \in \mathrm{SU}(2,\mathbb{C})$. Then the question is how to connect (5.1) and (5.3). We recall a classical result (See [**35**]).

PROPOSITION 5.1. *Suppose $y_1$ and $y_2$ are linearly independent solutions of (5.1), and set $f = \frac{y_2}{y_1}$. Then the Schwarz derivative*

$$\{f; z\} = -2(n(n+1)\wp(z) + B).$$

Therefore, the developing map $f$ is a ratio of two independent solution of (5.1), where $B$ is determined by (5.5). From this connection, we see that the monodromy group for $f$ is the *projective* monodromy group of the corresponding Lamé equation.

On one hand, we could compute the monodromy group for $f$ by the geometry of the mean field equation. We note that $f(z)$ is not necessary to be doubly periodic. Indeed, after a suitable normalization, $f$ satisfies

$$f(z+1) = e^{2\pi i\theta_1} f(z) \text{ and } f(z+\tau) = e^{2\pi i\theta_2} f(z), \text{ for } \theta_i \in \mathbb{R}. \tag{5.6}$$

There is a simple, but important, observation. If $f$ is a developing map satisfying (5.6), then $f_\lambda(z) = e^\lambda f(z)$, $\lambda \in \mathbb{R}$, also satisfies (5.6). Then plugging it into (5.4), the "rescaled" $u_\lambda$ defined by

$$u_\lambda(z) = \log \frac{8e^{2\lambda}|f'(z)|^2}{(1+e^{2\lambda}|f(z)|^2)^2},$$

is doubly periodic and satisfies the same mean field equation. Then we could easily see that as $\lambda \to \infty$, the blowup set of $u_\lambda$ consists of the zeros of $f$. Similarly, as $\lambda \to -\infty$, the blowup set of $u_\lambda$ consists of the poles of $f$. Clearly, by (5.6) the zeros or the poles of $f$ are well-defined as points in $E_\tau$. From the PDE result, any blowup set does not contain the origin 0 and has $n$ elements. By Theorem 2.1 (ii), both $a = (a_1, \ldots, a_n)$ and $b = (b_1, \ldots, b_n)$ satisfy (3.1) which implies $a, b \in Y^{(n)}(\tau)$ and both satisfy (3.4). Furthermore, we could prove $a = -b$. Clearly $\{a_1, \ldots, a_n\} \cap \{b_1, \ldots, b_n\} = \emptyset$. Thus $b = -a$ implies $\{a_1, \ldots, a_n\} \cap \{-a_1, \ldots, -a_n\} = \emptyset$, that is, $a$ is not a branch point of $Y^{(n)}(\tau)$ (see Theorem 3.2 (iv)). Let $\sigma_n(a) = r + s\tau$ for some $(r, s) \in \mathbb{R}^2$. In order to apply Theorem 4.6, we need to prove $(r, s) \notin \frac{1}{2}\mathbb{Z}^2$. Indeed, it was proved in [**6**]. Now we can apply Theorem 4.6 in §4, which implies that $Z^{(n)}_{r,s}(\tau) = 0$. This proves the sufficient part of the following result.

THEOREM 5.2. *Let $\tau \in \mathbb{H}$. Then (5.3) has a solution iff there exists a real pair $(r, s) \notin \frac{1}{2}\mathbb{Z}^2$ such that $Z^{(n)}_{r,s}(\tau) = 0$.*

PROOF. For the necessary part, we also use the sufficient part of Theorem 4.6: there exists $a = (a_1, \ldots, a_n) \in Y^{(n)}(\tau)$ such that $a$ satisfies (3.1) and $\sum_{i=1}^{n} a_i = r + s\tau$. Since $(r, s) \notin \frac{1}{2}\mathbb{Z}^2$, $a$ must be a non-branch point of $Y^{(n)}(\tau)$. Set

$$f_a(z) := \prod_{i=1}^{n} \exp \int_0^z \frac{\wp'(a_i)}{\wp(\omega) - \wp(a_i)} d\omega. \tag{5.7}$$

Then by a direct computation, we could prove $f_a(z)$ satisfies (5.6) for some real constant $\theta_1$ and $\theta_2$, and the function $u(z)$ defined by (5.4) satisfies the mean field equation. For a complete proof, we refer §5 and §6 of [**6**]. □

**5.2. The projective monodromy group of developing maps.** Recall the monodromy group for $f$ is a homomorphism $\bar{\rho}$ from $\pi_1(E_\tau)$ to PSL(2,$\mathbb{C}$). Hence the monodromy group is an Abelian group generated by $\bar{\rho}(\ell_i)$, $i = 1, 2$, where $\ell_i$ is the fundamental cycles of $E_\tau$ represented by the loop $[0, 1]$ and $[0, \tau]$ respectively. The identity (5.6) precisely says that

$$\bar{\rho}(L_1) = \pm \begin{pmatrix} e^{\pi i \theta_1} & 0 \\ 0 & e^{-\pi i \theta_1} \end{pmatrix} \text{ and } \bar{\rho}(L_2) = \pm \begin{pmatrix} e^{\pi i \theta_2} & 0 \\ 0 & e^{-\pi i \theta_2} \end{pmatrix}.$$

We claim

$$\theta_1 = -2s \text{ and } \theta_2 = 2r. \tag{5.8}$$

The claim (5.8) can be proved by (5.7) as follows. We recall the addition formula:

$$\frac{1}{2} \frac{\wp'(z)}{\wp(z) - \wp(a)} = \zeta(z + a) - \zeta(z) - \zeta(a). \tag{5.9}$$

We apply (5.9) to (5.7), and have

$$f(z) = e^{2z\sum_{i=1}^{n}\zeta(a_i)} \frac{\prod_{i=1}^{n}\sigma(z-a_i)}{\prod_{i=1}^{n}\sigma(z+a_i)}. \tag{5.10}$$

The transformation law

$$\sigma(z+\omega_i) = -e^{\eta_i(z+\frac{\omega_i}{2})}\sigma(z). \tag{5.11}$$

gives the transformation law for $f(z)$:

$$f(z+\omega_i) = f(z)\exp(2\omega_i \sum_{j=1}^{n}\zeta(a_j) - 2\eta_i \sum_{j=1}^{n} a_j).$$

Thus

$$\pi i\theta_1 = \sum_{j=1}^{n}\zeta(a_j) - \eta_1 \sum_{j=1}^{n} a_j = (\sum_{j=1}^{n}\zeta(a_j) - (\eta_1 r + s\eta_2)) - 2\pi i s,$$

$$\pi i\theta_2 = \tau\sum_{j=1}^{n}\zeta(a_j) - \eta_2(r+s\tau) = \tau(\sum_{j=1}^{n}\zeta(a_j) - (\eta_1 r + s\eta_2)) + 2\pi i r.$$

But (3.4) implies

$$\sum_{j=1}^{n}\zeta(a_j) = \eta_1 r + \eta_2 s \quad \text{if} \quad \sum a_j = r + s\tau, \ r, s \in \mathbb{R} \tag{5.12}$$

Hence (5.8) follows. So if the Lamé equation comes from the mean field equation, then the projective monodromy group is generated by

$$\pm\begin{pmatrix} e^{-2\pi i s} & 0 \\ 0 & e^{2\pi i s}\end{pmatrix} \text{ and } \pm\begin{pmatrix} e^{2\pi i r} & 0 \\ 0 & e^{-2\pi i r}\end{pmatrix},$$

where by (5.12), $(r, s)$ is determined by the zeros $\{z_1, \cdots, z_n\}$ of $f$.

**5.3. Monodromy group of integral Lamé equation.** Now we want to calculate the monodromy group for any Lamé equation (not necessarily related to mean field equation), (5.10) suggests that the fundamental solutions $y_1$ and $y_2$ should be given by

$$\begin{aligned} y_1(z) &= y_a(z) \doteqdot e^{z\sum_{i=1}^{n}\zeta(a_i)} \frac{\prod_{i=1}^{n}\sigma(z-a_i)}{\sigma(z)^n}, \text{ and} \\ y_2(z) &= y_{-a}(z) \doteqdot e^{-z\sum_{i=1}^{n}\zeta(a_i)} \frac{\prod_{i=1}^{n}\sigma(z+a_i)}{\sigma(z)^n}. \end{aligned} \tag{5.13}$$

Indeed, this is the so-called Hermite-Halphen Ansatz [**35**] in the literature for any integral Lamé equation (5.2), which does not necessarily come from the mean field equation. By a straightforward computation, we have the following theorem.

THEOREM 5.3. **[6]** *Let* $a = (a_1, \ldots, a_n) \in E_\tau^n$*. Then* $y_a(z)$ *satisfies (5.2) iff* $a \in Y^{(n)}(\tau)$*. Moreover,*

$$B = (2n-1)\sum_{i=1}^{n} \wp(a_i). \tag{5.14}$$

REMARK 5.4. If $a$ is not a branch point of $Y^{(n)}(\tau)$ then $y_a(z)$ and $y_{-a}(z)$ are two independent solutions of (5.2) with $B = (2n-1)\sum_{i=1}^{n} \wp(a_i)$. Then the monodromy group for (5.2) is generated by

$$\begin{pmatrix} e^{-2\pi i s} & 0 \\ 0 & e^{2\pi i s} \end{pmatrix} \text{ and } \begin{pmatrix} e^{2\pi i r} & 0 \\ 0 & e^{-2\pi i r} \end{pmatrix},$$

where $(r, s) \in \mathbb{C}^2 \backslash \frac{1}{2}\mathbb{Z}^2$ satisfies

$$\begin{aligned} \sum_{i=1}^{n} a_i &= r + s\tau, \text{ and} \\ \sum_{i=1}^{n} \zeta(a_i) &= r\eta_1 + s\eta_2. \end{aligned} \tag{5.15}$$

since the matrix $\begin{pmatrix} 1 & \tau \\ \eta_1 & \eta_2 \end{pmatrix}$ is invertible for any $\tau \in \mathbb{H}$, $(r, s)$ can always be solved. Note that $\sum_{i=1}^{n} a_i$ and $\sum_{i=1}^{n} \zeta(a_i)$ ate related to maps $\sigma_n(a)$ and $z_n(a)$ in §4.

We have only given a quick overview about the integral Lamé equation and its connection with the mean field equation. For the complete theory including the generalized Lamé equations, we refer **[6]**, **[8]** and **[27]**.

**5.4. Dahmen-Beukers conjecture.** In this subsection we would like to discuss a conjecture proposed by Dahmen and Beukers. See **[4, 15]**. Obviously, the Lamé equation possesses some symmetry from $SL(2,\mathbb{Z})$ action. For example, if we let $\tau' = \gamma \cdot \tau := \frac{a\tau+b}{c\tau+d}$ for some $\gamma = \begin{pmatrix} a & b \\ c & d \end{pmatrix} \in SL(2,\mathbb{Z})$ and make a scalar change of the independent variable $z \mapsto (c\tau+d)z$, then equation (5.2) is transformed to

$$y''(z) = \left(n(n+1)\wp(z|\tau') + (c\tau+d)^2 B\right) y(z) \text{ in } E_{\tau'}. \tag{5.16}$$

These scalar changes of variable induce a natural equivalence relation on the space of all Lamé equations; we call these two Lamé equations (5.2) and (5.16) *scalar equivalent.*

Equivalently, two Lamé equations of the algebraic form (5.1) are *scalar equivalent* if one can be transformed into the other by changing of variable $x \mapsto ax$ for some $a \in \mathbb{C}\backslash\{0\}$.

Conventionally, people are interested in (5.1) with algebraic solutions only, or equivalently with a finite monodromy group; see [**4, 15, 35**] and references therein. It is known that for an integral Lamé equation, if the monodromy group is finite, then the monodromy group $M$ of (5.1) is a dihedral group $D_N$ of order $2N$ for some $N \in \mathbb{N}_{\geq 3}$. Let $L_n(N)$ denote the number of Lamé equations (5.1) or (5.2) modulo scalar equivalence with monodromy group dihedral $D_N$. It is well known (cf. [**4, 15, 16**]) that $L_n(N)$ is *finite* for given $n \in \mathbb{N}$ and $N \in \mathbb{N}_{\geq 3}$. The issue is how to compute it. In [**15**] Dahmen and Beukers proposed an explicit formula about $L_n(N)$; see (5.22) below.

Denote the number of Lamé equations (5.1) modulo scalar equivalence with *projective* monodromy group dihedral $D_N$ by $PL_n(N)$. The well-known Klein theorem asserts that any second order Fuchsian differential equation on $\mathbb{CP}^1$ has finite projective monodromy group if and only if it is a pull-back of a hypergeometric equation belonging to the basic Schwarz list. Therefore, every integral Lamé equation (5.1) with projective monodromy group dihedral $D_N$ is such a pull-back by *Belyi functions* (see [**16, 26**]). By means of *Grothendieck correspondence* between *Belyi pairs* and *dessin d'enfants*, Litcanu [**16**] first showed how $PL_n(N)$ can be counted by using the combinatorics of dessins. Later by using this technique, Dahmen [**15**] obtained the explicit formula successfully:

$$PL_n(N) = \begin{cases} 0 & \textit{if } N \in \{1,2\}, \\ \frac{n(n+1)}{12}(\Psi(N) - 3\phi(N)) + \frac{2}{3}\varepsilon_n(N) & \textit{otherwise}, \end{cases} \tag{5.17}$$

where $\varepsilon_n(N)$ is defined in (5.24) below,

$$\begin{aligned} \phi(N) &:= \#\{k \in \mathbb{Z} | \gcd(k,N) = 1, 0 \leq k < N\} \\ &= N \prod_{p|N,\ p \text{ prime}} \left(1 - \frac{1}{p}\right) \end{aligned} \tag{5.18}$$

is the Euler function (set $\phi(N) = 0$ if $N \notin \mathbb{N}$), and

$$\Psi(N) := \#\mathcal{Q}(N) = N^2 \prod_{p|N,\ p \text{ prime}} \left(1 - \frac{1}{p^2}\right). \tag{5.19}$$

Here $\mathcal{Q}(N)$ is the set of $N$-torsion points of exact order $N$ defined by

$$\mathcal{Q}(N) := \left\{ \left(\frac{k_1}{N}, \frac{k_2}{N}\right) \middle| \gcd(k_1, k_2, N) = 1, 0 \leq k_1, k_2 < N \right\}. \tag{5.20}$$

However, differently from $PL_n(N)$, $L_n(N)$ can *not* be counted by means of dessins and seeking the explicit formula of $L_n(N)$ remains a long-standing challenging problem. It is not difficult to see that, when monodromy group

$M \simeq D_N$, then the projective monodromy group $PM \simeq D_{N/2}$ if $N$ is even and $PM \simeq D_N$ if $N$ is odd. Thus (cf. [**4**])

$$PL_n(N) = \left\{ \begin{array}{l} L_n(N) + L_n(2N) \text{ if } N \text{ is odd,} \\ L_n(2N) \text{ if } N \text{ is even.} \end{array} \right\}. \tag{5.21}$$

In 2003 Dahmen and Beukers proposed the following conjecture (see [**4**, Conjecture 73]):

CONJECTURE 5.5 (Dahmen-Beukers conjecture). *For every $n \in \mathbb{N}$ and $N \in \mathbb{N}_{\geq 3}$,*

$$\begin{aligned} L_n(N) = \frac{1}{2}\left(\frac{n(n+1)\Psi(N)}{24} - \left[a_n\phi(N) + b_n\phi\left(\frac{N}{2}\right)\right]\right) \\ +\frac{2}{3}\varepsilon_n(N), \end{aligned} \tag{5.22}$$

*where*

$$a_{2n} = a_{2n+1} = n(n+1)/2, \quad b_{2n} = b_{2n-1} = n^2, \tag{5.23}$$

*and*

$$\varepsilon_n(N) = \left\{ \begin{array}{ll} 1 & \text{if } N = 3 \text{ and } n \equiv 1 \bmod 3, \\ 0 & \text{otherwise.} \end{array} \right. \tag{5.24}$$

This conjecture was proved by Dahmen himself [**15, 16**] for two cases: (i) $n \in \mathbb{N}$ and $4|N$ (this follows directly from (5.17) and (5.21)); (ii) $n \in \{1, 2, 3\}$ and $N \in \mathbb{N}_{\geq 3}$. Recently, the case $n = 4$ and $N \in \mathbb{N}_{\geq 3}$ was proved by Y.-C. Chou [**27**, Appendix A]. As far as we know, the general case $n \geq 5$ and $4 \not| N$ still remains open. In this paper, we confirm Dahmen-Beukers conjecture completely. See §8.

**5.5. Generalized Lamé equation and the monodromy group.** Once we knew how to explicitly express the monodromy group for (5.2), it is natural to study the problem of isomonodromy by deforming the moduli parameter $\tau$. For this purpose, we consider the following generalized Lamé equation:

$$\frac{d^2y}{dz^2} = I(z)y, \tag{5.25}$$

where

$$\begin{aligned} I(z) = n(n+1)\wp(z) + \frac{3}{4}(\wp(z+p) + \wp(z-p)) \\ + A(\zeta(z+p) - \zeta(z-p)) + B, \ p \notin E_\tau[2], \end{aligned} \tag{5.26}$$

where $n \in \mathbb{N} \cup \{0\}$. Note that the local exponent at $p$ are $\frac{1}{2}$ and $-\frac{3}{2}$. Thus, the difference of the two local exponents is an integer. At this situation, the singular points $\pm p$ are called *apparent* if all solutions of (5.25) have no logarithmic singularity at $\pm p$. Throughout the paper, we always assume

(5.27) $\pm p$ *are apparent singularities of (5.25).*

Under the assumption of (5.27), the two constants $A$ and $B$ are not independent.

LEMMA 5.6 ([7]). *Suppose (5.27) holds. Then*

$$B = A^2 - \zeta(2p)A - \frac{3}{4}\wp(2p) - n(n+1)\wp(p).$$

REMARK 5.7. Let $\rho$ be the monodromy representation of (5.25):

$$\pi_1\left(E_\tau \backslash \{\pm p\}, *\right) \to \mathrm{SL}(2, \mathbb{C}).$$

Let $\gamma_\pm$ be the simple loop in $\pi_1(E_\tau \backslash \{\pm p\}, *)$ such that $\gamma_\pm$ encircles and $\pm p$ once. Then

$$\rho(\gamma_\pm) = -I_{2\times 2},$$

where $I_{2\times 2}$ is the identity matrix of two by two. Therefore by ignoring $\rho(\gamma_\pm)$, the monodromy representative is reduced to a homomorphism from $\pi_1(E_\tau)$ to $\mathrm{SL}(2, \mathbb{C})$, which implies the monodromy group of (5.25) is always Abelian. Hence the monodromy representation is always reducible.

Next we want to discuss how to compute the monodromy transformation $\rho(\ell_i)$, $i = 1, 2$, where $\ell_i$ are fundamental cycles in $\pi_1(E_\tau)$. Similar to (5.13), we consider the following ansatz for (5.25):

$$y_a(z) \doteqdot e^{\sum_{i=1}^{n+1} \zeta(a_i) z} \frac{\prod_{i=1}^{n+1} \sigma(z - a_i)}{\sigma(z)^n \sqrt{\sigma(z+p)\sigma(z-p)}}. \tag{5.28}$$

When $n = 0$, $a = a_1 \in E_\tau$ and the coefficient $A$ can be written in terms of $p$ and $a$.

THEOREM 5.8. [8] *Suppose the potential $I(z)$ of (5.25) is given by (5.26) with $n = 0$ and $p \notin E_\tau[2]$. Then $y_a$ of (5.28) is a solution of (5.25) iff*

$$A = \frac{1}{2}(\zeta(a+p) - \zeta(a-p) - \zeta(2p)). \tag{5.29}$$

*Moreover, the monodromy $\rho(\ell_i)$ is given by*

$$\rho(\ell_1) = \begin{pmatrix} e^{-2\pi i s} & 0 \\ 0 & e^{2\pi i s} \end{pmatrix}, \rho(\ell_2) = \begin{pmatrix} e^{2\pi i r} & 0 \\ 0 & e^{-2\pi i r} \end{pmatrix},$$

*where $(r, s)$ is determined by*

$$\begin{aligned} &r + s\tau = a, \\ &r\eta_1(\tau) + s\eta_2(\tau) = \zeta(a+p) + \zeta(a-p). \end{aligned} \tag{5.30}$$

For $n \geq 1$, the ansatz plays also important roles for calculating the monodromy representation.

THEOREM 5.9. *Let* $a = (a_1, \dots, a_n) \in E_\tau^{n+1}$, $[a_i] \notin \{\pm p\}$, *and* $n \geq 1$. *Then* $y_a$ *satisfies (5.25) iff*

$$\begin{aligned} &2\sum_{j\neq i}^{n+1}(\zeta(a_i - a_j) + \zeta(a_j) - \zeta(a_i)) \\ &\qquad = \zeta(a_i + p) + \zeta(a_i - p) - 2\zeta(a_i) \text{ and} \\ &\sum_{i=1}^{n+1}\zeta(a_i) = \frac{1}{2}\sum_{i=1}^{n+1}(\zeta(a_i + p) + \zeta(a_i - p)). \end{aligned} \tag{5.31}$$

*Moreover,* $A$ *satisfies*

$$A = \frac{1}{2}\sum_{i=1}^{n+1}(\zeta(a_i + p) - \zeta(a_i - p)) - \frac{1}{2}\zeta(2p). \tag{5.32}$$

Note that if $a$ satisfies (5.31), then $-a$ also satisfies (5.31). Clearly, $A$ also satisfies (5.32) with $-a$, that is, $y_{-a}$ is also a solution of (5.25) with the same coefficients $A$ and $B$. Moreover, it can be proved that either

$$\{a_1, \dots, a_{n+1}\} \cap \{-a_1, -a_2, \dots, -a_{n+1}\} = \emptyset \text{ or} \tag{4.33a}$$

$$\{a_1, \dots, a_{n+1}\} = \{-a_1, -a_2, \dots, -a_{n+1}\}. \tag{4.33b}$$

Let $(r, s) \in \mathbb{C}^2$ satisfy

$$\begin{cases} \sum_{i=1}^{n+1} a_i = r + s\tau \\ \sum_{i=1}^{n+1} \zeta(a_i) = r\eta_1(\tau) + s\eta_2(\tau). \end{cases} \tag{5.34}$$

Then $(r, s)$ also are related to the monodromy data of (5.7). Let $\rho$ be the monodromy representation of $\pi_1(E_\tau)$ to $\mathrm{SL}(2, \mathbb{C})$, and $\ell_i$, $i = 1, 2$, are the fundamental cycles of $E_\tau$.

THEOREM 5.10. [**8**] *Let* $a = (a_1, \dots, a_{n+1}) \in E_\tau^{n+1}$, $A$ *and* $B \in \mathbb{C}$ *such that (5.31), (5.32) and (4.33b) holds. Then GLE (5.25) is completely reducible, i.e.,* $y_a$ *and* $y_{-a}$ *are linearly independent and the monodromy representation is given by*

$$\rho(\ell_1)^* \begin{pmatrix} y_a \\ y_{-a} \end{pmatrix} = \begin{pmatrix} e^{2\pi i s} & 0 \\ 0 & e^{-2\pi i s} \end{pmatrix} \begin{pmatrix} y_a \\ y_{-a} \end{pmatrix},$$

$$\rho(\ell_2)^* \begin{pmatrix} y_a \\ y_{-a} \end{pmatrix} = \begin{pmatrix} e^{2\pi i r} & 0 \\ 0 & e^{-2\pi i r} \end{pmatrix} \begin{pmatrix} y_a \\ y_{-a} \end{pmatrix},$$

*where* $(r, s) \in \mathbb{C}^2$ *satisfies (5.34). Furthermore,* $(r, s) \notin \frac{1}{2}\mathbb{Z}^2$.

REMARK 5.11. The GLE (5.25) such that the hypothesis of Theorem 5.10 holds dose not cover all the possible $(A, B)$. In [**8**], we proved that the set of exceptional pair $(A, B)$ is a finite set. Indeed, the classification of GLE (5.25) is more complicated than the Lamé equation (5.2).

Similar to the Lamé equations, the GLE (5.25) also has a connection with the mean field equation,

$$\Delta u + e^u = 8\pi n\delta_0 + 4\pi(\delta_p + \delta_{-p}) \text{ in } E_\tau. \tag{5.35}$$

By the Liouville theorem, for any solution $u$ of (5.35), there exists an meromorphic function $f(z)$ in $\mathbb{C}$ such that (5.4) holds.

Assume $u(z)$ is an even solution. Then

$$\begin{aligned} u_{zz} - \frac{1}{2}u_z^2 =& -2(n(n+1)\wp(z) + \frac{3}{4}(\wp(z+p) + \wp(z-p))) \\ &+ A(\zeta(z+p) - \zeta(z-p) + B) \doteqdot I(z) \end{aligned} \tag{5.36}$$

and due to (5.4) the Schwarz derivative $\{f, z\}$ also satisfies

$$\{f, z\} = u_{zz} - \frac{1}{2}u_z^2 = -2I(z).$$

By this identity, Proposition 5.1 implies that the developing map $f$ can be written as a ratio of two independent solutions of (5.25), where $I(z) = u_{zz} - \frac{1}{2}u_z^2$.

PROPOSITION 5.12. [8] *Let GLE (5.25) come from the mean field equation (5.35). Then $\pm p$ are apparent singular points for (5.25).*

PROOF. Suppose that (5.25) has solutions with logarithmic singularity at $[p]$. Then the local monodromy matrix of (5.25) at $[p]$ is $\begin{pmatrix} -1 & D \\ 0 & -1 \end{pmatrix}$ for some $D \neq 0$. Since near $p$, $f(p + (z-p)e^{2\pi i})$ (i.e., the analytic continuation of $f$ along a loop encircling $p$ counter-clockwise) is also a developing map of the same $u$, there exists $M \in \mathrm{PSU}(2)$ such that $f(p + (z-p)e^{2\pi i}) = M \cdot f(z)$ holds near $p$. Because Proposition 5.1 shows that $f$ is a ratio of two linearly independent solutions of (5.25), there is $\gamma \in \mathrm{SL}(2, \mathbb{C})$ such that $\gamma \begin{pmatrix} -1 & D \\ 0 & -1 \end{pmatrix} \gamma^{-1} = \pm M$ (because $M = -M$ in PSU(2)). Since $\pm M \in \mathrm{PSU}(2)$ and $\gamma \in \mathrm{SL}(2, \mathbb{C})$, a direct computation gives $D = 0$, which yields a contradiction. □

After a normalization, $f$ satisfies also (5.6)

$$f(z+1) = e^{2\pi i\theta_1}f(z) \text{ and } f(z+\tau) = e^{2\pi i\theta_2}f(z), \ \theta_i \in \mathbb{R}.$$

The identity (5.6) shows that the projective monodromy representation of $f$ is unitary. In fact under this normalization, we can show

$$f = \frac{y_a(z)}{y_{-a}(z)},$$

where $y_a$ is given by (5.28). Again the unitary monodromy yields that the "rescaled" $u_\lambda$,

$$u_\lambda(z) = \log \frac{8e^{2\lambda}|f'(z)|^2}{(1 + e^{2\lambda}|f(z)|^2)^2},$$

is also a solution of (5.35). So the blowup set of $u_\lambda$ consists of the zeros of $f$ (the poles) as $\lambda \to +\infty$ ($\lambda \to -\infty$).

How many blowup points might $u_\lambda$ have? From the PDE result (see [**19**]), $u_\lambda$ never blows up at 0. But they might blow up at $p$ or $-p$. Here, we assume

(5.37) *The one parameter family of solutions $u_\lambda$ does not blow up at $\pm p$.*

Under the assumption (5.37), the blowup set consists of $n+1$ points. Hence $f$ has $n+1$ distinct simple zeros $a_1, \ldots, a_{n+1}$ in $E_\tau$, similarly $n+1$ distinct poles $b_1, \ldots, b_{n+1}$. Moreover, we could prove $0, \pm p \notin \{a_1, \ldots, a_{n+1}\} \cup \{b_1, \ldots, b_{n+1}\}$, if (5.36) holds. Moreover, we have

$$\{b_1, \ldots, b_{n+1}\} = \{-a_1, -a_2, \ldots, -a_{n+1}\}.$$

Since $a_1, \ldots, a_{n+1}$ are blowup points of $u_\lambda$, their location can be determined by a equation similar to (3.1) by the Pohozaev identity, which turns out to be (5.31). After that the coefficients $A$ and $B$ can be determined by (5.32) and Lemma 5.6 (by Proposition 5.12). Applying Theorem 5.10 we have

THEOREM 5.13. [**8**] *(5.35) has even solution such that (5.37) holds iff there exists $(a_1, \ldots, a_{n+1}) \in E_\tau^{n+1}$, $[a_i] \notin \{\pm p\}$ such that the hypothesis of Theorem 5.10 holds with $(r,s) \in \mathbb{R}^2 \backslash \frac{1}{2}\mathbb{Z}^2$, where $(r,s)$ is determined by (5.34).*

Now we consider the PDE (5.35) with the case $p = p(\tau)$, i,e, the vortex point $p$ is a holomorphic function $\tau$, such that the projective monodromy group of its developing map is a constant independent of $\tau$. This is an isomonodromic definition, and naturally it is related to the Painlevé equations, the main subject of the next section.

## 6. Painlevé VI equation

**6.1. Elliptic forms.** In literature, a nonlinear differential equation in one complex variable is said to possess the *Painlevé property* if its solutions have neither movable branch points nor movable essential singularities. For the class of second order differential equations

$$\lambda''(t) = F(t, \lambda, \lambda'), \quad t \in \mathbb{CP}^1$$

where $F(t, \lambda, \lambda')$ is meromorphic in $t$ and rational in both $\lambda$ and $\lambda'$, Painlevé (later completed by Gambier) obtained the classification of those nonlinear ODEs which possess the Painlevé property. They showed that there were fifty canonical equations of the form (6.1) with this property, up to Möbius transformations. Furthermore, of these fifty equations, forty-four are either integrable in terms of previously known functions (such as elliptic functions), equivalent to linear equations, or are reduced to one of six new nonlinear ODEs which define new transcendental functions (see e.g. [17]). These six nonlinear ODEs are called Painlevé equations. Among them, Painlevé VI is often considered to be the *master equation*, because others can be obtained from Painlevé VI by the *confluence*. Due to its connection with many different disciplines in mathematics and physics, Painlevé VI has been extensively studied in the past several decades. See [**5, 18, 19, 20, 23, 24, 25, 29, 31, 32, 33**] and the references therein.

Painlevé VI with four free parameters $(\alpha, \beta, \gamma, \delta)$ denoted by PVI$(\alpha, \beta, \gamma, \delta)$ can be written as

$$\begin{aligned}\frac{d^2\lambda}{dt^2} &= \frac{1}{2}\left(\frac{1}{\lambda}+\frac{1}{\lambda-1}+\frac{1}{\lambda-t}\right)\left(\frac{d\lambda}{dt}\right)^2 \\ &\quad -\left(\frac{1}{t}+\frac{1}{t-1}+\frac{1}{\lambda-t}\right)\left(\frac{d\lambda}{dt}\right) \\ &\quad +\frac{\lambda(\lambda-1)(\lambda-t)}{t^2(t-1)^2}\left[\alpha+\beta\frac{t}{\lambda^2}+\gamma\frac{t-1}{(\lambda-1)^2}+\delta\frac{t(t-1)}{(\lambda-t)^2}\right].\end{aligned} \tag{6.1}$$

An important fact to the PVI (6.1) says that PVI (6.1) possesses the *Painlevé property*. Notice that the branch points of PVI (6.1) which do not depend on any particular solutions, consist of $\{0, 1, \infty\}$ only. From the Painlevé property, any solution $\lambda(t)$ is a multi-valued meromorphic function in $\mathbb{CP}^1\backslash\{0, 1, \infty\}$. To avoid the multi-valueness of $\lambda(t)$, it is better to lift solutions of (6.1) to its universal covering. It is known that the universal covering of $\mathbb{CP}^1\backslash\{0, 1, \infty\}$ is $\mathbb{H}$. Then $t$ and solution $\lambda(t)$ can be lifted to $\tau$ and $p(\tau)$ respectively through the covering map by

$$t(\tau) = \frac{e_3(\tau) - e_1(\tau)}{e_2(\tau) - e_1(\tau)} \text{ and } \lambda(t) = \frac{\wp(p(\tau)|\tau) - e_1(\tau)}{e_2(\tau) - e_1(\tau)}. \tag{6.2}$$

Then $p(\tau)$ satisfies the following elliptic form

$$\frac{d^2 p(\tau)}{d\tau^2} = \frac{-1}{4\pi^2}\sum_{i=0}^{3}\alpha_i\wp'\left(p(\tau)+\frac{\omega_i}{2}|\tau\right), \tag{6.3}$$

where

$$(\alpha_0, \alpha_1, \alpha_2, \alpha_3) = (\alpha, -\beta, \gamma, \frac{1}{2}\delta).$$

This elliptic form was already known to Painlevé [**32**]. For a modern proof, see [**2, 30**]. The advantage of (6.3) is that $\wp(p(\tau)|\tau)$ is single-valued for $\tau \in \mathbb{H}$, although $p(\tau)$ has a branch point at those $\tau_0$ such that $p(\tau_0) \in E_{\tau_0}$[2].

**6.2. Isomonodromy and Hamiltonian systems.** Another important feature of (6.1) is that any solution $\lambda(t)$ (or $p(\tau)$) governs the isomonodromic deformation of some linear ODE (see [**19, 24**]). There are several different types of linear ODE associated with the Painlevé VI. Here we only discuss the second order ODE associated with (6.1).

Consider a Heun second order equation with five regular points $\{0, 1, \infty, t, \lambda\}$,

$$y''(x) + p_1(x)y'(x) + p_1(x)y(x) = 0, \tag{6.4}$$

with the following Riemann scheme

$$\begin{pmatrix} 0 & 1 & t & \infty & \lambda \\ 0 & 0 & 0 & \alpha & 0 \\ \theta_1 & \theta_2 & \theta_3 & \alpha+\theta_\infty & 2 \end{pmatrix}. \tag{6.5}$$

Here we always assume

$$\theta_i \notin \mathbb{Z},\ i=1,2,3,\infty \text{ and } \lambda \text{ is an apparent singular point.} \tag{6.6}$$

The classical Fuchsian relation implies that

$$\alpha = -\frac{1}{2}\{\sum_{i=1}^{3}\theta_i + \theta_\infty - 1\}. \tag{6.7}$$

The coefficients $p_1(x)$ and $p_2(x)$ are uniquely determined by the conditions (6.5) and (6.6) up to two constants $\mu$ and $K$:

$$\mu = \operatorname*{Res}_{x=\lambda} p_2(x) \text{ and } K = -\operatorname*{Res}_{x=\lambda} p_2(x). \tag{6.8}$$

Indeed, a straightforward computation shows

$$\begin{aligned} p_1(x) &= \sum_{i=1}^{3}\frac{1-\theta_i}{x-t_i} - \frac{1}{x-\lambda} \text{ and} \\ p_2(x) &= \frac{\kappa}{x(x-1)} - \frac{t(t-1)K}{x(x-1)(x-t)} + \frac{\lambda(\lambda-1)\mu}{x(x-1)(x-\lambda)}, \end{aligned} \tag{6.9}$$

where

$$t_1 = 0,\ t_2 = 1,\ t_3 = t,$$

and

$$\kappa = \frac{1}{4}[(\sum_{i=1}^{3}\theta_i - 1)^2 - \theta_\infty^2]. \tag{6.10}$$

Now suppose that Heun ODE (6.4) with the fixed Riemann scheme (6.5) keep the same monodromy representation while $t$ is deformed.

THEOREM 6.1. *Suppose (6.6). Then the monodromy representation of (6.4) with fixed Riemann scheme (6.5) is invariant under the deformation of $t$ iff $(\lambda(t), \mu(t))$ satisfies the Hamiltonian system*

$$\begin{cases} \frac{d\lambda(t)}{dt} &= \frac{\partial}{\partial\mu}K(\lambda,\mu,t), \\ \frac{d\mu(t)}{dt} &= -\frac{\partial}{\partial\lambda}K(\lambda,\mu,t), \end{cases} \tag{6.11}$$

*where*

$$\begin{aligned} K(\lambda,\mu,t) = \frac{1}{t(t-1)}[&\lambda(\lambda-1)(\lambda-t)\mu^2 - \{\theta_2(\lambda-1)(\lambda-t) \\ &+ \theta_3\lambda(\lambda-t) + (\theta_1-1)\lambda(\lambda-1)\}\mu + \kappa\lambda]. \end{aligned} \tag{6.12}$$

The Hamiltonian system (6.11) with $K$ given by (6.12) plays an important role in the application of the Painlevé VI. Indeed, this Hamiltonian system is equivalent to the Painlevé equation, as shown in the following theorem.

THEOREM 6.2. **[24]** *Suppose $\lambda(t)$ and $\mu(t)$ satisfies the Hamiltonian system (6.11) with the Hamiltonian $K(\lambda, \mu, t)$ iff $\lambda(t)$ is a solution of Painlevé VI (6.1).*

The parameters of the Hamiltonian system and the Painlevé VI has the relation:

$$(\alpha, \beta, \gamma, \delta) = \left(\tfrac{1}{2}\theta_4^2, -\tfrac{1}{2}\theta_1^2, \tfrac{1}{2}\theta_2^2, \tfrac{1}{2}\left(1-\theta_3^2\right)\right).$$

We remark that $\mu$ can be solved in terms of $\lambda$ and $\lambda'$ from the first equation of (6.11) (Note that $\frac{\partial K}{\partial \mu}$ is linear in $\mu$). Thus, the necessary part of Theorem 6.2 says that $\mu(t)$ satisfies the second equation of (6.11) if $\lambda(t)$ satisfies (6.1). The proof of Theorem 6.2 is a direct computation only. We note that the Painlevé VI (6.1) and the Hamiltonian system are equivalent without any condition. Nevertheless, the isomonodromy of (6.4) is equivalent to the Hamiltonian under the non-resonant condition $\theta_i \notin \mathbb{Z}$. We refer the readers to **[24]** for the excellent introduction of Painlevé equations and the proof of Theorem 6.1. and Theorem 6.2.

For the elliptic form (6.3) PVI($\alpha_i$), any solution $p(\tau)$ of (6.3) governs the isomonodromic deformation of a GLE:

$$y'' = I(z|\tau)y, \tag{6.13}$$

where

$$\begin{aligned} I(z|\tau) =& \sum n_i(n_i+1)\wp(z-\frac{\omega_i}{2}) + \frac{3}{4}(\wp(z+p)-\wp(z-p)) \\ &+ A(z)(\zeta(z+p)-\zeta(z-p)) + B, \end{aligned}$$

and the non-resonant condition is

$$n_i \notin \frac{1}{2} + \mathbb{Z}. \tag{6.14}$$

The Hamiltonian system is

$$\begin{cases} \frac{dp(\tau)}{d\tau} &= \frac{\partial \mathcal{K}}{\partial A}, \\ \frac{dA(\tau)}{d\tau} &= -\frac{\partial \mathcal{K}}{\partial p}, \end{cases} \tag{6.15}$$

where the Hamiltonian $\mathcal{K}$ is given by

$$\begin{aligned} \mathcal{K}(p, A, \tau) = \frac{-i}{4\pi}[A^2 &+ (2p\eta_1(\tau) - \zeta(2p|\tau))A - \frac{3}{4}\wp(2p|\tau) \\ &- \sum_{k=0}^{3} n_k(n_k+1)\wp(p+\frac{\omega_k}{2}|\tau)]. \end{aligned} \tag{6.16}$$

The equivalence of isomonodromy, the Hamiltonian and the elliptic form (6.4) are given in the following theorem:

THEOREM 6.3.

(i) *Suppose $n_i \notin \frac{1}{2} + \mathbb{Z}$, $i = 0, 1, 2, 3$. Then the monodromy representation of (6.13) is invariant under the deformation of $\tau$ iff $(p(\tau), A(\tau))$ satisfies the Hamiltonian system (6.15).*

(ii) *The pair* $(p(\tau), A(\tau))$ *satisfies the Hamiltonian system (6.15) iff* $p(\tau)$ *satisfies the elliptic form (6.3). Moreover the relation of parameters of (6.3) and (6.13) is*

$$\alpha_i = \frac{1}{2}(\frac{1}{2} + n_i)^2, \quad 0 \le i \le 3. \tag{6.17}$$

**6.3. Hitchin's theorem and applications.** The expression of the elliptic form (6.3) looks beautiful and more attractive than (6.1). For example, if $\alpha_i = 0, \forall i$, then $p(\tau)$ is linear in $\tau$, i.e., $p(\tau) = r + s\tau$. The corresponding $\lambda(t)$ (see (6.1)) is called the Picard solution. This is the first example of solutions with explicit expression. Another remarkable example is the Hitchin theorem [**22**] for PVI with the parameters: $\alpha_i = \frac{1}{8}$.

THEOREM 6.4 (Hitchin). [**22**] *For any* $(r, s) \in \mathbb{C}^2 \backslash \frac{1}{2}\mathbb{Z}^2$, *let* $p_{r,s}(\tau)$ *be given by*

$$\begin{aligned} \wp(p_{r,s}(\tau)|\tau) &= \wp(r + s\tau|\tau) \\ &\quad + \frac{\wp'(r + s\tau|\tau)}{2(\zeta(r + s\tau|\tau) - r\eta_1(\tau) - s\eta_2(\tau))}. \end{aligned} \tag{6.18}$$

*Then* $p_{r,s}(\tau)$ *is a solution to the elliptic form (6.3) with* $\alpha_i = \frac{1}{8}$, $i = 0, 1, 2, 3$.

There are several ways to prove the Hitchin theorem. One is to use Theorem 6.1 to find that the monodromy representation of (6.13) with $n = 0$ is generated by $\begin{pmatrix} e^{2\pi i s} & 0 \\ 0 & e^{-2\pi i s} \end{pmatrix}$ and $\begin{pmatrix} e^{2\pi i r} & 0 \\ 0 & e^{-2\pi i r} \end{pmatrix}$. This calculation of the monodromy representation has been done in §5. In (5.28), $y_{\pm a}(z)$ are two independent solutions of (5.10) iff $a \notin E_\tau[2]$,

$$\begin{aligned} A &= \frac{1}{2}\left[\zeta(p + a) - \zeta(p - a) - \zeta(2p)\right], \\ B &= A^2 - \zeta(2p)A - \frac{3}{4}\wp(2p). \end{aligned} \tag{6.19}$$

The monodromy data of (5.25) with (6.19) and $n = 0$ are provided by the pair (r,s) solving the $2 \times 2$ linear system

$$\begin{aligned} &r + s\tau = a, \\ &2(r\eta_1(\tau) + s\eta_2(\tau)) = \zeta(a + p) + \zeta(a - p). \end{aligned} \tag{6.20}$$

See Theorem 5.8.

Recall the addition theorem

$$\frac{\wp'(a)}{\wp(a) - \wp(p)} = \zeta(a + p) + \zeta(a - p) - 2\zeta(a).$$

PROOF OF HITCHIN THEOREM. Suppose $p_{r,s}(\tau)$ is a smooth branch given by (6.18). Without loss of generality, we may assume $p_{r,s}(\tau_0) \notin E_{\tau_0}[2]$. Then, $p_{r,s}(\tau)$ is uniquely determined by (6.18) and $p_{r,s}(\tau) \to p_{r,s}(\tau_0)$ as $\tau \to \tau_0$. Let $a(\tau) = r + s\tau$ and $(A(\tau), B(\tau))$ are given by (6.19). Let (6.13) be the GLE with $(p_{r,s}(\tau), A(\tau), B(\tau))$ and the monodromy data are given by $(\bar{r}, \bar{s})$ given by (6.20) with $a = a(\tau)$ and $p = p_{r,s}(\tau)$. In a small neighborhood of $\tau_0 \notin E_\tau[2]$, (6.18) and the addition theorem implies

$$\zeta(a(\tau)+p(\tau))+\zeta(a(\tau)-p(\tau))-2\zeta(a(\tau))=2(r\eta_1(\tau)+s\eta_2(\tau)-2\zeta(a(\tau)))$$

which implied

$$\zeta(a+p)+\zeta(a-p)=r\eta_1(\tau)+s\eta_2(\tau). \tag{6.21}$$

Together with $a(\tau)=r+s\tau$, it yields $(\bar{r},\bar{s})=(r,s)$. Therefore, the monodromy representation of GLE (6.13) with $(A(\tau),B(\tau))$ given by (6.19) is invariant under the deformation of $\tau$. By Theorem 6.3, $p_{r,s}(\tau)$ satisfies the Painlevé VI equation with $\alpha_i=\frac{1}{8},\forall i$. □

Conversely if $p(\tau)$ is a solution of PVI($\frac{1}{8}$) and assume the monodromy representation associated with GLE (6.13) is completely reducible, then our previous work [**10**] shows

$$\wp(p(\tau))=\wp(p_{r,s}(\tau)) \text{ for some } r,s\in\mathbb{C}\backslash\frac{1}{2}\mathbb{Z}^2.$$

Thus, the formula (6.18) gives all the completely reducible solutions of PVI($\frac{1}{8}$).

It is remarkable that the denominator of (6.18) is the Hecke form $Z_{r,s}(\tau)$, which is discussed in previous sections. The Painlevé property says that if $\tau_0$ is a pole of $\wp(p(\tau)|\tau)$, then $\tau_0$ is a simple pole. This Painlevé property implies

THEOREM 6.5. [**10**] *Suppose* $(r,s)\in\mathbb{C}^2\backslash\frac{1}{2}\mathbb{Z}^2$. *Then* $Z_{r,s}(\tau)$ *has only simple zeros.*

PROOF. First, we note that the situations $r+s\tau\in E_\tau[2]$ and $Z_{r,s}(\tau)=0$ can not occur simultaneously. If not, then there are $\tau_0$ and $m,n\in\mathbb{Z}$ such that $r+s\tau_0=m+n\tau_0+\frac{\omega}{2}$, where $\omega\in\Lambda_\tau$, and also $\zeta(r+s\tau_0)=r\eta_1(\tau_0)+s\eta_2(\tau_0)$. Without loss of generality, we might assume $\omega=\omega_1$. The other cases can be proved similarly.

The zero of denominator of (6.18) implies

$$\begin{aligned}\frac{1}{2}\eta_1(\tau_0)&=\zeta(\frac{\omega_1}{2})=\zeta((r-m)+(s-n)\tau_0)\\&=\zeta(r+s\tau_0)-m\eta_1(\tau_0)-n\eta_2(\tau_0)\\&=(r-m)\eta_1(\tau_0)+(s-n)\eta_2(\tau_0).\end{aligned}$$

Therefore, we have

$$\begin{aligned}(r-m-\frac{1}{2})+(s-n)\tau_0=0,\\(r-m-\frac{1}{2})\eta_1(\tau_0)+(s-n)\eta_2(\tau_0)=0,\end{aligned}$$

which implies $(r-m-\frac{1}{2})=0$ and $s=n$ because of the matrix $\begin{pmatrix}1&\tau\\\eta_1(\tau)&\eta_2(\tau)\end{pmatrix}$ is non-degenerate for any $\tau$ due to the Legendre relation. Obviously it contradicts to the assumption $(r,s)\notin\frac{1}{2}\mathbb{Z}^2$.

Now suppose $Z_{r,s}(\tau_0) = 0$, which implies $\wp(p(\tau_0)) = \infty$ by (6.18) because $\wp(r + s\tau_0) \neq 0$. Consider the transformation $\tau_0 \mapsto t_0$ via (6.2). Then by the Painlevé property, we know that $\lambda(t)$ has a pole at $t = t_0 \notin \{0, 1, \infty\}$. By substituting the local expansion of $\lambda(t)$ at $t = t_0$ into (6.1), it is easy to prove that the order of pole at $t = t_0$ is 1, which implies the zero of $Z_{r,s}$ at $\tau = \tau_0$ is simple. □

Theorem 6.5 is the first application of the Painlevé property. In the next subsection, we will give another application of this property.

**6.4. The smoothness of the degenerate curves.** Next we are going to discuss another application of PVI($\frac{1}{8}$). As we explained, (6.18) consists of only completely reducible solutions of PVI($\frac{1}{8}$). The other non-completely reducible solutions come from the solutions of Riccati equations.

THEOREM 6.6. [**10**] *There are four Riccati equations reduced from Painlevé VI with* $(\alpha, \beta, \delta, \gamma) = (\frac{1}{8}, -\frac{1}{8}, \frac{1}{8}, \frac{3}{8})$. *The complete list of solutions to these four Riccati equations are given by:*

$$\lambda_C(t) = \frac{\wp(p_C(\tau)|\tau) - e_1(\tau)}{e_2(\tau) - e_1(\tau)}, \quad t = \frac{e_3(\tau) - e_1(\tau)}{e_2(\tau) - e_1(\tau)},$$

*where* $p_C(\tau)$ *is given by the following (i) and (ii).*

*(i)* $\lambda_C(t)$ *solves the Riccati equation*

$$\lambda'(t) = -\frac{1}{2t(t-1)}(\lambda^2 - 2t\lambda + t)$$

*if and only if there exists constant* $C \in \mathbb{C} \cup \{\infty\}$ *such that*

$$\wp(p_C(\tau)|\tau) = \frac{\eta_2(\tau) - C\eta_1(\tau)}{C - \tau} = -\eta_1 + \frac{2\pi i}{\tau - C}, \tag{6.22}$$

$$\eta_2 = \tau\eta_1 - 2\pi i.$$

*(ii) For* $C \in \mathbb{C} \cup \{\infty\}$, *we let*

$$\wp(p_C(\tau)|\tau) = \frac{e_k(C\eta_1(\tau) - \eta_2(\tau)) + (\frac{g_2}{4} - 2e_k^2)(C - \tau)}{C\eta_1(\tau) - \eta_2(\tau) + e_k(C - \tau)}, \quad k = 1, 2, 3$$

*Then* $\lambda_C(t)$ *satisfies the Riccati equation*

$$\lambda'(t) = -\frac{1}{2t(t-1)}\left(\lambda^2 - 2t\lambda + t\right), \quad \textit{for } k = 1$$

$$\lambda'(t) = -\frac{1}{2t(t-1)}\left(\lambda^2 - t\right), \quad \textit{for } k = 2$$

$$\lambda'(t) = -\frac{1}{2t(t-1)}\left(\lambda^2 + 2(t-1)\lambda + t\right), \quad \textit{for } k = 3.$$

*Furthermore, such* $\lambda_C(t)$ *give all the solutions of these three Riccati equations in (ii).*

A simple observation is that $C = \infty$ in (6.22) gives that

$$\lambda_\infty(t) = -\frac{\eta_1(\tau) + e_1(\tau)}{e_2(\tau) - e_1(\tau)} \tag{6.23}$$

is a solution of $\mathrm{PVI}_{(\frac{1}{8},-\frac{1}{8},\frac{1}{8},\frac{3}{8})}$. Since $\lambda_\infty(t)$, $\lambda_\infty(t)-1$ and $\lambda_\infty(t)-t$ can have only simple zeros, a direct consequence is

THEOREM 6.7. [**10**] *For fixed $k \in \{1,2,3\}$, the followings hold:*

*(i) Any zero of $\eta_1(\tau) + e_k(\tau)$ must be simple.*

*(ii)*

$$\frac{\mathrm{d}}{\mathrm{d}\tau}\left((\eta_1(\tau) - e_k(\tau))^{-1}\right) \neq \frac{1}{2\pi i} \quad \textit{for any } \tau \in \mathbb{H}. \tag{6.24}$$

*(iii)* $\tau - \frac{2\pi i}{\eta_1(\tau)-e_k(\tau)} = \frac{\eta_2(\tau)-\tau e_k(\tau)}{\eta_1(\tau)-e_k(\tau)}$ *is a locally one-to-one map from $\mathbb{H}$ to $\mathbb{C}\cup\{\infty\}$.*

PROOF. Recall

$$t = t(\tau) = \frac{e_3(\tau) - e_1(\tau)}{e_2(\tau) - e_1(\tau)}.$$

Since $t'(\tau) \neq 0$ for all $\tau \in \mathbb{H}$, the assertion (i) follows readily from the fact that $\lambda_\infty(t)$ (for $k=1$), $\lambda_\infty(t)-1$ (for $k=2$) and $\lambda_\infty(t)-t$ (for $k=3$) can have only simple zeros.

For the assertion (ii), we note from the Legendre relation that $\lambda_C(t)$ defined by (6.22) is

$$\lambda_C(t) = \frac{\wp(p_C(\tau)|\tau) - e_1(\tau)}{e_2 - e_1(\tau)} = \frac{-\eta_1(\tau) - e_1(\tau) + \frac{2\pi i}{\tau - C}}{e_2(\tau) - e_1(\tau)}.$$

Fix any $\tau_0 \in \mathbb{H}$. If $\tau_0$ is a zero of $\eta_1 + e_1$, then by (i) or Painlevé property of $\lambda_\infty(t)$, $\tau_0$ is a simple pole. So $\tau - \frac{2\pi i}{\eta_1(\tau)-e_1(\tau)}$ is locally one-one near $\tau_0$. So it suffices to consider the case $\eta_1(\tau_0) + e_1(\tau_0) \neq 0$. Then by letting

$$C = \tau_0 - \frac{2\pi i}{\eta_1(\tau_0) - e_k(\tau_0)},$$

we see that $t_0 = t(\tau_0)$ is a zero of $\lambda_C(t)$. Since $\lambda_C(t)$ has only simple zeros, we have

$$\frac{\mathrm{d}}{\mathrm{d}\tau}\left(\eta_1(\tau) - e_1(\tau) - \frac{2\pi i}{\tau - C}\right)\bigg|_{\tau=\tau_0} \neq 0.$$

This, together with $\eta_1(\tau_0) - e_1(\tau_0) - \frac{2\pi i}{\tau_0 - C} = 0$, easily implies

$$\frac{\mathrm{d}}{\mathrm{d}\tau}\left((\eta_1(\tau) - e_k(\tau))^{-1}\right) \neq \frac{1}{2\pi i}.$$

This proves (6.24) for $k=1$. Similarly, by considering $\lambda_C(t)-1$ and $\lambda_C(t)-t$, we can prove (6.24) for $k=2,3$. This prove the assertion (ii).

Finally, using the Legendre relation leads to

$$\frac{\eta_2(\tau) - \tau e_k(\tau)}{\eta_1(\tau) - e_k(\tau)} = \tau - \frac{2\pi i}{\eta_1(\tau) - e_k(\tau)}.$$

Therefore, $\frac{\eta_2(\tau)-\tau e_k(\tau)}{\eta_1(\tau)-e_k(\tau)}$ is locally one-to-one. This completes the proof. □

We have already proved that the critical points of Green function $G(z|\tau)$ could bifurcate only at half periods. It is easy to see that the bifurcation could occur at $\frac{\omega_i}{2}$ iff the Hessian of $G(z|\tau)$ at $z = \frac{\omega_i}{2}$ vanishes. By a direct computation, we have

$$\begin{aligned}\det \mathrm{D}_z^2 G(\frac{\omega_k}{2}|\tau) &= -\frac{1}{4\pi^2 \mathrm{Im}\tau}|\eta_1(\tau)+e_k(\tau)|^2 \\ &\quad \mathrm{Im}\left\{\tau - \frac{2\pi i}{\eta_1(\tau)+e_k(\tau)}\right\}.\end{aligned} \tag{6.25}$$

Recall $C_k$ to be the curve of those $\tau \in F_2$ such that $\frac{\omega_k(\tau)}{2}$ is degenerate. Then Theorem 6.7 implies $C_k$ is smooth.

COROLLARY 6.8. [**10**] *The degenerate curve* $C_k$, $k = 1, 2, 3$, *is smooth.*

To conclude this section, we state another application of Theorem 6.7.

THEOREM 6.9. [**10**] *For any fundamental domain* $G$ *of* $\Gamma(2)$, $\eta_1(\tau) + e_k(\tau)$ *has at most a zero in* $F$.

PROOF. Note that if $e_k(\tau) + \eta_1(\tau) \neq 0$, then $\tau \in C_k$ if and only if $\mathrm{Im}\,(\tau - \frac{2\pi i}{e_k(\tau)+\eta_1(\tau)}) = 0$. Theorem 6.7 says that $e_k(\tau)+\eta_1(\tau)$ has only simple zeros, and any zero of $e_k(\tau) + \eta_1(\tau)$ lies on the curve $C_k$.

Clearly $e_k(\tau) + \eta_1(\tau)$ is not a modular form. However, if $\tau' = \gamma \cdot \tau$ for some $\gamma \in SL(2, \mathbb{Z})$, then $E_{\tau'}$ is conformally equivalent to $E_\tau$ and transforms $C_k$ to another degenerate curve $C_j$. Therefore, without loss of generality, we may assume $k = 1$.

From the map $z \to \frac{z}{c\tau+d}$, if $\gamma \in \Gamma_0(2) = \{\gamma \in \mathrm{SL}(2, \mathbb{Z}) \mid c \equiv 0 \pmod 2\}$, the image of $C_1(F_2)$ is mapped to $C_1(F_2')$ for another fundamental domain $F_2' := \gamma(F_2)$. For example, if $\gamma = TS^{-1}T^2S^{-1} = \begin{pmatrix} 1 & -1 \\ 2 & -1 \end{pmatrix}$, i.e.

$$\tau' = \gamma \cdot \tau = \frac{\tau - 1}{2\tau - 1},$$

then $F_2'$ is the domain bounded by 3 half circles:

$$F_2' = \{\tau \in \mathbb{H} \mid |\tau - \tfrac{1}{2}| \leq \tfrac{1}{2}, |\tau - \tfrac{1}{4}| \geq \tfrac{1}{4}, |\tau - \tfrac{3}{4}| \geq \tfrac{1}{4}\}.$$

Noting that the curve $\{\tau \mid |\tau - \frac{1}{2}| = \frac{1}{2}\}$ is invariant under $\gamma$, the curves $\overline{C_1(F_2)}$, $\overline{C_1(F_2')}$ bound a *simply connected domain* $\mathcal{D}$ in

$$F \doteqdot F_2 \cup F_2' = \{\tau \in \mathbb{H} \mid 0 \leq \mathrm{Re}\,\tau \leq 1, |\tau - \tfrac{1}{4}| \geq \tfrac{1}{4}, |\tau - \tfrac{3}{4}| \geq \tfrac{1}{4}\},$$

where $\overline{\mathcal{D}} \cap \mathbb{R} = \{0, 1\}$. Since $\Gamma_0(2) = \Gamma(2) \cup \gamma\Gamma(2)$, $F_2 \cup F_2'$ is also a fundamental domain of $\Gamma(2)$. Note that for any $\tau \in \mathcal{D}$, the half period $\frac{1}{2}$ is a minimum point of $G(z|\tau)$ in $E_\tau$, and (6.25) yields that $\frac{1}{2}$ is actually a non-degenerate critical point of $G(z|\tau)$. Therefore $\det D_z^2(\frac{1}{2}|\tau) > 0$ and then $\mathrm{Im}\{\tau - \frac{2\pi i}{\eta_1+e_1}\} > 0$ $\forall \tau \in D$. For simplicity of notation, we denote $\tau - \frac{2\pi i}{\eta_1(\tau)+e_1(\tau)}$ by $f(\tau)$. Then Theorem 6.7 implies $f(\tau)$ is a locally one-one map from $\mathcal{D}$ to the lower half plane $\mathbb{C}_- = \{\kappa \mid \mathrm{Im}\,\kappa < 0\}$.

Obviously, the map $f(\tau)$ maps $\overline{C_1(F_2)\cup C_1(F_2')}$ to the real axis. By [**28**, Theorem 1.6],

$$C_1(F_2)\cap\{\tau \mid \operatorname{Re}\tau=\tfrac{1}{2}\}=\{\tfrac{1}{2}+ib_1\}, \tag{6.26}$$

where $b_1\in(\frac{1}{2},\frac{\sqrt{3}}{2})$ is the unique zero of the increasing function in $b$

$$e_1+\eta_1-\frac{2\pi}{b} \tag{6.27}$$

along the vertical line $\frac{1}{2}+ib$. Similarly, $C_1(F_2')\cap\{\tau\mid \operatorname{Re}\tau=\frac{1}{2}\}=\{\frac{1}{2}+ib_0\}$ where $b_0\in(0,\frac{1}{2})$ is the unique zero of the increasing function $e_1+\eta_1$ along $\frac{1}{2}+ib$. Then

$$f(\tfrac{1}{2}+ib)=2(\tfrac{1}{2}+ib)-\frac{4\pi i}{e_1+\eta_1}=1+\frac{2bi}{e_1+\eta_1}\Big(e_1+\eta_1-\frac{2\pi}{b}\Big).$$

In particular, $f(\frac{1}{2}+ib_1)=1$, $f(\frac{1}{2}+ib_0)=1-i\infty$. We recall the classical transformation law for $\eta_1$:

$$\eta_1(\tau')=(c\tau+d)^2\eta_1(\tau)-2\pi ic(c\tau+d),$$

for $\tau'=\gamma\cdot\tau$, $\gamma\in SL(2,\mathbb{Z})$. In particular, $\eta_1(\frac{1}{\tau})=\tau^2\eta_1(\tau)-2\pi i\tau$. From those transformation, we have $\eta_1(\tau)+e_1(\tau)\to\infty$ when $\tau\in C_1(F_2)\cup C_1(F_2')$ tends to the boundary point 0 (resp. 1), we have that $f(\tau)\to 0$ (resp. 2). Therefore $f$ maps $\overline{C_1(F_0)}$ and $\overline{C_1(F_0')}$ onto $[0,2]$ and $\mathbb{R}\cup\{\infty\}\backslash(0,2)$ respectively. Then $f(\tau)=\infty$ has only one solution in $F$ at $\tau=\frac{1}{2}+ib_0$.

Now let $G$ be any fundamental domain of $\Gamma(2)$ and $G=\gamma\cdot F$. Let $p$ and $q$ are the cusps corresponding to 0 and 1 by $\gamma$. It is easy to see that from the transformation $\eta_1(\gamma\cdot\tau)=(c\tau+d)^2\eta_1(\tau)-2\pi ic(c\tau+d)$, we have $\eta_1(\tau)+e_1(\tau)\to\infty$ as $\tau\to p,q$. Thus this proves the theorem. □

REMARK 6.10. In the proof, we see that $f(\tau)$ maps $\bar{D}$ to $C_-\cup\mathbb{R}$ locally one-one. It is easy to see that the map is onto and globally one-one.

## 7. A generalization of the Hitchin theorem

**7.1. The statement.** In this section, we want to give a proof of Theorem 4.10. Recall the premodular form $Z_{r,s}^{(n)}(\tau)$ associated with the hyperelliptic curve $Y^{(n)}(\tau)$. For convenience, we use $Z_{r,s}^{(0)}(\tau)$ and $Z_{r,s}^{(1)}$ to denote 1 and $Z_{r,s}(\tau)$, respectively. Recall that in the last section, we use the Hitchin Theorem to obtain that $Z_{r,s}(\tau)$ satisfies the simple zeros property. The following theorem is a generalization of Hitchin theorem to Painlevé VI$(\alpha_i)$, where

$$\alpha_0=\frac{1}{2}(n+\frac{1}{2})^2 \text{ and } \alpha_i=\frac{1}{8},\quad 1\le i\le 3. \tag{7.1}$$

THEOREM 7.1. [**9**] *There exists a polynomial $P_{n+1}(X;\sigma)\in Q[g_2,g_3,\wp(\sigma),\wp'(\sigma),X]$ such that for any $(r,s)\in\mathbb{C}^2\backslash\frac{1}{2}\mathbb{Z}^2$, the following $p_{r,s}^{(n)}(\tau)$ sat-*

*isfies the Painlevé VI($\alpha_i$), where $\alpha_i$ is given by (7.1) and*

$$\wp(p_{r,s}^{(n)}(\tau)) = \wp(r+s\tau) + \frac{P_{n+1}\left(Z_{r,s}(\tau); r+s\tau\right)}{Z_{r,s}^{(n-1)}(\tau) Z_{r,s}^{(n+1)}(\tau)}. \tag{7.2}$$

*Furthermore, the followings holds:*

*(i) any two polynomials of $W^{(n-2)}(X;\sigma)$, $W^{(n)}(X;\sigma)$ and $P_{n+1}(X;\sigma)$ has no common roots for any $\sigma \notin E_\tau[2]$, where $W^{(n)}(\tau) \equiv 1$ if $n \leq 0$.*

*(ii) any two polynomials of $W^{(n)}(X;\sigma)$, $W^{(n+1)}(X,\sigma)$ and $W^{(n+2)}(X,\sigma)$ has no common roots for any $\sigma \notin E_\tau[2]$.*

*(iii) The monodromy group of the GLE associated with $p_{r,s}^{(n)}(\tau)$ is generated by $\begin{pmatrix} e^{2\pi i s} & 0 \\ 0 & e^{-2\pi i s} \end{pmatrix}$ and $\begin{pmatrix} e^{-2\pi i r} & 0 \\ 0 & e^{2\pi i r} \end{pmatrix}$. Conversely, any completely reducible solutions are given by (7.2).*

Due to (i), the denominator and numerator of the RHS in (7.2) can not vanish simultaneously. So we have

Corollary 7.2. *Suppose $Z_{r,s}^{(n)}(\tau_0) = 0$, then $\tau_0$ is a simple zero of $Z_{r,s}^{(n)}(\tau)$ if $(r,s) \in \mathbb{C}\backslash\frac{1}{2}\mathbb{Z}^2$ and $r+s\tau_0 \notin E_{\tau_0}[2]$.*

The proof of Theorem 7.1 is to apply the Okamoto transformation. The Okamoto transformation is a bi-rational transformation of the solutions of the Hamiltonian systems related to Painlevé VI($\frac{1}{8}$) and Painlevé VI($\alpha_i$) of (7.1). See [**23, 25, 31**]. The formula for this Okamoto transformation is very complicated. In the followings, we will briefly describe it.

**7.2. Okamoto transformation.** We recall that PVI is equivalent to the following Hamiltonian system

$$\frac{d\lambda(t)}{dt} = \frac{\partial K}{\partial \mu}, \quad \frac{d\mu(t)}{dt} = -\frac{\partial K}{\partial \lambda}, \tag{7.3}$$

where $K = K(\lambda, \mu, t)$ is given by

$$K = \frac{1}{t(t-1)} \left\{ \begin{array}{c} \lambda(\lambda-1)(\lambda-t)\mu^2 + \theta_0(\theta_0+\theta_4)(\lambda-t) \\ -\left[ \begin{array}{c} \theta_1(\lambda-1)(\lambda-t) + \theta_2\lambda(\lambda-t) \\ +(\theta_3-1)\lambda(\lambda-1) \end{array} \right] \mu \end{array} \right\}, \tag{7.4}$$

and the relation of parameters is given by

$$(\alpha,\beta,\gamma,\delta) = \left(\tfrac{1}{2}\theta_4^2, -\tfrac{1}{2}\theta_1^2, \tfrac{1}{2}\theta_2^2, \tfrac{1}{2}\left(1-\theta_3^2\right)\right), \tag{7.5}$$
$$\theta_0 := \tfrac{1}{2}(1-\theta_1-\theta_2-\theta_3-\theta_4),$$

$$\text{i.e.,} \quad 2\theta_0+\theta_1+\theta_2+\theta_3+\theta_4 = 1. \tag{7.6}$$

It is known that solutions of PVI with parameters

$$(\alpha,\beta,\gamma,\delta) = \left(\tfrac{1}{2}(n_0+\tfrac{1}{2})^2, \ -\tfrac{1}{2}(n_1+\tfrac{1}{2})^2, \ \tfrac{1}{2}(n_2+\tfrac{1}{2})^2, \right.$$
$$\left. \tfrac{1}{2} - \tfrac{1}{2}(n_3+\tfrac{1}{2})^2\right), \ n_k \in \mathbb{N}\cup\{0\} \text{ for all } k. \tag{7.7}$$

could be obtained from solutions of PVI$(\frac{1}{8}, \frac{-1}{8}, \frac{1}{8}, \frac{3}{8})$ (i.e., $n_k = 0$ for all $k$) via the Okamoto transformation ([**31**]).

First we recall the explicit form of the Okamoto transformations. By (7.5) and (7.6), it is convenient to think of the parameter space of PVI (7.1) (equivalently the Hamiltonian system (7.3)–(7.4)) as an affine space

$$\mathcal{K} = \{\theta = (\theta_0, \theta_1, \theta_2, \theta_3, \theta_4) \in \mathbb{C}^5 \ : \ 2\theta_0 + \theta_1 + \theta_2 + \theta_3 + \theta_4 = 1\}.$$

An Okamoto transformation $\sigma$ maps solutions $(\lambda(t), \mu(t), t)$ of the Hamiltonian system (7.3) with parameter $\theta \in \mathcal{K}$ to solutions $\sigma(\lambda)(t)$ $(\sigma(\lambda)(t), \sigma(\mu)(t), t)$ of (7.3)) with new parameter $\sigma(\theta) \in \mathcal{K}$. The list of the Okamoto transformations $\sigma_j (0 \le j \le 4)$ is given in the Table 1 (cf. [**33**]).

TABLE 1. Okamoto transformations

| | $\theta_0$ | $\theta_1$ | $\theta_2$ | $\theta_3$ | $\theta_4$ | $t$ | $\lambda$ | $\mu$ |
|---|---|---|---|---|---|---|---|---|
| $\sigma_0$ | $-\theta_0$ | $\theta_1+\theta_0$ | $\theta_2+\theta_0$ | $\theta_3+\theta_0$ | $\theta_4+\theta_0$ | $t$ | $\lambda+\frac{\theta_0}{\mu}$ | $\mu$ |
| $\sigma_1$ | $\theta_0+\theta_1$ | $-\theta_1$ | $\theta_2$ | $\theta_3$ | $\theta_4$ | $t$ | $\lambda$ | $\mu-\frac{\theta_1}{\lambda}$ |
| $\sigma_2$ | $\theta_0+\theta_2$ | $\theta_1$ | $-\theta_2$ | $\theta_3$ | $\theta_4$ | $t$ | $\lambda$ | $\mu-\frac{\theta_2}{\lambda-1}$ |
| $\sigma_3$ | $\theta_0+\theta_3$ | $\theta_1$ | $\theta_2$ | $-\theta_3$ | $\theta_4$ | $t$ | $\lambda$ | $\mu-\frac{\theta_3}{\lambda-t}$ |
| $\sigma_4$ | $\theta_0+\theta_4$ | $\theta_1$ | $\theta_2$ | $\theta_3$ | $-\theta_4$ | $t$ | $\lambda$ | $\mu$ |

For example, the $\theta$ parameter is $(-\frac{1}{2}, \frac{1}{2}, \frac{1}{2}, \frac{1}{2}, \frac{1}{2})$ for the Hitchin equation and is $(\frac{1}{2}, 0, 0, 0, 0)$ for the Picard equation (whose solutions are called Picard solutions). From Table 1, we see $\sigma_0$ maps $(\frac{1}{2}, 0, 0, 0, 0)$ to $(-\frac{1}{2}, \frac{1}{2}, \frac{1}{2}, \frac{1}{2}, \frac{1}{2})$. Indeed the inverse transformation of $\sigma_0$ is given by

$$\hat{\lambda}(t) = \lambda(t) - \frac{1}{2\mu(t)}, \ \mu(t) = \frac{t(t-1)\lambda' + \frac{1}{2}\lambda^2 - t(\lambda - \frac{1}{2})}{2\lambda(\lambda-1)(\lambda-t)}, \tag{7.8}$$

where $(\lambda(t), \mu(t))$ is a solution of the Hamiltonian system corresponding to the Hitchin equation, and $\hat{\lambda}(t)$ is the corresponding solution of the Picard equation.

SECOND PROOF OF THE HITCHIN THEOREM. As an application of the Okamoto transformation, we give a rigorous derivation of Picard solution from Hitchin's formula by the inverse transformation $\sigma_0$. First, we derive any Picard solution can be written as

$$\hat{\lambda}(t) = \hat{\lambda}_{r,s}(t) = \frac{\wp(r+s\tau|\tau) - e_1(\tau)}{e_2(\tau) - e_1(\tau)}, \quad t = \frac{e_3(\tau) - e_1(\tau)}{e_2(\tau) - e_1(\tau)}. \tag{7.9}$$

Recall the Picard solution is $p(\tau) = r + s\tau$. From (6.2), we have $\frac{t+1}{3} = \frac{-e_1(\tau)}{e_2(\tau)-e_1(\tau)}$, and it is well known that locally the inverse function of $t(\tau) = \frac{e_3(\tau)-e_1(\tau)}{e_2(\tau)-e_1(\tau)}$ can be expressed as

$$\tau = \tau(t) = i\frac{F(\frac{1}{2}, \frac{1}{2}, 1; t)}{F(\frac{1}{2}, \frac{1}{2}, 1; 1-t)} = \frac{\omega_2(t)}{\omega_1(t)},$$

where $F(\alpha,\beta,\gamma;t)$ is the hypergeometric function. Furthermore,

$$e_2(\tau)-e_1(\tau)=-\pi^2F(\tfrac{1}{2},\tfrac{1}{2},1;1-t)^2=\omega_1(t)^2.$$

Therefore, we easily deduce from (6.2) that

$$\hat{\lambda}(t)=\frac{\wp(r+s\tau\mid 1,\tau)}{\omega_1(t)^2}-\frac{e_1(\tau)}{e_2(\tau)-e_1(\tau)}=\frac{\wp(r+s\tau|\tau)-e_1(\tau)}{e_2(\tau)-e_1(\tau)},$$

which proves (7.9).

On the other hand, the above expression of Picard solutions can be obtained from the inverse of $\sigma_0$ as follows. Let $\lambda(t)=\lambda_{r,s}(t)$ given by Hitchin's formula (6.18), namely

$$\lambda(t)=\frac{\wp(p_{r,s}(\tau)|\tau)-e_1(\tau)}{e_2(\tau)-e_1(\tau)}=\frac{\wp(r+s\tau|\tau)-e_1(\tau)+\frac{\wp'(r+s\tau|\tau)}{2Z_{r,s}(\tau)}}{e_2(\tau)-e_1(\tau)}, \tag{7.10}$$

where $Z_{r,s}$ is the Hecke form. The important thing is that, by studying the isomonodromic deformation of GLE (5.25) with $n=0$, we proved in [**10**, Theorem 4.7] (without computing $\lambda'(t)$) that

$$\begin{aligned}\mu(t)&=\frac{e_2(\tau)-e_1(\tau)}{2(\wp(p_{r,s}(\tau)|\tau)-\wp(r+s\tau|\tau))}\\&=\frac{(e_2(\tau)-e_1(\tau))Z_{r,s}(\tau)}{\wp'(r+s\tau|\tau)}.\end{aligned} \tag{7.11}$$

We do not think that it is easy to obtain (7.11) via (7.10) and the second formula of (7.8). Substituting (7.10) and (7.11) into the first formula of (7.8), we immediately obtain the same formula (7.9) of Picard solution $\hat{\lambda}(t)$. □

These five transformations $\sigma_j$ $(0\le j\le 4)$, which satisfy $\sigma_j\circ\sigma_j=Id$, generate the affine Weyl group of type $D_4^{(1)}$:

$$W(D_4^{(1)})=\langle\sigma_0,\sigma_1,\sigma_2,\sigma_3,\sigma_4\rangle\,. \tag{7.12}$$

In $W(D_4^{(1)})$ there are four special transformations that interest to us very much. Define

$$\begin{aligned}\sigma_5&:=\sigma_0(\sigma_3\sigma_2\sigma_1\sigma_0)^2\sigma_4,\\ \sigma_6&:=\sigma_0(\sigma_4\sigma_2\sigma_1\sigma_0)^2\sigma_3,\\ \sigma_7&:=\sigma_0(\sigma_4\sigma_3\sigma_1\sigma_0)^2\sigma_2,\\ \sigma_8&:=\sigma_0(\sigma_4\sigma_3\sigma_2\sigma_0)^2\sigma_1.\end{aligned} \tag{7.13}$$

A straightforward computation shows

$$\begin{aligned}\sigma_5(\theta)&=(\theta_0-1,\theta_1,\theta_2,\theta_3,\theta_4+2),\\ \sigma_6(\theta)&=(\theta_0-1,\theta_1,\theta_2,\theta_3+2,\theta_4),\\ \sigma_7(\theta)&=(\theta_0-1,\theta_1,\theta_2+2,\theta_3,\theta_4),\\ \sigma_8(\theta)&=(\theta_0-1,\theta_1+2,\theta_2,\theta_3,\theta_4).\end{aligned} \tag{7.14}$$

Then for any $n_k \in \mathbb{N} \cup \{0\}$ for all $k$, it is easy to see that
(7.15)
$$(\theta_0, \theta_1, \theta_2, \theta_3, \theta_4) = \left( -\frac{1 + \sum_{k=0}^{3} n_k}{2}, n_1 + \tfrac{1}{2}, n_2 + \tfrac{1}{2}, n_3 + \tfrac{1}{2}, n_0 + \tfrac{1}{2} \right)$$

can be obtained from $(-\frac{1}{2}, \frac{1}{2}, \frac{1}{2}, \frac{1}{2}, \frac{1}{2})$ (which corresponds to PVI$(\frac{1}{8}, \frac{-1}{8}, \frac{1}{8}, \frac{3}{8})$) by using $\sigma_j$ with $1 \leq j \leq 8$. But for Theorem 7.1, we have $n_1 = n_2 = n_3 = 0$ and only need $\sigma_5$.

LEMMA 7.3. [9] *Let $(\lambda(t), \mu(t))$ be a solution of the Hamiltonian system (7.3) with parameter $\theta \in \mathcal{K}$. Then $(\tilde{\lambda}(t), \tilde{\mu}(t)) \doteqdot (\sigma_5(\lambda)(t), \sigma_5(\mu)(t))$, which is a solution of the Hamiltonian system (7.3) with new parameter*

$$\tilde{\theta} = (\tilde{\theta}_0, \tilde{\theta}_1, \tilde{\theta}_2, \tilde{\theta}_3, \tilde{\theta}_4) \doteqdot \sigma_5(\theta) = (\theta_0 - 1, \theta_1, \theta_2, \theta_3, \theta_4 + 2),$$

*are expressed as*

$$\tilde{\lambda}(t) = \hat{\lambda}(t) + \frac{1 - \theta_0}{\tilde{\mu}(t)}, \tag{7.16}$$

$$\tilde{\mu}(t) = \hat{\mu}(t) + \frac{\theta_0 + \theta_1 - 1}{\hat{\lambda}(t)} + \frac{\theta_0 + \theta_2 - 1}{\hat{\lambda}(t) - 1} + \frac{\theta_0 + \theta_3 - 1}{\hat{\lambda}(t) - t}, \tag{7.17}$$

*where*

$$\hat{\lambda}(t) = \bar{\lambda}(t) + \frac{1 + \theta_4}{\hat{\mu}(t)}, \qquad \bar{\lambda}(t) = \lambda(t) + \frac{\theta_0 + \theta_4}{\mu(t)}, \tag{7.18}$$

$$\hat{\mu}(t) = \mu(t) - \frac{\theta_0 + \theta_1 + \theta_4}{\bar{\lambda}(t)} - \frac{\theta_0 + \theta_2 + \theta_4}{\bar{\lambda}(t) - 1} - \frac{\theta_0 + \theta_3 + \theta_4}{\bar{\lambda}(t) - t}. \tag{7.19}$$

PROOF. The proof is just a straightforward computation via Table 1. □

Now we focus on the corresponding parameter $\theta^n$ by

$$\theta^n \doteqdot \left(-\tfrac{n+1}{2}, \tfrac{1}{2}, \tfrac{1}{2}, \tfrac{1}{2}, n + \tfrac{1}{2}\right), \quad n \in \mathbb{N} \cup \{0\}, \tag{7.20}$$

$$\text{i.e., } (\alpha, \beta, \gamma, \delta) = \left(\tfrac{1}{2}(n + \tfrac{1}{2})^2, \ -\tfrac{1}{8}, \ \tfrac{1}{8}, \ \tfrac{3}{8}\right), \ n \in \mathbb{N} \cup \{0\}. \tag{7.21}$$

Let $\sigma^{n-1,n} \in W(D_4^{(1)})$, $n \geq 1$, be any Okamoto transformation satisfying $\sigma^{n-1,n}(\theta^{n-1}) = \theta^n$. Recall that the Hitchin theorem shows that any completely reducible solution $p(\tau)$ of PVI$(\frac{1}{8})$ is given by

$$\wp(p(\tau)|\tau) = \wp(r + s\tau|\tau) + \frac{\wp'(r + s\tau)|\tau)}{2Z_{r,s}(\tau)}$$

with $(r, s) \in \mathbb{C}^2 \backslash \frac{1}{2}\mathbb{Z}^2$. Then by the transformation (6.2) and the Hamiltonian system (6.11), we have corresponding solution of the Hamiltonian system (6.11), denoted by $(\lambda_{r,s}^0(t), \mu_{r,s}^0(t))$. By induction, for any $(r, s) \in \mathbb{C}^2 \backslash \frac{1}{2}\mathbb{Z}^2$, we denote the corresponding completely reducible solution to PVI with parameter $\theta^n$ (i.e., (7.20) and (7.21)) by $\lambda_{r,s}^{(n)}(t)$ and $p_{r,s}^{(n)}(\tau)$, that is,

$$\left(\lambda_{r,s}^{(n)}(t), \mu_{r,s}^{(n)}(t)\right) = \sigma^{n-1,n}\left(\lambda_{r,s}^{(n-1)}(t), \mu_{r,s}^{(n-1)}(t)\right).$$

In the following, we usually omit the subscripts $r,s$ for convenience. The following result gives the explicit expression of $(\lambda^{(n)}(t),\mu^{(n)}(t))$ in terms of $(\lambda^{(n-1)}(t),\mu^{(n-1)}(t))$.

LEMMA 7.4. **[9]** *Under the above notations, for $n\geq 1$ there holds:*

$$\mu^{(n)}=\mu^{(n-1)}-\frac{n}{2}\left(\frac{1}{\lambda^{(n-1)}+\frac{n-1}{2\mu^{(n-1)}}}+\frac{1}{\lambda^{(n-1)}+\frac{n-1}{2\mu^{(n-1)}}-1}+\frac{1}{\lambda^{(n-1)}+\frac{n-1}{2\mu^{(n-1)}}-t}\right), \tag{7.22}$$

$$\lambda^{(n)}=\lambda^{(n-1)}+\frac{n-1}{2\mu^{(n-1)}}+\frac{n+1}{2\mu^{(n)}}. \tag{7.23}$$

We prove Theorem 7.1 by induction on $n$. Here (7.22) and (7.23) are our formula for the induction process. Nevertheless, the proof still contains many complicate compilations. See **[9]** for the complete proof.

We remark that in general the right hand side of 7.2 is very difficult to compute either by the method in **[27]** or the so-called Krichever's Ansatz. Our proof suggests the Okamoto transformation could be useful in such a computation. In the followings, we provide some examples of (7.2) by applying Okamoto transformation.

EXAMPLE 7.5. $a=a(\tau)=r+s\tau$, $Z^{(1)}_{r,s}(\tau)=\zeta(r+s\tau)-r\eta_1(\tau)-s\eta_2(\tau)$.

(i) For $n=2$, we have

$$Z^{(2)}_{r,s}(\tau)=k^3-3\wp(a)k-\wp'(a)$$
$$\wp(p^{(1)}_{r,s}(\tau))=\wp(r+s\tau)+\frac{3\wp'(a)k^2+(12\wp(a)^2-g_2)k+3\wp(a)\wp'(a)}{2Z^{(2)}_{r,s}(\tau)}.$$

(ii) For $n=3$, we have

$$Z^{(3)}_{r,s}(\tau;\sigma)=k^6-15\wp(\sigma)k^4-20\wp'(\sigma)k^3+\left(\frac{27}{4}g_2-15\wp(\sigma)^2\right)k^2-12\wp(\sigma)\wp'(\sigma)k-\frac{5}{4}\wp'(\sigma)^2$$

$$\wp(p^{(2)}_{r,s}(\tau))=\wp(r+s\tau)+\frac{28\wp'(\sigma)k^6+(288\wp(\sigma)^2-24g_2)k^5+300\wp(\sigma)\wp'(\sigma)k^4+1640\wp(\sigma)^3-88g_2\wp(\sigma)}{8Z^{(1)}_{r,s}(\tau;\sigma)Z^{(3)}_{r,s}(\tau;\sigma)},$$

where $\sigma=r+s\tau$.

In the next subsection, we will discuss the Okamoto transformation for not completely reducible solutions and its application. For the simplicity, we only consider the case $n=2$.

**7.3. Smoothness of degenerate curves for $n = 2$.** Now we come to compute the Hessian of $G^{(n)}(z)$ on $E_\tau^n$ at each branch point of $Y^{(n)}(\tau)$

$$G^{(n)}(z) = n\sum_{i=1}^{n} G(z_i) - \sum_{i<j} G(z_i - z_j), \quad z = (z_1, \ldots, z_n).$$

The calculation is highly non-trivial even for $n = 2$. Here we present the case $n = 2$ in order to show how it is related to Painlevé VI.

In section 3, we have shown that $Y^{(2)}(\tau)$ has 5 branch points: $(\frac{\omega_i}{2}, \frac{\omega_j}{2})$ and $(q_\pm, -q_\pm)$, where $\wp(q_\pm) = \pm\sqrt{\frac{g_2}{12}}$. The Hessian at those points are

$$\det\left(D^2G^{(2)}\left(\frac{\omega_i}{2}, \frac{\omega_j}{2}\right)\right) = \frac{4|3e_k\eta_1 + \frac{g_2}{2} - 3e_k^2|}{(2\pi)^4\mathrm{Im}\tau}\mathrm{Im}\left(\tau - \frac{6\pi i}{3e_k\eta_1 + \frac{g_2}{2} - 3e_k^2}\right),$$

and

$$\det\left(D^2G^{(2)}\left(q_\pm, -q_\pm\right)\right) = \frac{3g_2}{4\pi^4\mathrm{Im}\tau}|\wp(q_\pm) + \eta_1|^2\mathrm{Im}\left(\tau - \frac{2\pi i}{\wp(q_\pm) + \eta_1}\right).$$

We let $C_{ij}$ or $C_\pm$ denote the curve of $\tau$ such that $(\frac{\omega_i}{2}, \frac{\omega_j}{2})$ or $(q_\pm, -q_\pm)$ are degenerate critical points of $G^{(2)}(\tau)$.

Theorem 3.6 asserts that $C_{ij}$ or $C_\pm$ has no self-intersections. This is equivalent to the following result.

THEOREM 7.6. *The meromorphic functions*

$$\phi_\pm(\tau) := \tau - \frac{2\pi i}{\eta_1(\tau) \pm \sqrt{\frac{g_2(\tau)}{12}}}, \quad \textit{and}$$

$$\phi_k(\tau) := \tau - \frac{6\pi i e_k(\tau)}{3e_k(\tau)\eta_1(\tau) + \frac{g_2(\tau)}{2} - 3e_k(\tau)^2}.$$

*are locally 1-1.*

To prove Theorem 7.6, we have to express the non-completely reducible solutions of Painlevé$(\alpha_i)$, $\alpha = (\frac{9}{8}, \frac{1}{8}, \frac{1}{8}, \frac{1}{8})$.

THEOREM 7.7. *Suppose that $p^{(1)}(\tau)$ is a solution to the elliptic form*

$$\frac{d^2p(\tau)}{d\tau^2} = \frac{-1}{32\pi^2}\sum_{i=1}^{3}\wp'\left(p(\tau) + \frac{\omega_i}{2}\Big|\,\tau\right) + \frac{-9}{32\pi^2}\wp'\left(p(\tau)\,|\tau\right). \tag{7.24}$$

*Then $p^{(1)}(\tau)$ is a non completely reducible solution if and only if there exists constant $C \in \mathbb{C} \cup \{\infty\}$ such that either*

$$\begin{aligned}&\wp(p^{(1)}(\tau)|\tau)\\ &= \frac{-4(C\eta_1 - \eta_2)^3 - g_2(C\eta_1 - \eta_2)(C-\tau)^2 + 2g_3(C-\tau)^3}{(C-\tau)[12(C\eta_1 - \eta_2)^2 - g_2(C-\tau)^2]},\end{aligned} \tag{7.25}$$

*or*

$$\wp(p^{(1)}(\tau)|\tau) = \frac{(\frac{g_2}{2} - 3e_k^2)(C\eta_1 - \eta_2) + \frac{g_2}{4}e_k(C - \tau)}{3e_k(C\eta_1 - \eta_2) + (\frac{g_2}{2} - 3e_k^2)(C - \tau)} \tag{7.26}$$

*with* $k \in \{1, 2, 3\}$.

Remark that if $C = \infty$, then formulae (7.25)–(7.26) turn to be

$$\wp(p^{(1)}(\tau)|\tau) = \frac{-4\eta_1^3 - g_2\eta_1 + 2g_3}{12\eta_1^2 - g_2},$$

$$\wp(p^{(1)}(\tau)|\tau) = \frac{(\frac{g_2}{2} - 3e_k^2)\eta_1 + \frac{g_2}{4}e_k}{3e_k\eta_1 + \frac{g_2}{2} - 3e_k^2},$$

respectively.

For $C \in \mathbb{C} \cup \{\infty\}$ and $k \in \{1, 2, 3\}$, we define functions from $\mathbb{H}$ to $\mathbb{C} \cup \{\infty\}$:

$$f_C(\tau) \doteqdot \begin{cases} 12(C\eta_1(\tau) - \eta_2(\tau))^2 - g_2(\tau)(C - \tau)^2 & \text{if } C \neq \infty, \\ 12\eta_1(\tau)^2 - g_2(\tau) & \text{if } \quad C = \infty, \end{cases} \tag{7.27}$$

$$f_{k,C}(\tau) \doteqdot \begin{cases} 3e_k(\tau)(C\eta_1(\tau) - \eta_2(\tau)) \\ +(\frac{g_2(\tau)}{2} - 3e_k(\tau)^2)(C - \tau) & \text{if } \quad C \neq \infty, \\ 3e_k(\tau)\eta_1(\tau) + \frac{g_2(\tau)}{2} - 3e_k(\tau)^2 & \text{if } \quad C = \infty, \end{cases} \tag{7.28}$$

$$\phi_\pm(\tau) \doteqdot \tau - \frac{2\pi i}{\eta_1(\tau) \pm \sqrt{g_2(\tau)/12}}, \tag{7.29}$$

$$\phi_k(\tau) \doteqdot \tau - \frac{6\pi i e_k(\tau)}{3e_k(\tau)\eta_1(\tau) + \frac{g_2(\tau)}{2} - 3e_k(\tau)^2} = \tau - \frac{6\pi i e_k(\tau)}{f_{k,\infty}(\tau)}. \tag{7.30}$$

Notice that $\phi_\pm(\tau)$ have branch points at $\tau \in \mathfrak{S}$, where

$$\mathfrak{S} \doteqdot \left\{ \frac{ae^{\pi i/3} + b}{ce^{\pi i/3} + d} \middle| \begin{pmatrix} a & b \\ c & d \end{pmatrix} \in SL(2, \mathbb{Z}) \right\},$$

because $g_2(\tau) = 0$ if and only if $\tau \in \mathfrak{S}$. Remark that

$$\phi_\pm(\tau) = \tau - \frac{2\pi i}{\eta_1(\tau)} = \frac{\eta_2(\tau)}{\eta_1(\tau)} \notin \mathbb{R} \cup \{\infty\}, \quad \forall \tau \in \mathfrak{S}. \tag{7.31}$$

THEOREM 7.8. *For* $C \in \mathbb{CP}^1$ *and* $k \in \{1, 2, 3\}$, *let* $f_C(\tau)$, $f_{k,C}(\tau)$, $\phi_\pm(\tau)$ *and* $\phi_k(\tau)$ *be defined in (7.27)–(7.30).*

(i) *Any of* $f_C(\tau)$ *and* $f_{k,C}(\tau)$ *has simple zeros only in* $\mathbb{H}$.

(ii) $\phi'_\pm(\tau) \neq 0$ *for any* $\tau \in \mathbb{H}\backslash\mathfrak{S}$. *In particular,* $\phi_\pm(\tau)$ *has only simple zeros in* $\mathbb{H}$ *and is locally one-to-one from* $\mathbb{H}\backslash\mathfrak{S}$ *to* $\mathbb{C} \cup \{\infty\}$.

(iii) *For* $k \in \{1, 2, 3\}$, $\phi'_k(\tau) \neq 0$ *for any* $\tau \in \mathbb{H}$. *In particular,* $\phi_k(\tau)$ *has only simple zeros in* $\mathbb{H}$ *and is locally one-to-one from* $\mathbb{H}$ *to* $\mathbb{C} \cup \{\infty\}$.

Theorem 7.6 is a direct consequence of Theorem 7.8.

PROOF. (i) Recalling

$$\lambda^{(1)}(t) = \frac{\wp(p^{(1)}(\tau)|\tau) - e_1(\tau)}{e_2(\tau) - e_1(\tau)}, \quad t = t(\tau) = \frac{e_3(\tau) - e_1(\tau)}{e_2(\tau) - e_1(\tau)},$$

Theorem 7.7 imply that for any zero $\tau_0 \in \mathbb{H}$ of $f_C(\tau)$ or $f_{k,C}(\tau)$, $t(\tau_0) \in \mathbb{C}\backslash\{0,1\}$ is a pole of the corresponding solution $\lambda^{(1)}(t)$ to $\mathrm{PVI}(\frac{9}{8}, \frac{-1}{8}, \frac{1}{8}, \frac{3}{8})$. Since $t'(\tau) \neq 0$ for all $\tau \in \mathbb{H}$ and any pole $t_0 \in \mathbb{C}\backslash\{0,1\}$ of $\lambda^{(1)}(t)$ is a simple pole, the assertion (i) follows readily.

(ii) Consider function $\phi_+(\tau)$. Fix any $\tau_0 \in \mathbb{H}\backslash\mathfrak{S}$, then $g_2(\tau_0) \neq 0$. If $\eta_1(\tau_0) + \sqrt{g_2(\tau_0)/12} = 0$, then $\tau_0$ is a simple zero of $f_\infty(\tau)$ of (7.27). Thus $\phi_+(\tau)$ is locally 1-1 near $\tau_0$. It suffices to consider $\eta_1(\tau_0) + \sqrt{g_2(\tau_0)/12} \neq 0$. Then by letting

$$C \doteqdot \phi_+(\tau_0) = \tau_0 - \frac{2\pi i}{\eta_1(\tau_0) + \sqrt{g_2(\tau_0)/12}}, \tag{7.32}$$

we see from $\tau\eta_1 - \eta_2 = 2\pi i$ that $\tau_0$ is a zero of

$$\begin{aligned}\psi(\tau) &\doteqdot 2\pi i + (C - \tau)\left(\eta_1(\tau) + \sqrt{g_2/12}\right) \\ &= C\eta_1(\tau) - \eta_2(\tau) + \sqrt{g_2/12}(C - \tau),\end{aligned}$$

which implies that $\tau_0$ is a zero of

$$f_C(\tau) = 12\left[C\eta_1(\tau) - \eta_2(\tau) - \sqrt{g_2/12}(C - \tau)\right]\psi(\tau).$$

Then (i) gives $\psi'(\tau_0) \neq 0$, i.e.,

$$\eta_1(\tau_0) + \sqrt{g_2(\tau_0)/12} + (\tau_0 - C)\frac{d}{d\tau}\left(\eta_1(\tau) + \sqrt{g_2/12}\right)\Big|_{\tau=\tau_0} \neq 0.$$

Together with (7.32) and (7.29), we easily obtain $\phi_+'(\tau_0) \neq 0$. Similarly we can prove $\phi_-'(\tau) \neq 0$ for any $\tau \in \mathbb{H}\backslash\mathfrak{S}$. This proves (ii).

(iii) Now we consider function $\phi_k(\tau)$ defined in (7.30). Fix $k \in \{1,2,3\}$ and any $\tau_0 \in \mathbb{H}$. If $f_{k,\infty}(\tau_0) = 0$, then it is easy to see that $e_k(\tau_0) \neq 0$ and so $\frac{d}{d\tau}(\frac{e_k(\tau)}{f_{k,\infty}(\tau)})|_{\tau=\tau_0} = \infty$, i.e., $\phi_k'(\tau_0) \neq 0$. It suffices to consider $f_{k,\infty}(\tau_0) \neq 0$. Then by letting

$$C \doteqdot \phi_k(\tau_0) = \tau_0 - \frac{6\pi i e_k(\tau_0)}{f_{k,\infty}(\tau_0)}, \tag{7.33}$$

it follows from $\tau\eta_1 - \eta_2 = 2\pi i$ that $f_{k,C}(\tau_0) = 0$, where $f_{k,C}(\tau)$ is defined in (7.28), i.e.,

$$f_{k,C}(\tau) = 6\pi i e_k(\tau) + (C - \tau)f_{k,\infty}(\tau).$$

Again (i) gives $f_{k,C}'(\tau_0) \neq 0$, i.e.,

$$6\pi i e_k'(\tau_0) - f_{k,\infty}(\tau_0) + (C - \tau_0)f_{k,\infty}'(\tau_0) \neq 0.$$

Together with (7.33) and (7.30), we easily obtain $\phi_k'(\tau_0) \neq 0$. This completes the proof. □

## 8. Application

**8.1. Zeros of Eisenstein series of weight 1.** In this section, we discuss the zeros of the Eisenstein series of weight 1, as an application of Hitchin theorem. Consider

$$Z_{(N)}(\tau) = \prod_{(r,s)\in Q_N} Z_{r,s}(\tau),$$

where $Q_N$ is the set of $N$-torsion. $Z_{(N)}(\tau)$ is a modular form of weight $|Q_N|$ w.r.t $\mathrm{SL}(2,\mathbb{Z})$. It is easy to write

$$Z_{(N)}(\tau) = C_N\Delta(\tau)^m(\ell_{1,N}(j))^2 \tag{8.1}$$

for some monic polynomial $\ell_{1,N}$ of $j$ and nonzero constant $C_N$, where $j(z)$ is the invariant function under $\mathrm{SL}(2,\mathbb{Z})$ and $m \in \mathbb{N}$.

THEOREM 8.1. **[10]** *For any $N \geq 5$ with $N \neq 6$, the monic polynomial $\ell_{1,N}(j)$ determined by (8.1) has rational coefficients and satisfies*

*(i) for any zero $j_0$ of $\ell_{1,N}(j)$, there is an algebraic solution $\lambda_{r,s}(t)$, $(r,s) \in Q_N$, such that $j_0 = j(\tau_0)$, where $t_0 = t(\tau_0)$ satisfies $\lambda_{r,s}(t_0) = \infty$. Conversely, for any algebraic solution $\lambda_{r,s}(t), (r,s) \in Q_N$, if $\lambda_{r,s}(t_0) = \infty$ for some $t_0 = t(\tau_0)$, then $j_0 = j(\tau_0)$ is a zero of $\ell_{1,N}(j)$.*

*(ii) $\ell_{1,N}(j)$ has distinct roots.*

*(iii) for any $N_1 \neq N_2$, $\ell_{1,N_1}(j)$ and $\ell_{1,N_2}(j)$ have no common zeros.*

*(iv)*

$$\deg \ell_{1,N} = \begin{cases} \frac{|Q_N|}{24} & \text{if } N \text{ is odd,} \\ \frac{|Q_N|}{24} - \frac{1}{2}\varphi(\frac{N}{2}) & \text{if } N \text{ is even,} \end{cases}$$

*where $\varphi(\cdot)$ is the Euler function.*

Recall the elementary formulae

$$|Q_N| = N^2 \prod_{p|N, p\text{prime}} \left(1 - \frac{1}{p^2}\right), \quad \varphi(N) = N \prod_{p|N, p\text{prime}} \left(1 - \frac{1}{p}\right).$$

Define

$$J_N^- = \{(r,s) \in Q_N | 2r + s = 1 \text{ and } \frac{1}{3} < s < \frac{1}{2}\},$$
$$J_N^+ = \{(r,s) \in Q_N | 2r + s = 1 \text{ and } 0 < s < \frac{1}{3}\}.$$

Then we have the following interesting result.

THEOREM 8.2. *For any $N \geq 5$ with $N \neq 6$, $\ell_N(j)$ has exactly $\#J_N^+$ real zeros in (0, 1728) and exactly $\#J_N^-$ real zeros in $(-\infty, 0)$. Furthermore, $\ell_N(j)$ has no zeros in $\{0\} \cup [1728, +\infty)$.*

It is very difficult to compute the polynomial $\ell_{1,N}(j)$ even for small $N$. To simplify the computation, we use a summation of solutions of the Painlevé

VI, instead of the product $Z_{(N)}$ as follows. Fix $N \geq 5$ with $N \neq 6$. Define $J(N)$ to be the zero set of $\ell_{1,N}(j)$:

$$J(N) := \{j(\tau)|Z_{r,s}(\tau) = 0 \text{ for some } (r,s) \in Q_N\} = \{j_k | 1 \leq k \leq \deg \ell_{1,N}\}.$$

Recall that $\lambda_{r,s}(t)$ is a solution of Painlevé VI$(\frac{1}{8}, -\frac{1}{8}, \frac{1}{8}, \frac{3}{8})$. Instead of considering the product like $Z_{(N)}(\tau)$, we consider the summation of $\lambda_{r,s}(t)$ with $(r,s) \in Q'_N$:

$$y_N(t) := \sum_{(r,s)\in Q'_N} \lambda_{r,s}(t) = \frac{1}{2} \sum_{(r,s)\in Q_N} \lambda_{r,s}(t),$$

where

$$Q'_N = \left\{ (r,s) \in Q_N \;\middle|\; \begin{array}{l} r < \frac{1}{2} \text{ if } s = 0; \\ s < \frac{1}{2} \text{ if } r = 0; \\ s \leq \frac{1}{2} \text{ if } r \neq 0, s \neq 0. \end{array} \right\}$$

Obviously, $y_N(t)$ is a rational function.

For $j_k \in J(N)$, since $j_k \notin \{0, 1728\}$ by Theorem 8.2, there are exactly six different $t$'s which satisfies $j(t) = j_k$. We fix a $t_k \in \mathbb{C}$ such that $j(t_k) = j_k$, then $\Xi(t_k)$ gives precisely these six different $t$'s. Therefore, we conclude that

$$\bigcup_{k=1}^{\deg \ell_{1,N}} \Xi(t_k) = \bigcup_{k=1}^{\deg \ell_{1,N}} \left\{ t_k, 1 - t_k, \frac{1}{t_k}, 1 - \frac{1}{t_k}, \frac{1}{1-t_k}, \frac{t_k}{t_k - 1} \right\}$$

gives precisely all the poles of $y_N(t)$.

From the above argument, we have

$$y_N(t) = -\sum_{k=1}^{\deg \ell_{1,N}} \sum_{a\in\Xi(t_k)} \frac{2a(a-1)}{t-a} + Ct + D,$$

where $C$, $D$ are two constants that can be easily determined.

We go back to the problem of computing $J(N)$ (or $\ell_{1,N}(j)$). The key observation is that the coefficients of the Taylor expression of $y_N(t)$ at $t = 0$ are expressed in terms of $j_k \in J(N)$. In general, $J(N)$ can be determined by $y_N^{(2L)}(0)$, the $2L$-th derivatives, with $1 \leq L \leq \deg \ell_{1,N}$. On the other hand, we can compute the Taylor expansion of $y_N(t)$ at $t = 0$ up to the term $t^{2 \deg \ell_N}$ by using Mathematica. In this way $J(N)$ can be computed. Here are some examples:

$$J(3) = \{0\},\ J(5) = \{\frac{5 \cdot 2^{12}}{3^5}\},\ J(8) = \{\frac{207646}{3^8}\}. \tag{8.2}$$

$N = 7$, we have

$$\ell_{1,7} = j^2 + \frac{2^{12} \cdot 37001}{3^2 \cdot 5^7} j - \frac{2^{24} \cdot 571787}{3^7 \cdot 5^7},$$

and

$$J(7) = \{\frac{2^{11}}{5^7 \cdot 3^4}(-333009 \pm 175519\sqrt{21})\}. \tag{8.3}$$

For $N = 9$, the polynomial is

$$\begin{aligned}\ell_{1,9}(j) = j^3 + \frac{861913910400000000}{78815638671875}j^2 + \frac{19885648112869441536}{78815638671875}j \\ - \frac{72057122256042716037\,12}{78815638671875}.\end{aligned} \tag{8.4}$$

So $J(9) = \{a, b, \bar{b}\}$, where $a \in \mathbb{R}$ and $b \notin \mathbb{R}$. Numerically,

$$a \approx 186.3, \ b \approx -639.9 + 285.0\sqrt{-1}. \tag{8.5}$$

It seems that except for $N = 3$, all elements in $J(N)$ are not *algebraic integers*.

We ask the following question: Is any real root $j$ of $\ell_{1,N}$, $N \geq 5$, not an algebraic integer? The affirmative answer implies the corresponding $\tau$ is transcendental by a classical result of Schneider.

**8.2. The proof of Dahmen-Beukers' conjecture.** The second application is the proof of the conjecture of Dahmen-Beukers. To prove Dahmen-Beukers conjecture, we define

$$M_{n,N}(\tau) := \prod_{(r,s)\in\mathcal{Q}(N)} Z_{r,s}^{(n)}(\tau). \tag{8.6}$$

Clearly $M_{n,N}(\tau)$ is a modular form of weight $\frac{n(n+1)}{2}\Psi(N)$ with respect to $SL(2,\mathbb{Z})$, where $\Psi(N)$ is given by (5.19). Set

$$U_n(N) = \frac{1}{2}\left(\frac{n(n+1)\Psi(N)}{24} - v_\infty\left(M_{n,N}(\tau)\right)\right) + \frac{2}{3}\varepsilon_n(N), \tag{8.7}$$

where $v_\infty(M_{n,N}(\tau))$ denotes the vanishing order of $M_{n,N}(\tau)$ at infinity. Then the crucial result in both Dahmen's and Chou's proof of the conjecture for $n \in \{1,2,3,4\}$ is

LEMMA 8.3. [**4**] *There holds $L_n(N) \leq U_n(N)$. Furthermore, $L_n(N) = U_n(N)$ if and only if $Z_{r,s}^{(n)}(\tau)$ has only simple zeros in $\mathbb{H}$ $(r,s) \in \mathcal{Q}_N$.*

See [**4**, Lemma 65] for the proof. For $n \in \{1,2,3,4\}$, the explicit expressions of $Z_{r,s}^{(n)}(\tau)$ are known; for example, see §7. Hence the vanishing order $v_\infty(M_{n,N}(\tau))$ of $M_{n,N}(\tau)$ at $\tau = \infty$ can be calculated explicitly:

$$v_\infty(M_{n,N}(\tau)) = a_n\phi(N) + b_n\phi\left(\frac{N}{2}\right) \tag{8.8}$$

for $n \in \{1,2,3,4\}$, where $a_n$ and $b_n$ are given by (5.23). Then Lemma 8.3 yields

$$\begin{aligned}L_n(N) &\leq U_n(N) \\ &= \frac{1}{2}\left(\frac{n(n+1)\Psi(N)}{24} - \left(a_n\phi(N) + b_n\phi\left(\frac{N}{2}\right)\right)\right) + \frac{2}{3}\varepsilon_n(N).\end{aligned}$$

This, together with (5.17) and (5.21), easily implies that for odd $N$,

$$PL_n(N) = L_n(N) + L_n(2N) \leq U_n(N) + U_n(2N) = PL_n(N),$$

so $L_n(N) = U_n(N)$ and $L_n(2N) = U_n(2N)$ if $N$ is odd. If $4|N$ then $L_n(N) = U_n(N)$ follows directly from (5.17) and (5.21). This proves Dahmen-Beukers conjecture for $n \in \{1, 2, 3, 4\}$. A consequence of this proof and Lemma 8.3 is that for $n \in \{1, 2, 3, 4\}$, $Z_{r,s}^{(n)}(\tau)$ has at most simple zeros in $\mathbb{H}$ if $(r, s) \in \mathcal{Q}_N$.

From the discussion above, the identity (8.8) is a crucial step to confirm Dahmen-Beukers conjecture. Our second main result is

THEOREM 8.4. [9] *For any $n \in \mathbb{N}$ and $N \in \mathbb{N}_{\geq 3}$, $v_\infty(M_{n,N}(\tau)) = a_n\phi(N) + b_n\phi(N/2)$ holds. Consequently, Dahmen-Beukers conjecture holds true.*

REMARK 8.5. As pointed out above, in both Dahmen's and Chou's proof of the conjecture for $n \in \{1, 2, 3, 4\}$, since the explicit expression of $Z_{r,s}^{(n)}(\tau)$ is known, they could prove the identity (8.8) first and then obtained the simple zero property of $Z_{r,s}^{(n)}(\tau)$ as a byproduct. However, this approach can not apply for general $n \geq 5$ because it is impossible to write down the explicit expression of $Z_{r,s}^{(n)}(\tau)$ for general $n \geq 5$. In this paper, we exploit an opposite idea, namely we prove the simple zero property of $Z_{r,s}^{(n)}(\tau)$ (i.e. Corollary 7.2) first, and then apply it to prove the identity (8.8)! Indeed, we need to prove that pre-modular forms $Z_{r,s}^{(n)}(\tau)$ possess the following properties:

(P-1) For any $N \in \mathbb{N}_{\geq 3}$ and $(r, s) \in \mathcal{Q}(N)$, $Z_{r,s}^{(n)}(\tau)$ has only simple zeros in $\mathbb{H}$;

(P-2) For any $N \in \mathbb{N}_{\geq 3}$, $v_\infty(Z_{r,s}^{(n)}(\tau)) = 0$ for any $(r, s) \in \mathcal{Q}(N)$ with $s \notin \{0, \frac{1}{2}\}$;

(P-3) There exist $\tilde{a}_n, \tilde{b}_n \in \mathbb{N} \cup \{0\}$ independent of $N$ such that

$$v_\infty(Z_{r,0}^{(n)}(\tau)) = \tilde{a}_n \text{ and } v_\infty(Z_{r,\frac{1}{2}}^{(n)}(\tau)) = \frac{\tilde{b}_n}{2}$$

for any $N \in \mathbb{N}_{\geq 3}$ and $(r, s) \in \mathcal{Q}(N)$ with $s \in \{0, \frac{1}{2}\}$;

(P-4) $\tilde{a}_n = a_n$ and $\tilde{b}_n = b_n$.

Obviously, to prove any one of (P-1)–(P-4) is not simple at all. However, it is unexpected that Painlevé VI can be applied to prove those properties.

We would like to study the asymptotics of pre-modular forms $Z_{r,s}^{(n)}(\tau)$ as $\tau \to \infty$. As discussed earlier, these asymptotics are the most important steps towards the proof of properties (P-2)–(P-4) in Remark 8.5. We only need to consider $\tau \in F_2$, where $F_2$ is a fundamental domain of $\Gamma(2)$ defined by

$$F_2 := \{\tau \in \mathbb{H} \mid 0 \leq \operatorname{Re} \tau < 2,\ |\tau - 1/2| \geq 1/2,\ |\tau - 3/2| > 1/2\}.$$

The first theorem is concerned with $\operatorname{Re} s \in (0, \frac{1}{2})$, which is a stronger version of property (P-2) in Remark 8.5.

THEOREM 8.6. [**9**] *Given any $n \geq 1$ and $(r,s) \in \mathbb{C}^2\backslash\frac{1}{2}\mathbb{Z}^2$ such that* $\mathrm{Re}\, s \in (0, \frac{1}{2})$, *there holds*

$$Z^{(n)}_{r,s}(\tau) = \check{Z}^{(n)}(s) + o(1) \text{ as } F_2 \ni \tau \to \infty,$$

*where $\check{Z}^{(n)}(s)$ is a polynomial of $s$ given as follows:*

(1) *if $n = 2m+1$ with $m \in \mathbb{N} \cup \{0\}$,*

$$\begin{aligned}\check{Z}^{(n)}(s) =&(2\pi)^{2m(m+1)} [\pi i(2s-1)]^{m+1} \prod_{k=0}^{m-1} \Big[ (s+k)\left(s+k+\frac{1}{2}\right) \\ &(s-k-1)\left(s-k-\frac{3}{2}\right)\Big]^{m-k};\end{aligned}$$

(2) *if $n = 2m$ with $m \in \mathbb{N}$,*

$$\begin{aligned}\check{Z}^{(n)}(s) =&(-1)^{m^2}(2\pi)^{2m^2} [\pi i(2s-1)]^m s^m (s-1)^m \prod_{k=0}^{m-2} \Big[ \left(s+k+\frac{1}{2}\right) \\ &(s+k+1)\left(s-k-\frac{3}{2}\right)(s-k-2)\Big]^{m-1-k}.\end{aligned}$$

*In particular,*

$$\lim_{F_2 \ni \tau \to \infty} Z^{(n)}_{r,s}(\tau) \neq 0 \text{ as long as } \mathrm{Re}\, s \in (0,1/2) \cup (1/2,1).$$

The final result deals with $s \in \{0, \frac{1}{2}\}$, which confirms properties (P-3)–(P-4) in Remark 8.5. Denote $q = e^{2\pi i \tau}$. Clearly $q \to 0$ as $F_2 \ni \tau \to \infty$.

THEOREM 8.7. [**9**] *Given any $n \geq 1$ and $(r,s) \in \mathbb{C}^2\backslash\frac{1}{2}\mathbb{Z}^2$ such that* $s \in \{0, \frac{1}{2}\}$, *there hold*

$$\begin{aligned} Z^{(n)}_{r,0}(\tau) &= \alpha^{(n)}_0(r) q^{a_n} + O(|q|^{a_n+1}), \\ Z^{(n)}_{r,\frac{1}{2}}(\tau) &= \beta^{(n)}_0(r) q^{\frac{b_n}{2}} + O(|q|^{\frac{b_n+1}{2}}), \end{aligned}$$

*where both $\alpha^{(n)}_0(r)$ and $\beta^{(n)}_0(r)$ have no rational zeros in $(0,1/2) \cup (1/2,1)$. Here $(a_n, b_n)$ is given by (5.23).*

We conclude the introduction by the following remark. To prove Theorem 8.7, it suffices for us to study the asymptotics of $Z^{(n)}_{\frac{1}{4},0}(\tau)$. In [**9**], we could prove

$$Z^{(n)}_{\frac{1}{4},0}(\tau) = (-1)^{C_n} 16^{a_n} \pi^{\frac{n(n+1)}{2}} \prod_{k=1}^{n-1} (2k+1)^{n-k} q^{a_n} + O(|q|^{a_n+1}), \tag{8.9}$$

where

$$C_n = \begin{cases} \frac{n^2-1}{4} \text{ if } n \text{ is odd,} \\ \frac{n^2}{4} \text{ if } n \text{ is even.} \end{cases} \tag{8.10}$$

The interesting thing is that *every* odd positive integer will appear in the coefficient of the leading term $q^{a_n}$ as $n$ increases. This phenomena shows that pre-modular form $Z_{r,s}^{(n)}(\tau)$ should possess many interesting unknown properties that are worthy to study in a future work.

## References

[1] L.V. Ahlfors; *Complex analysis, An introduction to the theory of analytic functions one complex variable, third edition.* McGraw-Hill Higher Education, 1978

[2] M.V. Babich and L.A. Bordag; *The elliptic form of the sixth Painlevé equation.* Preprint NT Z25/1997, Leipzig (1997).

[3] W. Bergweiler and A. Eremenko; *Green's function and anti-holomorphic dynamics on a torus.* Proc. Amer. Math. Soc. **144** (2016), no.7, 2911–2922.

[4] F. Beukers and Waall; *Lamé equations with algebraic solutions.* J. Differ. Equ. **197** (2004), 1–25.

[5] P. Boalch; *From Klein to Painlevé via Fourier, Laplace and Jimbo.* Proc. Lond. Math. Soc. **90** (2005), 167–208.

[6] C.L. Chai, C.S. Lin and C.L. Wang; *Mean field equations, Hyperelliptic curves, and Modular forms: I.* Cambridge J. Math. **3** (2015), no. 1–2, 127–274.

[7] Z. Chen, T.J. Kuo and C.S. Lin; *Hamiltonian system for the elliptic form of Painlevé VI equation.* Preprint 2015. arXiv:1506.06545v1 [math. AG] 2015.

[8] Z. Chen, T.J. Kuo and C.S. Lin; *Mean field equation, Isomonodromic deformation and Painlevé VI equation.* Preprint.

[9] Z. Chen, T.J. Kuo and C.S. Lin; *On the geometry of generalized Lamé equations, in preparation.*

[10] Z. Chen, T.J. Kuo, C.S. Lin and C.L. Wang; *Green function, Painlevé VI equation, and Eisenstein series of weight one.* Preprint 2015. http://www.math.ntu.edu.tw/dragon/works.htm

[11] C.C. Chen and C.S. Lin; *Sharp estimates for solutions of multi-bubbles in compact Riemann surfaces.* Comm. Pure Appl. Math. **55** (2002), 728–771.

[12] C.C. Chen and C.S. Lin; *Topological degree for a mean field equation on Riemann surfaces.* Comm. Pure Appl. Math. **56** (2003), 1667–1727.

[13] C.C. Chen and C.S. Lin; *Mean field equation of Liouville type with singular data: Topological degree.* Comm. Pure Appl. Math. **68** (2015), 887–947.

[14] C.C. Chen, C.S. Lin and G. Wang; *Concentration phenomenon of two-vortex solutions in a Chen-Simons model.* Ann. Scuola Norm. Sup. Pisa **3** (2004), 367–379.

[15] S. Dahmen; *Counting integral Lamé equations with finite monodromy by means of modular forms.* Master Thesis, Utrecht University 2003.

[16] S. Dahmen; *Counting integral Lamé equations by means of dessins d'enfants.* Trans. Amer. Math. Soc. **359** (2007), 909–922.

[17] F. Diamond and J. Shurman; *A first course in modular forms.* Springer, 2005.

[18] B. Dubrovin and M. Mazzocco; *Monodromy of certain Painlevé-VI transcendents and reflection groups.* Invent. Math. **141** (2000), 55–147.

[19] R. Fuchs; *Über lineare homogene Differentialgleichungen zweiter Ordnung mit drei im Endlichen gelegenen wesentlich singulären Stellen.* Math. Ann. **63** (1907), 301–321.

[20] D. Guzzetti; *The elliptic representation of the general Painlevé VI equation.* Comm. Pure Appl. Math. **55** (2002), 1280–1363.

[21] E. Hecke; *Zur Theorie der elliptischen Modulfunctionen.* Math. Ann. **97** (1926), 210–242.

[22] N.J. Hitchin; *Twistor spaces, Einstein metrics and isomonodromic deformations.* J. Differ. Geom. **42** (1995), 30–112.

[23] M. Inaba, K. Iwasaki and M. Saito; *Bäcklund transformations of the sixth Painlevé equation in terms of Riemann-Hilbert correspondence.* Inter. Math. Res. Not. **1** (2004), 1–30.

[24] K. Iwasaki, H. Kimura, S. Shimomura and M. Yoshida; *From Gauss to Painlevé: A Modern Theory of Special Functions.* Aspects of Mathematics, **16**, 1991, Springer.

[25] M. Jimbo; *Monodromy problem and the boundary condition for some Painlevé equations.* Publ. Res. Inst. Math. Sci. **18** (1982), 1137-1161.

[26] R. Liţcanu; *Lamé operators with finite monodromy – a combinatorial approach.* J. Differ. Equ. **207** (2004), 93–116.

[27] C.S. Lin and C.L. Wang; *Mean field equations, Hyperelliptic curves, and Modular forms: II.* arXiv:1502.03295v1 [math. AP] 2015.

[28] C.S. Lin and C.L. Wang; *Elliptic functions, Green functions and the mean field equations on tori.* Annals of Math. **172** (2010), 911–954.

[29] O. Lisovyy and Y. Tykhyy; *Algebraic solutions of the sixth Painlevé equation.* J. Geom. Phys. **85** (2014), 124–163.

[30] Y. Manin; *Sixth Painlevé quation, universal elliptic curve, and mirror of* $\mathbb{P}^2$. Amer. Math. Soc. Transl. (2), **186** (1998), 131–151.

[31] K. Okamoto; *Studies on the Painlevé equations. I. Sixth Painlevé equation* $P_{VI}$. Ann. Mat. Pura Appl. **146** (1986), 337–381.

[32] P. Painlevé; *Sur les équations différentialles du second ordre à points critiques fixes.* C. R. Acad. Sic. Paris Sér. **143** (1906), 1111–1117.

[33] T. Tsudo, K. Okamoto and Sakai; *Finding transformations of the Painlevé equations.* Math. Ann. **331** (2005), 713–738.

[34] A.V.D. Waall; *Lamé Equations with Finite Monodromy.* PhD Thesis, Utrecht University, 2002.

[35] E. Whittaker and G. Watson; *A course of modern analysis.* Cambridge University Press, 1996.

TAIDA INSTITUTE FOR MATHEMATICAL SCIENCES, NATIONAL TAIWAN UNIVERSITY, TAIPEI 10617, TAIWAN

*E-mail address*: cslin@math.ntu.edu.tw

Made in United States
North Haven, CT
28 February 2026

89275744R00109